The Cambrian Explosion

THE CONSTRUCTION OF ANIMAL BIODIVERSITY

DOUGLAS H. ERWIN

Department of Paleobiology
National Museum of Natural History
Washington, DC
and
Santa Fe Institute
Santa Fe, NM

JAMES W. VALENTINE

Department of Integrative Biology
University of California, Berkeley

ROBERTS AND COMPANY
Greenwood Village, Colorado

The Cambrian Explosion: **The Construction of Animal Biodiversity**

Roberts and Company Publishers, Inc.
4950 South Yosemite Street, F2 #197
Greenwood Village, CO 80111 USA
Tel: (303) 221-3325
Fax: (303) 221-3326
Web: www.roberts-publishers.com
Email: info@roberts-publishers.com

Publisher: Ben Roberts
Production management: Kathi Townes at TECHarts
Artists: Quade Paul at Echo Medical Media and Tom Webster at Lineworks
Copyeditor: Kathleen Lafferty
Proofreader: Dawn Hall
Designer: Kathi Townes
Composition: TECHarts

Front cover: Artist's reconstruction of one of the more unusual animals of the Cambrian explosion, *Opabinia regalis,* fossil representatives of which have been recovered from the Burgess Shale deposits in British Columbia, Canada. The animal shown here was about 5 inches long. Artwork by Quade Paul.

ISBN: 978-1-936221-03-5

Library of Congress Cataloging-in-Publication Data

Douglas, Erwin.

The Cambrian explosion : the construction of animal biodiversity / Douglas H. Erwin, Department of Paleobiology, National Museum of Natural History, Washington, DC and Santa Fe Institute, Santa Fe, NM, James W. Valentine, Department of Integrative Biology, University of California, Berkeley.

pages cm

ISBN 978-1-936221-03-5 (hardback)

1. Evolutionary paleobiology. 2. Paleontology--Cambrian. 3. Biodiversity–Climatic factors. I. James, Valentine. II. Title.

QE721.2.E85D68 2013

560'.1723--dc23

2012036689

10 9 8 7 6 5 4 3 2

Contents

Preface

The Cambrian explosion is one of the major evolutionary transitions in the history of life. From a meager and only dimly perceived beginning from single-celled ancestors nearly 800 million years ago (Ma), early animals spent over 200 million years in which their evolutionary changes chiefly involved the development of different cell types in relatively simple organisms, but patterned by increasingly sophisticated genetic programs. Beginning about 579 Ma, during the Ediacaran period, many novel clades of animals began to appear, radiated into distinctive forms, thus producing an increasingly diverse family tree of early animals. Unfortunately, only some of these lineages are represented by fossils. Finally, during a relatively narrow time span in the early Cambrian (after about 542 Ma), there is an explosive appearance of many different, morphologically distinctive fossils, including representatives of most of the major animal groups that are alive today.

In 2007 there was an international meeting of molecular developmental biologists at Moorea, in French Polynesia, sponsored by the Center for Integrative Genomics at Berkeley. By that time many of the features of the gene regulatory systems that mediate the development of complex animal body plans from fertilized eggs had been demonstrated in model organisms, but the evolution of these developmental pathways was still unclear. We are both paleontologists but without laboratory experience in developmental research, yet we were invited to give talks at the meeting about the early fossil record of animal body plans. After our presentations Nicole King, whose lab at Berkeley was pioneering the study of regulatory genes in unicellular forms, asked us to suggest a book about the Cambrian explosion for her graduate students. We had no answer, but her question clearly indicated a need. It has taken us a while, but here is our attempt at such a book.

In *The Cambrian Explosion: The Construction of Animal Biodiversity*, we examine our current understanding of early animal evolution from a variety of perspectives, including details of the fossil record, changes in geochemical cycles, the information revealed by the use of morphological and molecular data to establish the evolutionary relationships among the major animal groups, and the startling new insights into the evolution of development produced by comparative studies of modern animal groups. The Cambrian explosion, however, provides a more important window into understanding the processes that generate the biodiversity on our planet and, in particular, the evolutionary mechanisms responsible for those processes. So, we delve into how our understanding of evolutionary change has been affected by studies of the Cambrian explosion, and we discuss the very large problems that remain and indicate how they are being addressed. The range of scientific disciplines involved in reconstructing the history of this major evolutionary transition is broad and often involves specialized approaches within particular disciplines. Consequently, we must provide some of the details of the appropriate methods and findings to support our conclusions.

Chapter 1 is an introduction to the subject. Chapter 2 establishes the temporal framework from about 635 Ma to about 500 Ma, discussing the various techniques of radiometric dating and correlation that geologists have used to establish the temporal order of events in the geologic record. With this framework in hand, chapter 3 then focuses on the physical environmental contexts—from global glaciations to rapid shifts in the cycling of carbon, sulfur, and other elements through the oceans and atmosphere—that accompanied the origin and early rise of animals. Many of our arguments require an understanding of the evolutionary relationships between various groups of animals. Such evolutionary trees are based on rigorous phylogenetic studies of both morphologic and molecular data, as we review in chapter 4. Studies of DNA can potentially provide insight into the timing of divergences of different groups through a technique known as a molecular clock, and a discussion of the conflicting results of various molecular clock studies concludes the chapter.

In chapter 5, we consider the initial phase of animal origins during the Ediacaran period (635–542 Ma), the geologic interval just before the Cambrian that includes an enigmatic menagerie of soft-bodied forms known as the Ediacaran macrofauna. Animals do not just leave behind impressions of their bodies, however. We can learn a great deal about the complexity and behavior of animals from the tracks, trails, and burrows they leave in the sediment as well. From the record of such trace fossils we have learned that the burrowing activity of animals increased rapidly from the Ediacaran into the Cambrian, destroying microbially bound sediments that had been an important feature of late Neoproterozoic seafloors. In chapter 6, we turn to the Cambrian explosion itself and review the exceptionally preserved soft-bodied assemblages of the Chengjiang and the Burgess Shale faunas. These assemblages provide insights into the startling breadth of the explosion that we could not recover from normally preserved elements of the fossil record.

In the case of the Cambrian explosion, the great evolutionary puzzle is in establishing the relative importance of three pieces: (1) changes in the environment, which were unusual by today's standards; (2) the appearance of developmental innovations that permitted the appearance of the progressively more complex animal body plans found through the Ediacaran and Early Cambrian; and (3) the construction of the ecological relationships that allowed these new animals to succeed. We consider the ecological issues in chapter 7 and the developmental components in chapters 8 and 9. Finally, in chapter 10, we bring all these pieces together to integrate the patterns and processes involved and to address what the Cambrian explosion tells us about the nature of the evolutionary process.

Acknowledgments

In writing *The Cambrian Explosion: The Construction of Animal Biodiversity*, we have benefited from discussions, comments, and critiques from many colleagues. We are particularly indebted to readers of the entire volume: Nick Butterfield, Ronald Jenner, Marc Laflamme, Kevin Peterson, Sarah Tweedt, and Shuhai Xiao. Their perceptive comments greatly improved the content and structure of the final book and saved us from a number of important errors or misinterpretations. We also received detailed comments on individual chapters from Loren Babcock, Soren Jensen, Marc Laflamme, and Adam Maloof. During the preparation of the manuscript and our research on this interval, we greatly benefited from discussions with Sam Bowring, Jean-Bernard Caron, Eric Davidson, Bernard Degnan, Bill DiMichele, Galen Halverson, Paul Hoffman, Nicole King, David Krakauer, Marc Laflamme, Michael Levine, Charles Marshall, Simon Conway Morris, Guy Narbonne, Kevin Peterson, John Odling-Smee, Daniel Rokhsar, Mansi Srivastava, Martin Stein, and Roger Summons. Sarah Tweedt was primarily responsible for the compilation of the first-occurrence data presented in the appendix, and Marc Laflamme produced many of the photographic plates of the Ediacaran fauna in chapter 5. At the UC Berkeley Museum of Paleontology Chris Mejia and David Smith provided aid in reproducing illustrations. The final book was greatly improved through copyediting by Kathleen Lafferty, proofreading by Dawn Hall, rendering of the line art by Tom Webster, and the incredible artwork of Quade Paul. Kathi Townes did a magnificent job of design and production of the book, under occasionally trying circumstances. The contributions of two people were particularly significant: Ben Roberts was enormously patient and supportive during the long gestation of this volume, and we particularly want to acknowledge the support of Wendy Wiswall at the National Museum of Natural History, without whose support this book would never have been completed.

Research such as ours would not be possible without support from a number of funding sources, which we gratefully acknowledge, including NASA's National Astrobiology Institute (MIT Node) and the Santa Fe Institute.

DOUGLAS H. ERWIN
JAMES W. VALENTINE

PART ONE

ꕥ

Historical and Geological Settings

CHAPTER ONE

Introduction

Life arose on Earth more than three and a half billion years ago (3.5 gigayears, or Ga, ago; 1 Ga = 1 billion years) with the evolution of the first living cells. Soon (geologically speaking), three major, single-celled evolutionary lineages became established: the Eubacteria ("true" bacteria), the Archaea, and the stem Eukaryota (the lineage that now includes plants, animals, and fungi). The cells of each of these domains are very different from one another: Eubacteria and Archaea have their DNA dispersed throughout the cell, whereas in Eukaryota the DNA is enclosed within a membrane-bound nucleus. These single-celled forms diversified broadly, inventing a wide array of biochemical specializations. Although the lineage that gave rise to eukaryotes seems to be as ancient as the Eubacteria and Archaea, the earliest eukaryotic cells now recognized in the fossil record date to about 1.8 Ga, and it seems unlikely that they evolved much before 2 Ga. The cell lineages that gave rise to the Eukaryota are unknown. These early organisms were able to transfer genetic material between lineages, so the phylogenetic histories of their genes—their family trees—do not branch in a tree-like manner but instead involve many cross-links, greatly complicating the reconstruction of their phylogenies (Rivera and Lake 2004). Lateral gene transfers occur in animals as well, but at much lower frequencies and chiefly among early lineages.

The evolutionary changes since the origin of life have been accompanied by extensive changes in environmental conditions, some caused by purely physical and chemical processes and others by interactions of organisms with the atmosphere, oceans, and crustal materials. For example, the initial rise of oxygen in the oceans and atmosphere about 2.4 Ga was caused by the evolution of oxygenic photosynthesis that probably became important about 2.7 Ga. This process entrained a complex series of oxidation reactions with elements such as iron and uranium, and it eventually led to the spread of low levels of free oxygen. The evolutionary history of life has been sensitive to such changes.

The earliest fossil eukaryotic cells arose through the acquisition of symbionts by a host cell (Embley and Martin 2006; Margulis 1970). Organelles such as mitochondria and the chloroplasts began as small cells that lived within a host cell symbiotically—for mutual benefits—and that became incorporated as obligate endosymbionts. They eventually evolved into cellular structures, organelles that function within eukaryotic cells much like organs function in animal bodies. Not all organelles arose from symbiosis; some simply evolved to function within their present clades. All organisms that are large, complex, or multicellular or that have a diversity of structures are eukaryotes; evidently, the structure of the eukaryotic genome is an advantage in achieving a certain kind of multicellularity. It may be that the evolution of the traits that permitted the eukaryote host cell to acquire symbionts as permanent organelles, making the host's cells multicellular in a sense, underlay the ability of eukaryotes to eventually form bodies composed of many differentiated cell types (Awramik and Valentine 1985).

Today, some two dozen major eukaryotic groups have bodies composed of more than one cell, but few have progressed beyond the stage of an association of essentially identical cell types (Buss 1987; Knoll 2011). Eukaryotes include protistan colonies and various algae that have many cells, but there is no evidence that any of these groups has ever achieved the developmental control required to produce more complex morphologic patterns. Multicellular algae and fungi have only a few cell types, whereas other eukaryotic lineages are multicellular but exhibit none of the hierarchical structure of differentiation seen in plants and animals. At least eight different groups of these multicellular eukaryotes arose well before animals finally evolved sometime more than 750 million years ago (Ma). Complex multicellularity involves a hierarchical structure of differentiated cell types, tissues, organs, and the regionally differentiated structures found in animals and vascular land plants. As we will see in chapter 3, there is good evidence that the environmental settings of mid-Proterozoic time would have inhibited the evolutionary success of those groups even if they had appeared.

Multicellularity is a generative evolutionary innovation in the sense that it provides the basis for two additional important evolutionary steps: greater body size and increased division of labor among differentiated body parts. Greater size quite literally changes the nature of the world experienced by organisms. Most single cells in the sea live in a world in which their motion is dominated by the viscosity of the water rather than by the inertia of their bodies (as expressed by the Reynolds number[1]). Body size is a multiplier of inertia, and most multicellular organisms are large enough that they cross the boundary into a world where inertial forces become important. At such larger sizes, most organisms evolved new ways of locomotion and feeding, facilitated by the specialization of cells, tissues, organs, and differentiated body parts. Such division of labor is evident even in sponges, the earliest metazoan group, but becomes far more pronounced in more complex animals.

Choanoflagellates, a unicellular group that feeds on bacteria and other minute food items, are the closest living relatives of animals. The earliest animals likely fed in a similar manner, but evolved larger, multicellular feeding chambers and were able to capture more food, to support their increased body masses, from larger volumes of water. From this fairly unpromising beginning arose all the rich diversity of the animal kingdom. Humanity owes a special debt to sponges.

Some 120 million to 170 million years after the origin of sponges, the scrappy fossil record improved with a bang, geologically speaking. Following a prelude of a diverse suite of enigmatic, soft-bodied organisms beginning about 579 Ma, a great variety and abundance of animal fossils appear in deposits dating from a geologically brief interval between about 530 to 520 Ma, early in the Cambrian period. During this time, nearly all the major living animal groups (phyla) that have skeletons first appeared as fossils (at least one appeared earlier). Surprisingly, a number of those localities have yielded fossils that preserve details of complex organs at the tissue level, such as eyes, guts, and appendages. In addition, several groups that were entirely soft-bodied and thus could be preserved only under unusual circumstances also first appear in those faunas. Because many of those fossils represent complex groups such as vertebrates (the subgroup of the phylum Chordata to which humans belong) and arthropods, it seems likely that all or nearly all the major phylum-level groups of living animals, including many small soft-bodied groups that we do not actually find as fossils, had appeared by the end of the early Cambrian. This geologically abrupt and spectacular record of early animal life is called the Cambrian explosion. The explosion tells us a great deal about ecological and evolutionary history and, even more importantly, about the many processes involved in evolutionary change.

Although the earliest, preexplosion history of animals is not well recorded by fossils, it can be pieced together from a fragmentary fossil record and from knowledge of the morphologies and genomes of animals whose body plans evolved during that remote period, especially living sponges, cnidarians, and primitive bilaterian groups. The rest is history, one that can be reconstructed from the fossil record, the comparative study of living metazoans, the record of evolution preserved in the genomes of living groups, and the study of the ecological and evolutionary processes that shaped the living fauna.

The reality of the Cambrian explosion has been questioned by a number of scientists. One line of argument has been that because the fossil record is incomplete, the absence of earlier animal fossils is not evidence that such forms were not present in earlier faunas. Therefore, it has been argued that the explosion is more apparent than real and simply reflects unusual conditions of fossil preservation. Another frequent criticism is that because evolution is assumed to proceed by rather gradual change and not by jumps, the rapid appearance of such a diverse, novel fauna could not have arisen in such a short period of time as the explosion interval, almost a geological instant. Arguing from negative evidence is clearly dangerous.

Because unusual claims require unusual evidence, such concerns are entirely legitimate. The unique Cambrian fossil assemblages were revealed in large part due to studies of the Burgess Shale fauna discovered in 1909 by paleontologist Charles Walcott (fig. 1.1) but finally brought to full light by Harry Whittington and his colleagues beginning in the 1970s. Even older, spectacular Cambrian faunas from China were discovered in 1984 by Hou Xian-guang. The evolutionary history behind these assemblages was not well understood, however. A long, hidden history of animal evolution seemed possible. Taken at face value, the geologically abrupt appearance of Cambrian faunas with exceptional preservation suggested the possibility that they represented a singular burst of evolution, but the processes and mechanisms were elusive.

Although there is truth to some of the objections, they have not diminished the magnitude or importance of the explosion. A long history of metazoan evolution did precede the Cambrian, perhaps by 200 million years or more. This long history was unknown when the explosion faunas were first described but is being gradually revealed by comparative studies of the rates of molecular sequence divergence (so-called molecular clocks), by studies of molecular fossils or biomarkers and by a growing knowledge of the fossil record. The only animals present during most of the preexplosion interval were sponges and architecturally simple organisms built of sheets of tissues. A few tens of millions of years before the explosion, however, small organisms that had body plans designed for locomotion on the seafloor—bilaterian-grade forms—evolved. Their evolution may have been enabled by an increase in oxygen levels that permitted an expansion in metabolic activities, although oxygen levels were still quite low by modern standards (1–10% of present atmospheric levels). A continuing increase in oxygen levels may have permitted the evolution of larger-bodied architectures that arose from varied members of the chiefly worm-like bilaterian fauna. Solutions to some of the biomechanical problems posed by those larger bodies commonly involved the evolution of either tough organic or even mineralized skeletons. It was this round of metazoan evolution that produced the fossils of the Cambrian explosion. Thus, explosion fossils did have a metazoan ancestry stretching back well before the Cambrian into the Neoproterozoic, but the earlier faunas did not consist of numerous, large, complicated body plans.

Several lines of evidence are consistent with the reality of the Cambrian explosion. First, metazoan fossils are not the only fossil groups to suggest an unusual period of evolutionary activity during the early and middle Cambrian. The rise in fossil metazoan diversity is closely tracked by an increased

Figure 1.1 Paleontologist Charles Walcott, the fourth secretary of the Smithsonian Institution, discovered the Burgess Shale Fauna in August of 1909. Walcott is shown with three other men excavating the quarry, in British Columbia, Canada. Walcott is in the center of the picture with hand on hip. Photograph courtesy of the Smithsonian Institution Archives.

diversity of organic-walled microfossils known as acritarchs as well as by an increase in the diversity and complexity of trace fossils—the signs of animal activity such as trails or burrows. Each of those fossil types is subject to unique preservational requirements; thus, they represent independent metrics of diversity change. The similarity in their patterns suggests a general expansion of biodiversity, not just among early animals but among many other groups as well. Finally, many of the changes in preservation are a consequence of the biotic innovations (Butterfield 2003). For example, a revolutionary change in the sedimentary environment—from microbially stabilized sediments during the Ediacaran to biologically churned sediments as larger, more active animals appeared—occurred during the early Cambrian. Thus, the quality of fossil preservation in some settings may actually have declined from the Ediacaran into the Cambrian, the opposite of what has sometimes been claimed, yet we find a rich and widespread explosion of fauna.

The Cambrian explosion is also correlated with changes in the amount of oxygen in the oceans, with the construction of animal-dominated marine ecosystems, and with the expansion of the developmental processes—leading from egg to adult—that underpin the ability of animal genomes to generate the morphologies of more complex animals. These three elements—changes in the physical environment, the establishment of new ecological relationships, and the evolution of developmental systems—form the changes that are most critical for understanding the explosion. A central theme of this book is the exploration of the contributions of each of these elements and particularly of the interactions between them. Many accounts of the explosion focus on only one, or sometimes two, parts of this triad as the primary driver for this extraordinary episode of evolutionary innovation. In our view, the early diversification of animals was not simply a response to a changing environment or to the acquisition of a particular new adaptation or to the invention of new types of development, but to interactions among all three.

The subtitle of this book, *The Construction of Animal Biodiversity*, captures a second theme: the importance of building the networks that mediate the interactions. Networks exist between the physical environment and the biota to affect geochemical cycles, between various species to construct marine ecosystems, and among genes and cells within diversifying animal lineages as the developmental process evolved. Each of these theaters of evolutionary change involved the formation of new interaction networks, and in many ways the Cambrian explosion is dominated by the issue of network dynamics. Take just one example that we will explore later in more detail: the oxygenation of the oceans. Precambrian oceans were largely anoxic, with unhealthy doses of sulfur and iron. Their conversion to the well-oxygenated oceans of today set the stage for the diversification of marine animals. The oxygen originally was generated by algal photosynthesis, but it is only a necessary precondition. What we need to know is how the change worked. Although many purely geological and geochemical scenarios have been proposed, there is also evidence of a significant role for the activity of animals. For example, it is possible that the action of sponges and their allies in sequestering carbon in the sediments may have been critical in the oxygen buildup in ocean waters in the late Neoproterozoic (Sperling, Pisani, and Peterson 2007). If this hypothesis is correct, it exemplifies the contributions of the animals in the building of an environment that permitted their own diversification through the development of ecological interaction networks.

Increased genetic and developmental interactions were also critical to the formation of new animal body plans. By the time a branch of advanced sponges gave rise to more complex animals, their genomes comprised genes whose products could interact with regulatory elements in a coordinated network. Network interactions were critical to the spatial and temporal patterning of gene expression, to the formation of new cell types, and to the generation of a hierarchical morphology of tissues and organs. The evolving lineages could begin to adapt to different regions within the rich mosaic of conditions they encountered across the environmental landscape, diverging and specializing to diversify into an array of body forms.

A third theme of this book is the tension between the nature of explanations for major evolutionary transitions in general and that of the Cambrian explosion in particular. In each of the three trajectories of change explored in this book—of the physical environment, of ecological interactions, and of the growth of developmental interactions—some workers have favored explanations that are consistent with how processes work today. Others, though, interpret the evidence to suggest that the world of the Ediacaran and Cambrian operated in very different ways or at least produced very different effects than what similar changes would produce in the modern world. For example, some geochemists have suggested that the carbon cycle involved in the late Ediacaran operated in very different ways from today and that only by reconstructing the ancient dynamic of the carbon cycle can we understand the increase in oxygen levels of the time. Geologists describe such settings as "nonanalog conditions." Much warmer or colder climates, more extensive continental seas, and widespread ocean anoxia are examples of such conditions. Having recognized the occurrence of nonanalog conditions, the challenge becomes to understand whether the processes that produced them also differ from those operating in the modern world.

As geologists, we view this tension as a debate over the extent to which uniformitarian explanations can be applied to understand the Cambrian explosion. Uniformitarianism is often described as the concept, most forcefully advocated by Charles Lyell in his *Principles of Geology,* that "the present is the key to the past" (Lyell 1830). Lyell argued that study of geological processes operating today provides the most scientific approach to understanding past geological events. Uniformitarianism has two components. Methodological uniformitarianism is simply the uncontroversial assumption that scientific laws are invariant through time and space. This concept is so fundamental to all sciences that it generally goes unremarked. Lyell, though, also made a further claim about substantive uniformitarianism: that the rates and processes of geological change have been invariant through time (Gould 1965). Few of Lyell's contemporaries agreed with him (Rudwick 2008). Today, geologists recognize that the rates of geological processes have varied considerably through the history of Earth and that many processes have operated in the past that may not be readily studied today.

Whether uniformitarian explanations can be appropriately applied to understanding events of the Ediacaran and Cambrian will arise in several chapters of this book. Although it has not usually been framed this way, we will see that debates over the nature of the geochemical evidence, the processes involved in the construction of Ediacaran and Cambrian ecological assemblages, and the processes of change in developmental evolution in early metazoans all involve differences of opinion as to whether a uniformitarian approach is appropriate (Erwin 2011).

The nature of appropriate explanations is particularly evident in the final theme of the book: the implications that the Cambrian explosion has for understanding evolution and, in particular, for the dichotomy between microevolution and macroevolution. If our theoretical notions do not explain the fossil patterns or are contradicted by them, the theory is either incorrect or is applicable only to special cases. Stephen Jay Gould employed the animals of the Burgess Shale and the early Cambrian radiation in his book *Wonderful Life* (Gould 1989) to advance his own view of evolutionary change. Gould argued persuasively for the importance of contingency—dependence on preceding events—in the history of life. Many other evolutionary biologists have also addressed issues raised by these events. One important concern has been whether the microevolutionary patterns commonly studied in modern organisms by evolutionary biologists are sufficient to understand and explain the events of the Cambrian or whether evolutionary theory needs to be expanded to include a more diverse set of macroevolutionary processes. We strongly hold to the latter position.

In general, microevolution treats changes within populations and species, underpinned by the natural selection of genetic variation that arises through mutation or recombination within the genome. These genetic changes arise within individuals but are promulgated over time within populations and species, depending on the advantage, disadvantage, or neutrality of the changes with respect to the relative reproductive success of the individuals that carry them; in other words, they are scrutinized by selection. Sometimes, this evolutionary mode is characterized as evolution by change in gene frequency. Microevolutionary change often produces new species when different populations of a species are isolated genetically, or nearly so, such that each pursues a separate pathway of genetic change and they become distinct species; in animals, it usually means that they can no longer exchange genes. Macroevolution, by contrast, involves the study of what happens in evolution beyond the mechanisms of the formation of species. Some species, for example, are founders of major clades that encompass millions of species that occupy a wide range of ecological occupations, whereas other species are merely found in minor branches of life's tree with rather similar ecologies or simply become extinct without issue (other patterns are not uncommon). Each of the species with those very different evolutionary outcomes arose through microevolutionary processes, yet there is obviously more to be said about their evolution, which forms the topic of macroevolution. Some macroevolutionary studies focus on the waxing and waning of clades through space and time and on the causes of their relative abilities to expand, to resist extinction, to deploy ecologically, and to generally prosper or not. Other studies focus on the rise of evolutionary novelties within some branches that produce novel body plans and, in some cases, many "subplans," as in the Arthropoda. In yet other branches, some rich in species and some not, only a single, narrow range of body plan morphology occurs, as in the priapulids (see chap. 4). Clearly, the results of all speciation events are not equal. These two macroevolutionary areas—relative richness and relative novelty—are clearly related, with differences in body plans being responsible for some differences in branching patterns in the tree of life. The change from studying microevolution to macroevolution involves a hierarchical step (Erwin 2000; Jablonski 2007) that is important because it moves the focus of interest from processes that affect individuals within species to those that affect species within higher-order groups. Thus, the move from micro to macro forms a discontinuity. Novel features arise within lineages, just as do changes leading to speciations, but the subsequent behavior of the groups with respect to evolutionary rates, diversifications, extinctions, and ecological and geographic ranges must be studied among lineages. It is in work on the origin of novelties that explanations for the Cambrian explosion are now emerging. Since the 1990s, there has been a revolution in our understanding of the mechanisms governing the development of animals and how these mechanisms evolve.

Here, then, is a perfect scientific challenge: to unravel events of basic importance to our understanding of the history of life and of the processes that underlay it, set in oceans of the remote past and obscured by far more than half a billion years of subsequent evolution of both the environment and the biota. With a fragmentary and often mysterious fossil record, combined with such information as can be gleaned from the rock sequences in which the fossils are preserved, the Cambrian explosion was a major transition in the history of life, and it plays a critical role in evaluating our theories and understanding of the processes of evolution. What could be more appealing?

NOTE

1. The Reynolds number can be expressed as $Re = lU/v$, where l represents the size of the organism as a function of some linear dimension, U is the velocity of a fluid medium relative to the organism, and v is the kinematic viscosity of the fluid (the ratio of dynamic viscosity to density). See S. Vogel (1994) for a most readable account of the Reynolds number and its consequences.

CHAPTER TWO

The Geological Context of Ediacaran and Cambrian Events

In 2004, geologists celebrated the establishment of the first new geological period in 113 years: the Ediacaran. This admittedly arcane event was the culmination of two decades of work by an international group of earth scientists, continuing a tradition that stretches back to the beginning of modern geology. The geologic timescale is largely a residue of nineteenth-century geology that was constructed from a desire to develop a global synthesis of geologic history through the examination of individual outcrops of rock. Geologists compiled information from individual outcrops into regional frameworks and eventually synthesized them into a global picture. This process required the ability to correlate rocks among different areas, and for this task fossils were the most accurate tool then available. Some fossil taxa are geographically widespread, easily preserved and identified, and existed for only a brief interval of geologic time. Such fossils proved critical in correlating between different outcrops of rock locally and regionally and eventually between different continents.

When the geologic timescale was being developed during the 1800s, the most ancient known signs of life were the skeletons of marine invertebrate animals, including trilobites, a long-extinct arthropod group. Trilobites were assigned to the Cambrian Period, established in 1835 by English geologist Adam Sedgwick based on rocks in northern Wales. The Cambrian and all younger periods were assigned to the Phanerozoic Era, the era during which there were signs of life. Since the 1950s, paleontologists have recognized that life originated far earlier than the Cambrian, probably about 3.5 Ga. Microbial fossils have been found well back into the Archean Eon (which dates between 3.8 and

2.4 Ga), although they did not become abundant until the Proterozoic. The earliest eukaryotic cells (those that have DNA encapsulated in a nucleus) appear by 1.8 Ga, and fossils of multicellular algae date to about 1.2 Ga. The discovery of soft-bodied fossils in a sequence of rocks older than the Cambrian, now called the Ediacaran Period, showed that macroscopic animal life was not strictly limited to the Phanerozoic Era as had been widely believed. Further, it is now known that skeletonized fossils are not limited to Cambrian and younger periods of geologic time.

The establishment of the Ediacaran Period was based on the distinctive fossils and geological events leading up to the explosion of observable animal life that characterizes the early Cambrian period (Knoll et al. 2004). Geologists recognized the unique nature of the Ediacaran interval in the 1960s, and several workers had formally proposed names for new periods and systems older than the Cambrian, but none of those units had received the imprimatur of the International Union of Geological Sciences. Establishing a new geologic period requires far more than just discovering distinctive fossils. Paleontologists had to build a convincing case that these fossils were indeed older than the Cambrian and that rocks of the same age could be reliably correlated globally. The goal, however, was not simply to add another period to the geologic timescale; rather, it was to establish a temporal framework for understanding the geological and biological events associated with the early diversification of animals. This chapter focuses on the construction of this temporal framework through the use of radiometric dates to establish absolute ages and on other modern stratigraphic methods to correlate events from one locality to events in other regions of the world. These techniques allow geologists to better understand the tempo and mode of events as they unfold over time and across the globe.

STRATIGRAPHY AND CORRELATIONS

By convention, the bases of intervals of the geologic timescale are defined at a specific point in a particular outcrop of rock, generally at the first (i.e., earliest) appearance of a diagnostic fossil or at a distinctive geologic marker. This point establishes a reference standard for correlation of sequences in other parts of the world. The Ediacaran Period is defined as beginning with the deposition of an unusual carbonate sequence, known as a cap carbonate, which lies immediately above glaciogenic rocks that outcrop along Enorama Creek in the Flinders Ranges of South Australia (fig. 2.1). These glacial deposits and the overlying carbonates are widespread and may have been nearly global in original extent; they are products of a climatic event known as the Marinoan glaciation. The broad extent of these deposits makes them an excellent marker

Figure 2.1 The base of the Ediacaran System at Enorama Creek, Australia. The boundary is defined here above the glacial deposits of the Marinoan glaciation. Australian geologist Jim Gehling is pointing at the boundary. Photograph by Doug Erwin.

horizon for correlation to rocks of similar age on different continents. Moreover, the carbonate rocks record a distinctive shift in carbon isotopes associated with the end of this glaciation (discussed in chap. 3). To date, almost all the fossil evidence for animal life has been found in rocks younger than the Marinoan event. Rocks of the Ediacaran document a dynamic and complex period of Earth history, stretching from the close of the Marinoan glacial event to the base of the Cambrian, that includes enormous swings in the global carbon cycle, at least one additional glaciation, and the conversion of the oceans from a sulfur-rich brew to something approaching the better oxygenated world we have today.

Within the Ediacaran, correlations are largely based on a combination of stratigraphy, radiometric dates, and fluctuations in carbon and strontium isotopes. Despite many efforts, fossils have not yet proven useful in further subdividing this interval to the same level of resolution as is possible in Phanerozoic periods. Most Ediacaran animals were soft bodied and had a low potential for preservation (see chap. 5). One major group of Ediacaran microfossils with tough organic walls, known as acritarchs, are more easily preserved and have been useful in correlation. A very distinctive group of acritarchs, known as the Doushantuo-Pertatataka assemblage, is found in early Ediacaran rocks of

Australia, south China, India, Siberia, and Svalbard. This group is characterized by its relatively large size and morphological complexity. As is often the case with attempts to use Ediacaran fossils for correlation, however, there is an ongoing controversy over the age of this assemblage. Some evidence suggests that it appeared immediately after the Marinoan glaciation at 635 Ma, as indicated by studies in south China, whereas in Australia it first appears somewhere near 580 Ma (Grey, Walter, and Calver 2003; Zhou et al. 2007). Until such major problems are resolved, it will be difficult to use fossils to subdivide or correlate events within the Ediacaran period.

Significant changes in sea level generate characteristic architectures in sedimentary deposits. When these changes are global (e.g., owing to seawater being locked up in glacial ice or released by glacial melting) rather than being caused by regional tectonic events, the changes in stratigraphic architecture produce globally correlative signatures known as sequence boundaries. Because of the glacial episodes and other events during the Ediacaran, correlation using sequence boundaries has proven to be useful, particularly when combined with other stratigraphic data. For example, in south China, sequence stratigraphy has allowed correlations to be made across the entire region, and it is likely that these correlations will be extended to other continents (fig. 2.2; Zhu, Zhang, and Yang 2007).

Unlike the Ediacaran, the base of the Cambrian is defined by a fossil, but not by a distinctive body fossil such as a trilobite. In fact, the base of the Cambrian as now defined is almost 20 million years older than the first trilobite occurrences. In 1994, after almost a decade of study of localities in northern Siberia, central China, and Canada, an international consortium of geologists settled on the first appearance of a trace fossil known as *Treptichnus pedum* in the Chapel Island Formation in the town of Fortune on the Burin Peninsula, Newfoundland, Canada, as the marker for the base of the Cambrian (Brasier, Cowie, and Taylor 1994; Landing 1994). The animal whose activity formed *T. pedum* produced a distinctive burrow just beneath the seafloor, a burrow that could have been made by a priapulid or a priapulid-like worm (Vannier et al. 2010). It is one of the earliest burrow types that penetrated vertically into the sediment. Earlier trace fossils represent the displacement of sediment as organisms moved horizontally along or just beneath the seafloor. Consequently, *T. pedum* represents an easily identifiable marker that indicates an increase in the complexity of organismal behavior at the base of the Cambrian, which is one reason it was selected to mark the boundary. This selection has been controversial, however, and not just because the locality once served as the town dump.

Trace fossils (tracks, burrows, and similar sedimentary evidence of animals) tend to be less diagnostic than body fossils, so they had not previously been considered sufficiently diagnostic to qualify as markers for defining a geologic unit's boundaries. In the Phanerozoic, the distribution of particular trace fossils is often limited to specific environments, while useful markers should have a broad ecological range. Further, many different organisms construct similar burrows, so some burrow types have very long geological ranges and are not very useful for correlation. Neither of these problems seems to hold for *T. pedum*, however, so this trace fossil appears to be a reasonably reliable marker. Furthermore, body fossils during the Cambrian have problems of their own that have made them unreliable for correlation in some settings. For example, trilobites, the iconic fossils of the Cambrian, tend to have very endemic (local) distributions in the early Cambrian. This provincial pattern produces regions with distinctive assemblages of trilobites, making it difficult to use them in correlating among different continents. By the mid-Cambrian, trilobites have broader ranges and become more reliable for long-distance correlations. As described later, a diverse assemblage of shelly fossils appears in the earliest Cambrian, but these fossils also tend to have limited distributions as well as the additional shortcoming of being largely restricted to shallow-water settings. Given that all other proposed boundary horizons had their own serious problems, the Newfoundland boundary with *T. pedum* got the nod (Landing 1994).

The effect of this selection was to make the base of the Cambrian some 20 million years older than it was in the days of Sedgwick and, of course, of Charles Darwin as well. (In the interim, the base had already been lowered by about 7 million years when earlier appearances of durably skeletonized fossils were discovered.) This expansion of the beginning of the Cambrian into a much earlier time shifted a lot of interesting events into the early Cambrian, incorporating a progressive dampening of changes in carbon isotopes (chap. 3), a growing diversity and abundance of trace fossils (chap. 5), and an increase in the variety of small shelly fossils (chap. 6). On the other hand, rather perversely, the great "Cambrian explosion" no longer occurs at the actual base of the Cambrian, but significantly later.

Historically, the Cambrian was subdivided into Early, Middle, and Late epochs. Unlike most latter parts of the timescale, however, there has been little agreement on their subdivisions, partly because of difficulties in finding globally useful fossils as biostratigraphic markers. For several decades, a regional stratigraphy that was developed in Siberia served as the standard, with the Nemakit-Dal'dynian, Tommotian, Atdabanian, Botomian, and Toyonian stages, from oldest to youngest, serving as the paradigmatic subdivisions of the classic Early Cambrian system, but it has never been easy to correlate between the Siberian sections and rocks of similar age elsewhere in the world. Many of the important sections, including those in Siberia, are incomplete at critical horizons. This lack of record has long been apparent to those of us fortunate enough to have visited these localities, but the extent of the discon-

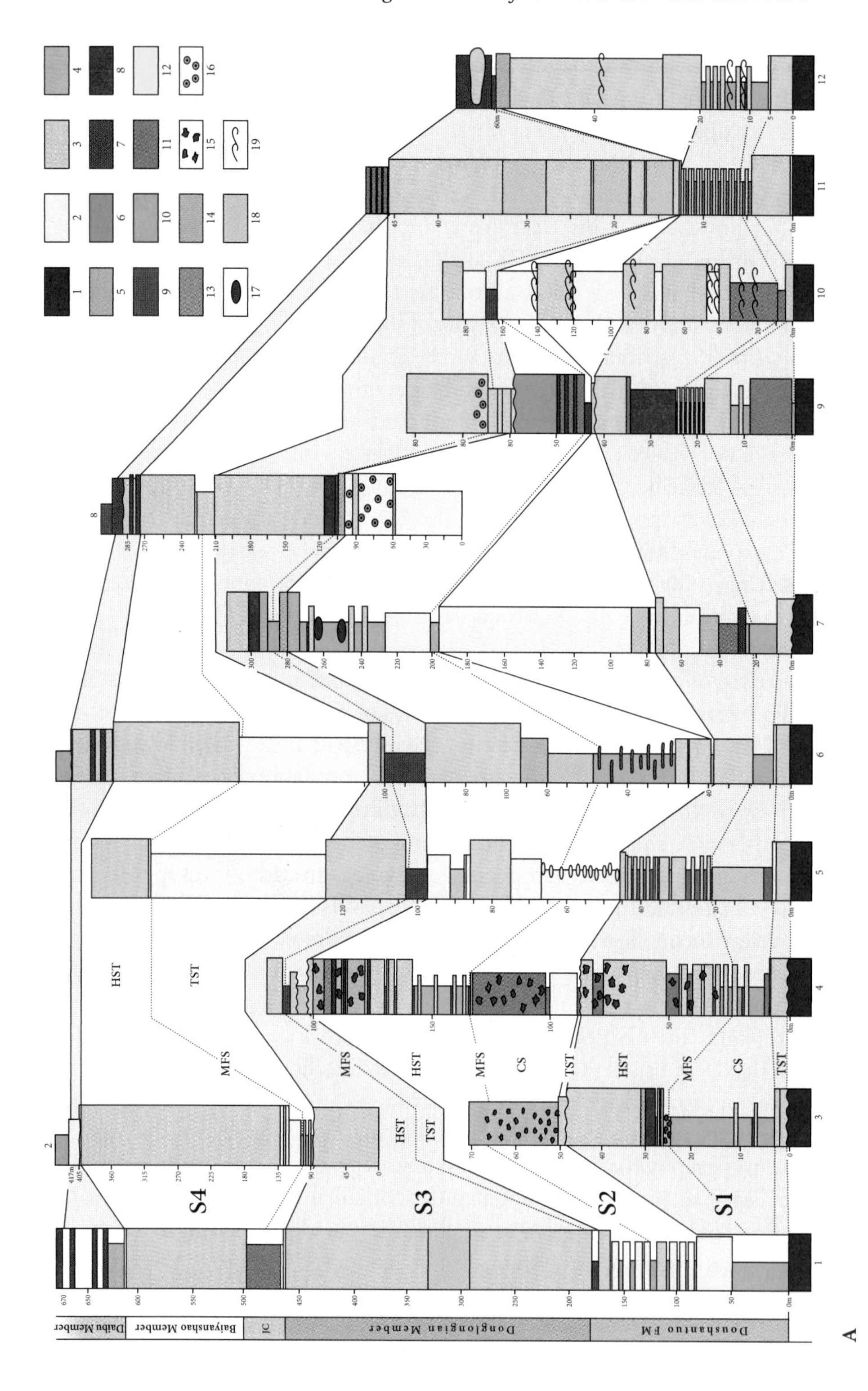
A
Doushantuo FM
Dongliongian Member
JC
Baiyanshao Member
Daibu Member
S1
S2
S3
S4
HST
TST
MFS
CS

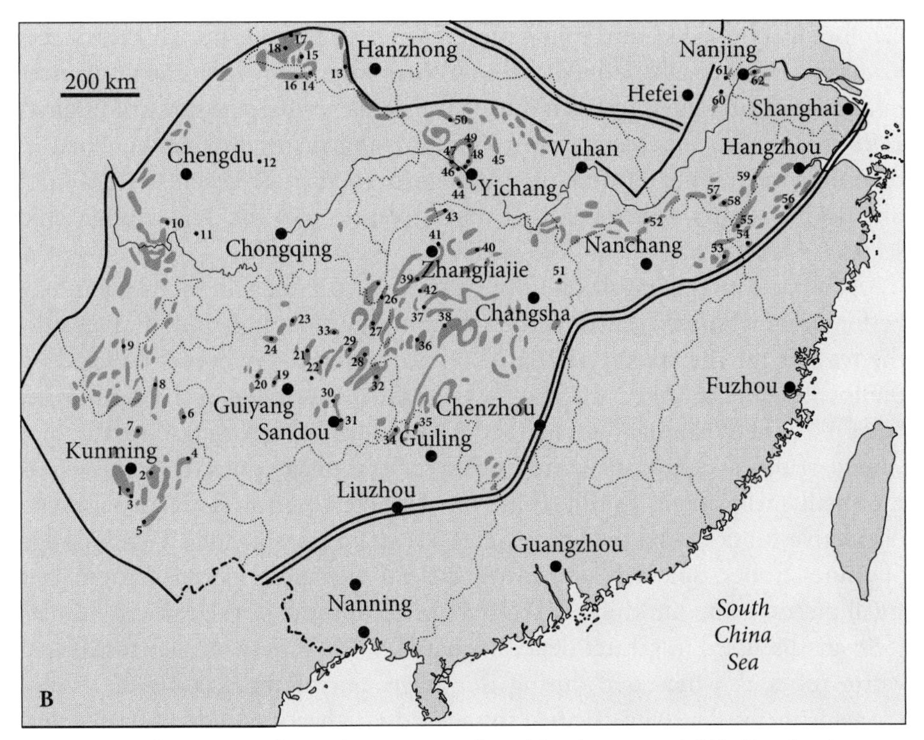

Figure 2.2 The stratigraphic architecture produced by the rise and fall of sea level is a critical tool for correlation. This figure illustrates sequence stratigraphic correlations across south China from localities in Yunnan Province on the left north to near Chengdu.

(A) (left) Outcrops of the Ediacaran sequences and locations of reference sections on the Yangtze Block of south China: (1) Wangjiawan, Jinning County, Yunnan; (2) Dongdahe, Chengjiang County, Yunnan; (3) Xiaoshiqiao, Yuxi City, Yunnan; (4) Wenquan, Qujin City, Yunnan; (5) Dazhai, Jianshui County, Yunnan; (6) Deze, Qujin City, Yunnan; (7) Cuihua, Lucquan County, Yunnan; (8) Dahai, Huize County, Yunnan; (9) Dacaohe, Puge County, Sichuan; (10) Hongchunping, Emeishan, Sichuan; (11) Drilling Core 15, Weiyuan County, Sichuan; (12) Deep Drilling Core at Longnüsi, Wusheng County, Sichaun.

(B) (above) The geographic distribution of localities and key cities across the Yangtze Block. Dark brown shading marks outcrops of Ediacaran and Cambrian rocks. From Zhu, Zhang, and Yang (2007).

formities has become more obvious with the development of detailed carbon isotope records (Zhu, Zhang, and Yang 2007b). In the past few years, a new stratigraphic framework (fig. 2.3) has emerged from the Cambrian Stratigraphic Subcommission of the International Commission on Stratigraphy. This framework represents a consensus of active Cambrian stratigraphers and employs a fourfold rather than the traditional threefold subdivision, with the

Cambrian subdivided into four series and ten stages, but not all these units have yet been formally defined. Because there were many important advances in understanding the Cambrian during the decades when the threefold framework was in use, much important literature and many texts are couched in that old stratigraphic terminology. Therefore, we include it here, with a brief discussion, as an aid in translating that literature into the new stratigraphic scheme.

As difficult as the transition from a threefold to a fourfold subdivision may be for those who are accustomed to the traditional stratigraphic framework, the reasons for the change are based on the evolutionary events revealed by the fossil sequence (fig. 2.4). The rocks of the lowest stage of the Cambrian, Stage 1 or the Fortunian, are marked by a gradually increasing diversity of a variety of tubes, cones, shells, and other skeletal fossils collectively known as the small shelly fossils (SSF). These fossils increase in diversity to form the distinctive biota of the overlying Stage 2, where they include a wide range of spines, tubes, spicules, and other skeletal elements, many of them just small pieces of the multiplated skeletal armor of some of the earliest animals (SSF are discussed in greater detail in chap. 6). The variety of trace fossils and worm tubes also increased during this stage. Furthermore, a rapidly evolving assortment of heavily calcified sponges, the archaeocyathids, appeared and diversified. The base of Stage 2 will probably be formally defined by the first appearance of either an SSF or an archaeocyathid, but that decision has not yet been made. The base of Stage 3 is marked by the first appearance of an early trilobite genus. At about this time there is also an increase in burrowing as larger worms and other invertebrates proliferated, which in turn caused changes in the nature of the seafloor. In addition, a fauna with truly marvelous preservation of soft-bodied forms, the Chengjiang biota of south China, indicates that the Cambrian explosion was well under way (chap. 6). The base of Stage 4 is defined at the first appearance of either of two distinctive trilobites (*Ollenellus* or *Redlichia*). Formal definitions of the boundaries of some of the remaining subdivision are currently under study. The correlation of stages between different regions has been made much easier by the recent discovery of wide swings in the carbon isotope record recorded in rocks of this interval (Brasier and Sukhov 1998; Halverson et al. 2006, 2010; Maloof et al. 2005; Maloof, Porter, et al. 2010; Zhu, Babcock, and Peng 2007). These stages take

Figure 2.3 (right) Stratigraphic divisions of the Ediacaran and Cambrian, showing both the newly established fourfold stages and series of the Cambrian and, for comparison, the more traditional Siberian stages to the right. Although not evident at this scale, none of the traditional Siberian biostratigraphic zones match the new stratigraphic scheme precisely.

Millions of years before present	Eon	Period	Series	Stage		Siberian stages
	Phanerozoic	Cambrian	Furongian Series	Stage 10	Late	
				Jiangshanian		
				Pabian		
500			Series 3	Guzhangian	Middle	
				Drumian		
				Stage 5		
			Series 2	Stage 4	Early	Botomian
520				Stage 3		Atdabanian
			Terrenuvian	Stage 2		Tommotian
540				Fortunian		Nemakit-Daldynian
560	Proterozoic	Ediacaran				
580			Gaskiers Glaciation			
600						
620						
640			Marinoan Glaciation			

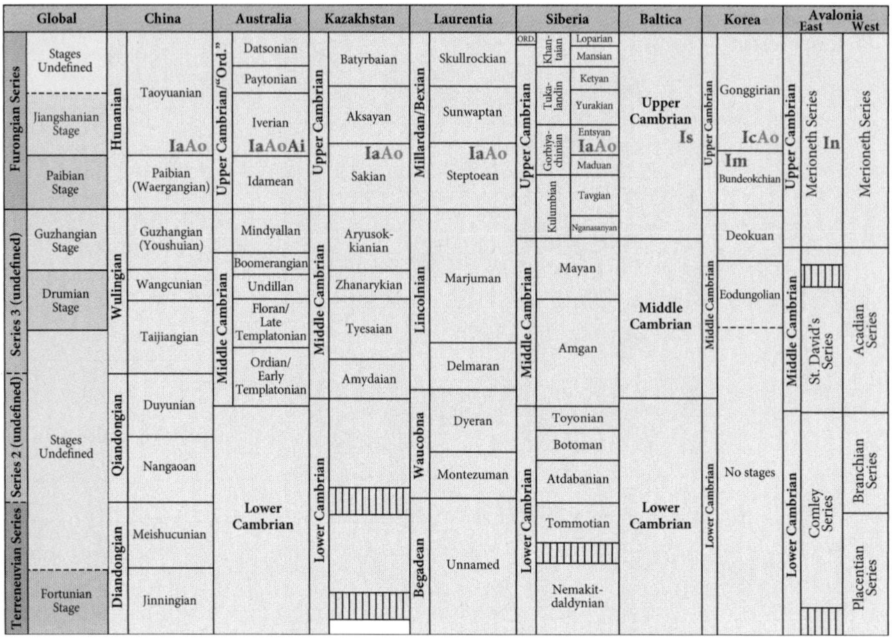

Region	Series / subdivisions (top to bottom)	Stages / units (top to bottom)
Global	Furongian Series; Series 3 (undefined); Series 2 (undefined); Terreneuvian Series	Stages Undefined; Jiangshanian Stage; Paibian Stage; Guzhangian Stage; Drumian Stage; Stages Undefined; Fortunian Stage
China	Hunanian; Wulingian; Qiandongian; Diandongian	Taoyuanian (IaAo); Paibian (Waergangian); Guzhangian (Youshuian); Wangcunian; Taijiangian; Duyunian; Nangaoan; Meishucunian; Jinningian
Australia	Upper Cambrian/“Ord.”; Middle Cambrian; Lower Cambrian	Datsonian; Paytonian; Iverian (IaAoAi); Idamean; Mindyallan; Boomerangian; Undillan; Floran/ Late Templatonian; Ordian/ Early Templatonian; Lower Cambrian
Kazakhstan	Upper Cambrian; Middle Cambrian; Lower Cambrian	Batyrbaian; Aksayan; Sakian (IaAo); Aryusok- kianian; Zhanarykian; Tyesaian; Amydaian
Laurentia	Millardan/Bexian; Lincolnian; Waucobna; Begadean	Skullrockian; Sunwaptan; Steptoean (IaAo); Marjuman; Delmaran; Dyeran; Montezuman; Unnamed
Siberia	ORD.; Upper Cambrian (Khan- taian; Tuka- landin; Gorbiya- chinian; Kulumbian); Middle Cambrian; Lower Cambrian	Loparian; Mansian; Ketyan; Yurakian; Entsyan (IaAo); Maduan; Tavgian; Nganasanyan; Mayan; Amgan; Toyonian; Botoman; Atdabanian; Tommotian; Nemakit- daldynian
Baltica		Upper Cambrian (Is); Middle Cambrian; Lower Cambrian
Korea	Upper Cambrian; Middle Cambrian; Lower Cambrian	Gonggirian (IcAo); Bundeokchian (Im); Deokuan; Eodungolian; No stages
Avalonia East	Upper Cambrian; Middle Cambrian; Lower Cambrian	Merioneth Series (In); St. David’s Series; Comley Series
Avalonia West		Merioneth Series; Acadian Series; Branchian Series; Placentian Series

Figure 2.4 Correlation chart of the Cambrian showing the proposed global chronostratigraphic stages compared with regional usage in major areas of the world (modified from Peng, Babcock, Robison, et al. 2004). In this revised stratigraphic framework, the base of Stage 2 corresponds to neither the base of the Tommotian in Siberia nor the Meishucunian in south China, although it is close to the latter. The Meishucunian of south China is roughly equivalent to the Tommotian, plus possibly the upper part of the Nemakit-Daldynian. The last three stages of the Siberian “Early Cambrian” (the Atdabanian, Botomian, and Toyonian) are roughly equivalent to Series 2 (Stages 3 and 4) in the new zonation. Series 3 and 4 of the new framework correspond roughly to the later part of the traditional Middle Cambrian and Upper Cambrian, respectively (Babcock and Peng 2007; Zhu, Babcock, and Peng 2007). Chart compiled from numerous sources, summarized principally in Geyer and Shergold (2000) and Peng, Babock and Robison et al. (2004).

us into the middle of the Cambrian, where the diversity of trilobites and their broad geographic distribution make correlations much more straightforward.

TIME

Improved stratigraphic correlations using fossils establishes the *relative* ordering of events through the late Neoproterozoic and Cambrian but provides little sense of how much *absolute* time was involved. Darwin took a try at estimating the duration of time since the Cretaceous by calculating erosion rates for some

Early Cretaceous rocks (rather overestimating the elapsed time, incidentally). Other geologists and physicists attempted to establish the absolute duration of geologic events by measuring the cooling rates of iron and extrapolating to the cooling of an originally molten Earth. Development of more reliable measurements became possible with the discovery of radioactivity in 1896. The analysis of rocks that contain radioactive minerals provides a kind of chronometer to gauge when the minerals were formed. Today, such radiometric clocks have provided a framework of fairly high temporal resolution.

One of the most accurate ways to estimate the distribution of time in sequences of fossiliferous rocks is by dating interbedded volcanic ash beds using the decay of radioactive isotopes of uranium, or other minerals. Many ash beds generated from silica-rich volcanic eruptions contain microscopic (100–300 microns) crystals of the mineral zircon. When zircon crystals (fig. 2.5) form in a magma chamber of molten rock, the lattice of atoms in the mineral incorporates a small amount of uranium, but virtually no lead. Uranium has several different isotopes (recall that isotopes of an element have the same number of protons in the nucleus but differ in the number of neutrons). Uranium-238 and uranium-235 release α and β particles through nuclear fission, decaying to form lead-206 and lead-207, respectively. Geochronologists thus describe each of these uranium isotopes as a parent and the appropriate lead isotope as a daughter. Each radioactive system has a characteristic rate of decay that can be measured in the laboratory and that is expressed as a half-life: the length of time required for half of the parent isotope to decay to the daughter product. So, knowing the half-life and the amount of parent and daughter present today, it is relatively simple to calculate the time at which minerals hosting the radioactive elements were formed. One of the great advantages of the uranium-lead (U-Pb) decay system is that the two different decay series provide an internal check on the results from a single crystal because they should both give the same age. By precisely determining the amount of these uranium and lead isotopic decay series in a crystal and measuring many dozens of crystals, geochronologists can now determine the date of the volcanic eruption that produced the ash bed to an accuracy of 0.01%. Ash produced during large volcanic eruptions, particularly when the ash cloud is high enough to enter the stratosphere, can be distributed across an entire continent and sometimes even more broadly. As the resulting ash beds become compacted into rock, they can form lithologically distinctive, widely distributed beds and have been used by stratigraphers as marker beds for more than a century. With the advent of radiometric dating, these ash beds now provide a reliable basis for age dating. Our understanding of the distribution of events through the Ediacaran and Cambrian has been revolutionized by the

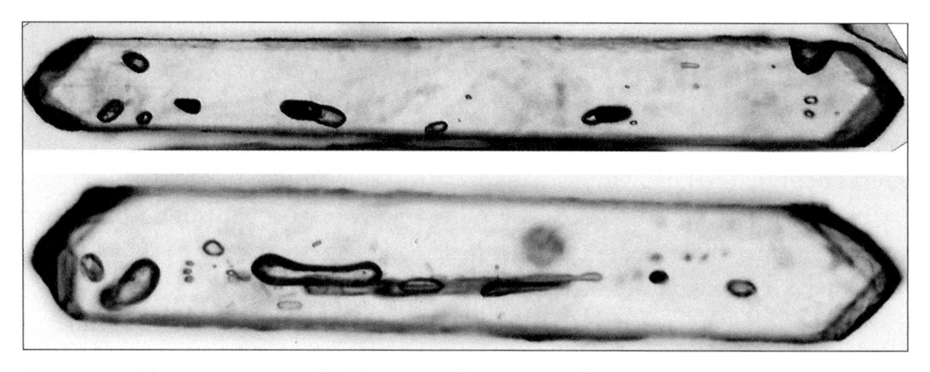

Figure 2.5 Two zircon crystals. Photograph courtesy of Sam Bowring, MIT.

introduction of these high-resolution radiometric dating techniques (Condon and Bowring 2011).

Volcanic rocks are not as common in the rock record as we would like, however. Because they are almost never present where they would be most useful, the ages generated by geochronologists generally have to be integrated with data from fossils (through biostratigraphy) and stable isotopes, such as carbon and strontium (through chemostratigaphy; see chap. 3), to produce a unified stratigraphic framework. Despite these challenges, studies of the ages of volcanic ash beds found in various localities around the world have produced radiometric dates of unprecedented accuracy, leading to a well-resolved timescale for the Ediacaran and Cambrian (fig. 2.6). We now have a refined temporal framework and thus a far better understanding of the ages of Ediacaran and Cambrian events and boundaries than was true just a few years ago (fig. 2.7). The age of the end of the Marinoan glaciation has been dated to about 635 Ma in both Namibia and south China; therefore, we assume that it represents a global deglaciation (Condon et al. 2005). Marinoan glacial deposits in Australia provided the lower boundary of the Ediacaran period, as shown in figure 2.1. It is one of several glacial periods through the late Neoproterozoic, including a later event known as the Gaskiers glaciation that is dated at 584–582 Ma (Bowring et al. 2003). In the 1980s, Ediacaran fossils were believed to only occur in rocks tens of millions of years older than the Cambrian, but Bowring's radiometric dates showed that in Namibia the Ediacara fauna persisted until the very base of the Cambrian. The age of the oldest clearly demonstrable bilaterian fossils was established by dating an ash

Figure 2.6 (right) Radiometric dates of the Ediacaran and Cambrian. Notice the absence of radiometric dates from the Ediacaran between 579 and 555 Ma. Superscripts refer to references: (1) Condon et al. 2005; (2) Amthor et al. 2003; (3) Martin et al. 2000; (4) Grotzinger et al. 1995; (5) Landing et al. 2000; (6) Maloof, Porter, et al. 2010; (7) Bowring et al. 1993; (8) Landing et al. 1998.

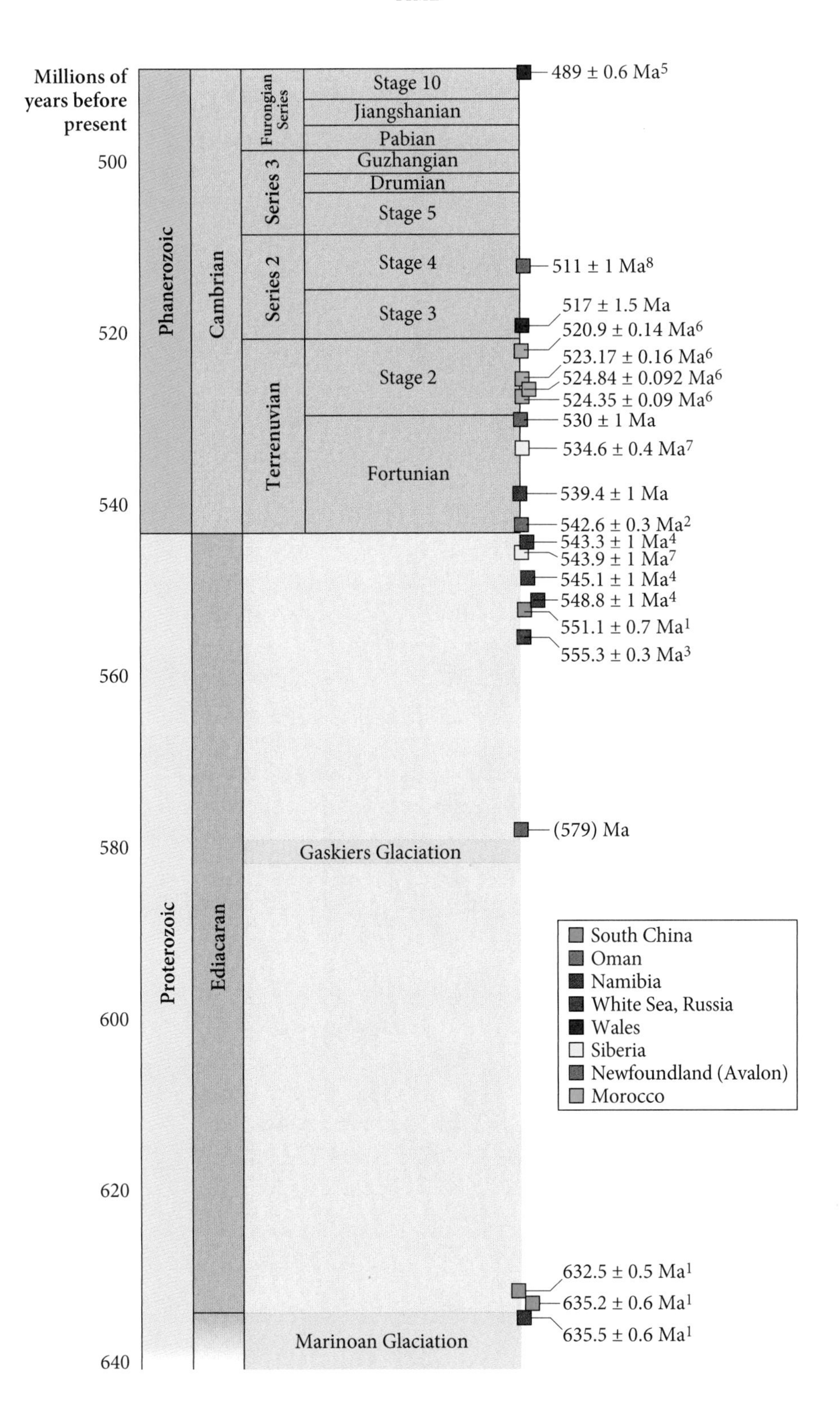
Millions of years before present
500
520
540
560
580
600
620
640
Phanerozoic
Proterozoic
Cambrian
Ediacaran
Furongian Series
Series 3
Series 2
Terrenuvian
Stage 10
Jiangshanian
Pabian
Guzhangian
Drumian
Stage 5
Stage 4
Stage 3
Stage 2
Fortunian
Gaskiers Glaciation
Marinoan Glaciation
489 ± 0.6 Ma5
511 ± 1 Ma8
517 ± 1.5 Ma
520.9 ± 0.14 Ma6
523.17 ± 0.16 Ma6
524.84 ± 0.092 Ma6
524.35 ± 0.09 Ma6
530 ± 1 Ma
534.6 ± 0.4 Ma7
539.4 ± 1 Ma
542.6 ± 0.3 Ma2
543.3 ± 1 Ma4
543.9 ± 1 Ma7
545.1 ± 1 Ma4
548.8 ± 1 Ma4
551.1 ± 0.7 Ma1
555.3 ± 0.3 Ma3
(579) Ma
632.5 ± 0.5 Ma1
635.2 ± 0.6 Ma1
635.5 ± 0.6 Ma1
South China
Oman
Namibia
White Sea, Russia
Wales
Siberia
Newfoundland (Avalon)
Morocco

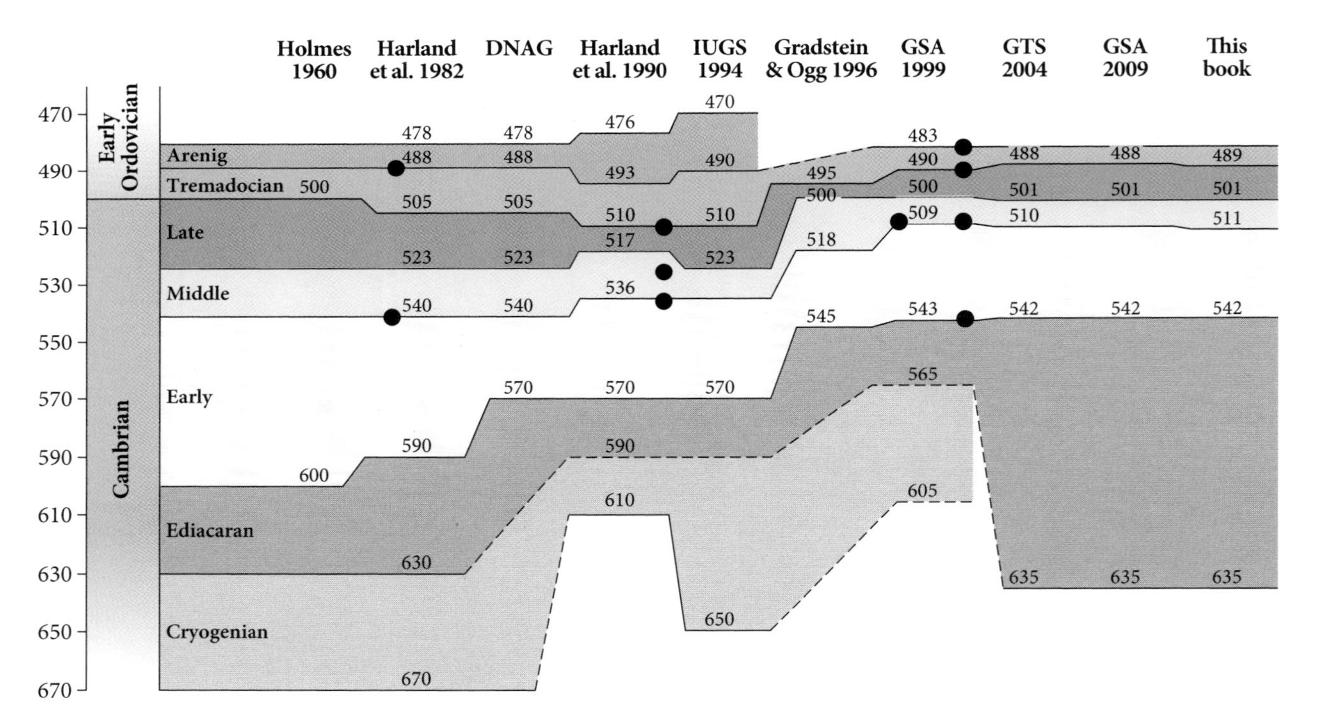

Figure 2.7 Estimates for the age of boundaries through the Ediacaran and Cambrian have changed considerably over time as a consequence of better definition of the boundaries and higher-resolution radiometric methods. In 1937, Arthur Holmes used radiometric data to place the base of the Cambrian at 470 Ma, subsequently extending it to 600 Ma in 1960. By 1992, the best estimates for the base of the Cambrian were that it was younger than about 560 Ma and older than about 525 Ma (Bowring et al. 1993). Most of the boundaries have now been formally defined by stratigraphers, and, as shown in figure 2.3, there is an increasing wealth of radiometric dates. Error estimates have been left off for clarity but have decreased from 10 million years to 20 million years for Holmes (1960) to a few hundred thousand years today. Abbreviations: DNAG: Decade of North American Geology: Palmer (1983); GSA 1999: Geological Society of America (1999); GSA 2009: Walker and J. W. Geissman (2009); Geological Time Scale (GTS) 2004: Gradstein et al. 2004; IUGS: Salvador (1994).

bed along the White Sea in northern Russia at 555 Ma (Martin et al. 2000). Samples from rock sequences in Siberia established the base of the Tommotian Stage to be about 524 Ma (Maloof, Porter, et al. 2010) and the base of the Botomian Stage, at the end of the early Cambrian, to be about 517 Ma. Ash beds interbedded with diagnostic fossils helped establish that the duration of the Cambrian is far shorter than previously thought and established a temporal framework in which to understand the duration of the biological events (Amthor et al. 2003; Bowring et al. 1993; Condon and Bowring 2011; Grotzinger et al. 1995; Isachsen et al. 1994). These radiometric results also constrained the pattern of change in carbon isotopes.

There are still a number of outstanding geochronological problems that have not been resolved. For example, there is a long span of time through the mid-Ediacaran (577–555 Ma) with only a single radiometric date, and, as we will see in the next chapter, this time span includes the Shuram carbon isotope excursion, a critical event in the history of the carbon cycle. Did this isotopic anomaly last for 1 million years or 10 million years? We do not know, but the difference matters a great deal for understanding the oxygenation of the oceans. In addition, we have no dates on the ages of the Ediacaran fossils in Australia, and there is a continuing dispute over the age of the Doushantuo fossils (chap. 5) in south China. The durations of some of the glacial episodes also need sorting out, and, as the error estimates of various dates are refined, we will be better able to understand the rates at which many early evolutionary events occurred.

OTHER CORRELATION TECHNIQUES

Geologists have employed still other methods of correlation in studying Ediacaran and Cambrian rocks. One of the most useful methods exploits the episodic reversals in Earth's magnetic polarity. Volcanic rocks (and some other deposits) record the magnetic pole sign and direction when they form: as the rock cools, magnetic minerals will align with the magnetic field, preserving a record of the magnetic pole sign and direction of that time. As long as the rocks are not heated too much during later burial or by subsequent magmatic events, they will preserve that record indefinitely. The polarity of Earth's magnetic field reverses episodically, and geophysicists can recover this reversal history by measuring the polarity of the rocks. Magnetic reversals are sufficiently common during most of the past 600 million years that they provide a sort of bar code: match the sequence of reversals in one region to sequences of rocks from another, and the result is a potential correlation. In practice, the magnetic reversal record rarely provides unambiguous correlations, so confirmation through fossil biostratigraphy, radiometric geochronology, or shifts in carbon isotopes are required as well. The sequence of reversals of Earth's magnetic field during the Ediacaran and Cambrian has been intensively investigated, and the resulting information has aided greatly in intercontinental correlations. Fluctuations in chemical signals, particularly changes in carbon isotopes, provide a useful chemostratigraphic tool. Such studies reveal far more about geochemical cycles in general, so we reserve their discussion for the next chapter.

The development of this high-resolution geologic timescale and the correlation of geologic sections from every continent through biostratigraphy, chemostratigraphy, and magnetostratigraphy have been essential to any detailed understanding of rates of geological and evolutionary change. With the framework in hand, we can now calculate such rates and, for example, determine how closely various evolutionary events correlate with changes in the physical environment that may have been causally related.

CHAPTER THREE

The Environmental Context of the Ediacaran and Early Cambrian

Some of the most dramatic environmental changes of the past 2 billion years occurred during late Neoproterozoic and early Cambrian times. The late Neoproterozoic world was, quite literally, a different Earth: atmospheric oxygen levels were much lower than today, and the deep oceans were oxygen poor but iron rich, perhaps similar to the noxious brew deep in the present Black Sea. During the Ediacaran, carbon and sulfur isotopes and other geochemical proxies provide evidence of growing oxygenation of the oceans. Geologists continue to debate the extent of this oxygenation and whether it reflects purely geological processes or was facilitated by biological innovations. These geochemical changes overlap with a tectonic reorganization growing out of the breakup of a supercontinent known as Rodinia about 750 Ma. Several glaciations were so extensive and were accompanied by such puzzling geological features—in particular, the formation of tropical glaciers at sea level—that they continue to challenge the ingenuity of geologists. By the early Cambrian, however, those changes had generated an environmental revolution: the oceans had become oxygenated to a depth of at least several hundred meters, ferrugious (iron-rich) ocean waters had retreated to become only regional or local phenomena, and the carbon cycle, which had been characterized by large fluctuations, began to settle into a pattern more similar to that of the Phanerozoic.

Much of the recent focus in a search for environmental triggers for the Cambrian explosion has been on increased oxygen levels (Knoll 2003b). The energy released from oxygen bonds provides, through aerobic respiration, about an order of magnitude more energy for the same amount of food than does anaerobic metabolism. Consequently, oxygen is critical to the origin and diversification of metabolically active, complex organisms (see Raymond and Segre 2006). Oxygen concentrations also limit the maximum size of organisms without active circulatory systems. Small organisms distribute oxygen via diffusion, with about 0.1% of the present atmospheric level of oxygen (21%) being required for millimeter-sized organisms. For centimeter-sized animals with a closed circulatory system, levels of 0.1 to 1% of the present atmospheric level are required (Berkner and Marshall 1964, 1965; Catling et al. 2005). Of course, there is no guarantee that any such animals will actually evolve simply because an appropriate threshold oxygen level has been reached. Nevertheless, the relationship among the energy available from oxygenic respiration, the oxygen requirements implied by the large size and mobility of early metazoans, and the suggestion, conveyed by geochemical proxies, of an increase in oxygen levels in the sea during the late Neoproterozoic has long fueled speculations that increased oxygenation drove the Cambrian explosion.

Determining causality can be very difficult in events that happened more than half a billion years ago. Correlation of events in time is not evidence of a causal connection, and we must carefully distinguish factors that may have triggered an event from those that merely allowed or facilitated a biological response. Distinguishing environmental from biological drivers is particularly challenging in the case of the changes in ocean chemistry during the Ediacaran and Cambrian. For years, geologists assumed that the biological changes were a response to geological and chemical changes in the physical environment. Indeed, major research efforts were focused on the extent to which specific environmental changes facilitated an evolutionary response. Recently, this question has been inverted, and it now appears possible that many of the geochemical changes may have been driven by the ecological and environmental effects of animals.

We begin this chapter with the environmental context of the late Neoproterozoic and Cambrian, including the configuration of the continents and the effects of plate tectonics; the pervasive climatic and environmental changes during the Neoproterozoic that were triggered by widespread glaciations; and the dynamics of carbon and sulfur isotopes, which serve as critical environmental proxies. This foundation leads to a discussion of the implications of the geochemical and other environmental data in understanding possible increases in atmospheric oxygen and their effects on the diversification of animals. A critical issue is whether oxygen levels in the ocean increased significantly during the Ediacaran. If so, was the increase primarily caused by geological processes or by biological events?

A note of caution is necessary. This subject is an active and indeed highly contentious area of research, particularly with respect to the history and causes of oxygenation of marine waters. The various geochemical proxies are not currently providing an unambiguous picture of events during the Ediacaran and Cambrian. At the end of this chapter, we provide our own views of what we believe was occurring during this time, but we must warn the reader that many of our colleagues may disagree with some of our perspectives. As with any area of active scientific inquiry, the models used to explain the available data evolve as new data become available, as do the opinions of those performing the research.

CONTINENTAL CONTEXT DURING THE EDIACARAN AND CAMBRIAN

The distribution of the continents across the globe influences patterns of chemical weathering of the continents, atmospheric and oceanic circulation, and planetary albedo (the reflectivity of Earth) and thus places first-order boundary conditions on the range of climatic conditions. At the advent of the theory of plate tectonics, Valentine and Moores (1970) suggested that a breakup of a Neoproterozoic supercontinent formed shallow, nutrient-rich seas and facilitated the diversification of animals. There is now much greater evidence for such a supercontinent, Rodinia (Russian for "motherland"), during the Neoproterozoic.

Continental reconstructions require the integration of a wide range of often conflicting information. Plant and animal fossils can provide crucial clues not only to the age of the rocks but also to the biogeography of the time: similar species on continents that are now widely separated are likely to reflect easier dispersal and therefore greater proximity between the two regions. Such fossil evidence is poor and of little utility during the Ediacaran, however, and as noted in chapter 2, even in the early Cambrian most trilobite lineages seem to have had quite restricted distributions. Today, the primary tool for continental reconstructions is the record of the magnetic field recovered from volcanic rocks, discussed in chapter 2, from which it is possible to determine the latitude and the direction of the magnetic pole when the rocks were originally formed. Magnetic minerals in volcanic rocks erupting in Europe and North America would point to the current magnetic pole today, but because the direction to the pole is different in each continent, they would point in different directions. When geologists track the paleomagnetic record of each continent back through the past 100 million years, they find that the movements of those two continents have been independent. In contrast, some 230 Ma, Europe and North America were connected, before the Atlantic Ocean began to form, and volcanic rocks in each area show the same motion for each continent. The relative movements of these ancient pole positions through time allow geologists to reconstruct the differential movement of continents; at times, continental blocks broke into smaller blocks, and at other times, they collided and united. Determining the past positions of the continents and their movements becomes increasingly difficult proceeding further into the past, and considerable uncertainty remains about reconstructions of continental positions during the Neoproterozoic and Cambrian.

On balance, the geologic evidence favors the formation of the supercontinent of Rodinia before 1 Ga, followed by breakup, beginning by the time of the Sturtian glaciation (717 Ma), into an array of smaller continents by Ediacaran time (Halverson et al. 2009). Africa, South America, and Asia eventually collided to form the large and long-lived supercontinent of Gondwana in the late Neoproterozoic (fig. 3.1) (Z. Li et al. 2008; McKerrow, Scotese, and Brasier 1992; Meert and Lieberman 2004, 2008; Piper 2007; Powell and Pisarevsky 2002). Most evidence indicates that the continents were largely in low latitudes during the interval of Neoproterozoic glaciations, from about 850 to 635 Ma and earlier, which is called the Cryogenian. As continents have a higher albedo (reflectivity) than the open ocean, with more continents in low latitudes, the Earth's albedo would increase, producing a reasonably moderate climate. Tropical conditions increase the weathering of continental rocks (granites and the like), lowering atmospheric carbon dioxide levels through a negative feedback process that controls long-term climate patterns on Earth (fig. 3.2).

Continental configurations during the Neoproterozoic indicate that large continental areas were exposed to subaerial weathering and that extensive sedimentary basins formed along the margins of many continents, allowing the deposition of sediments rich in organic carbon. The combination of increased burial of organic carbon and increased continental weathering would have reduced atmospheric carbon dioxide (CO_2) and set the stage for climatic cooling and, eventually, the onset of glaciation (Brasier and Lindsay 2001; Halverson et al. 2009; Meert and Lieberman 2008; Squire et al. 2006).

NEOPROTEROZOIC GLACIATIONS

Three widespread Neoproterozoic glacial epochs are well documented by glacial debris such as tillites (fine-grained glacially derived sediment) and dropstones (isolated boulders entrained in glaciers that are eventually deposited as glacial sediments when the ice melts). These glacial intervals are the Sturtian glaciations between 717 and 665 Ma, the Marinoan glaciation between 645 and 635 Ma, and the short-lived Gaskiers glaciation between 583 and 584 Ma.[1] Geologists have long puzzled over these deposits because many glacial

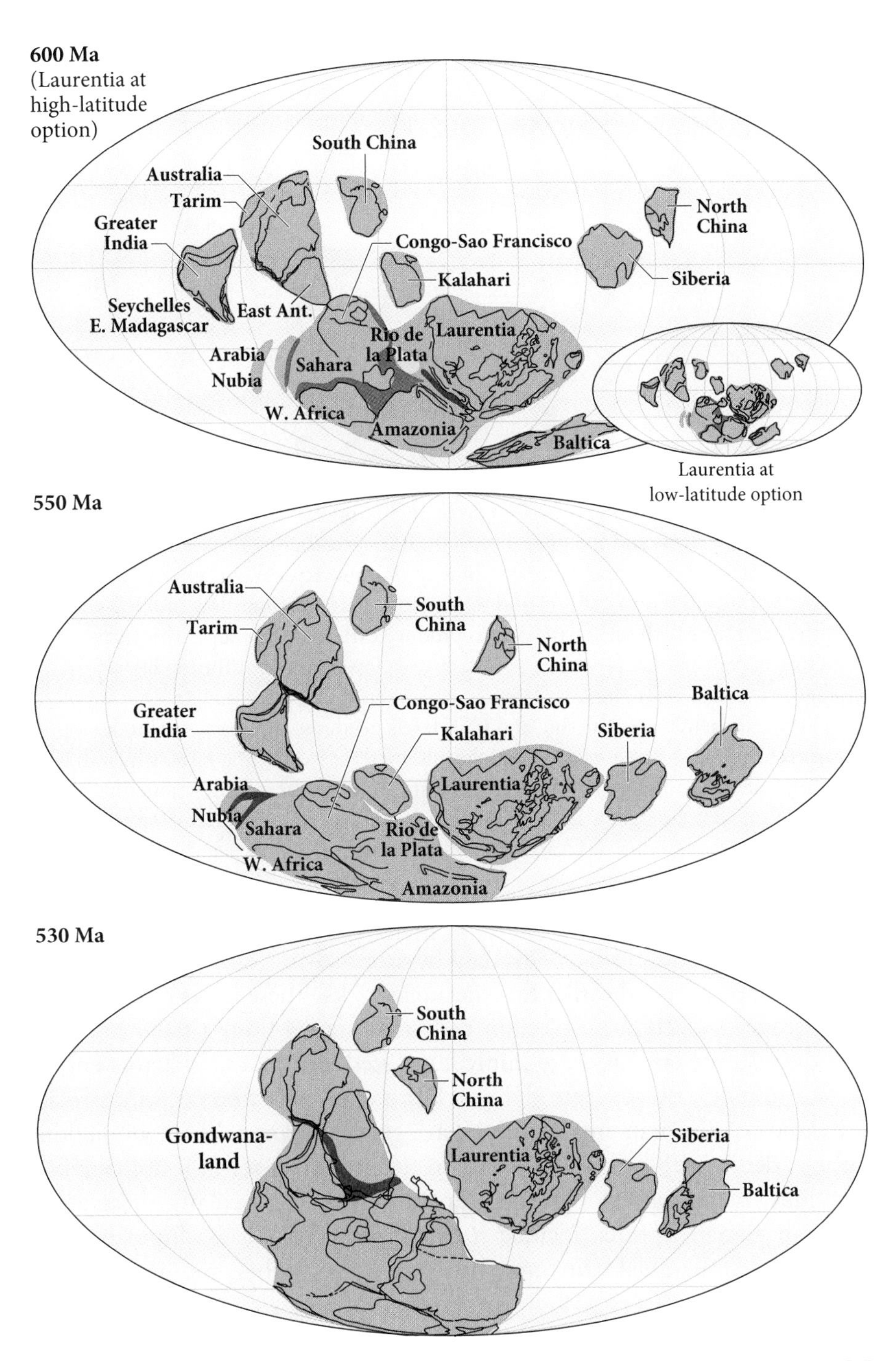

Figure 3.1 Paleocontinental reconstructions from 600, 550, and 530 Ma. From Z. X. Li et al. (2008).

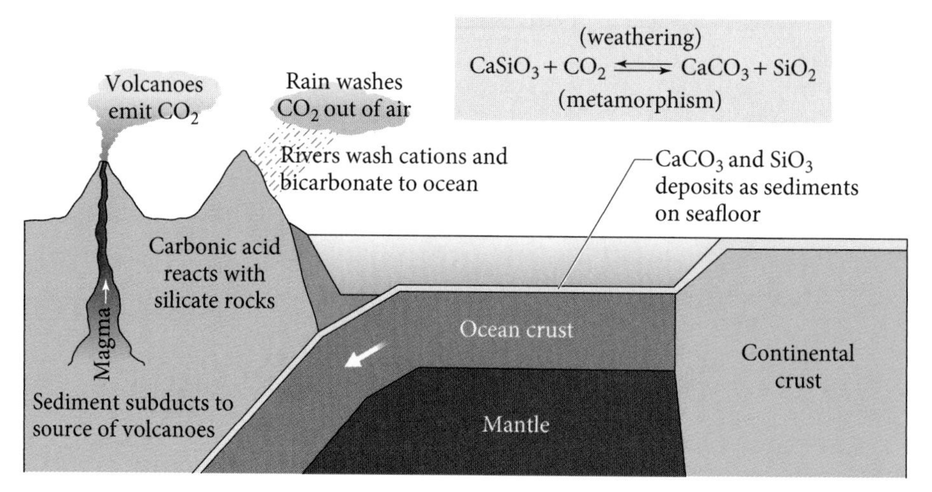

Figure 3.2 Schematic drawing of carbon cycle, including the size of the major organic and inorganic reservoirs and the importance of silicate weathering for the carbon cycle. In the simplest sense, CO_2 and water combine to produce acids that chemically weather silicate rocks into silica plus calcium and magnesium ions. The ions flow into the sea, where they combine with carbonate ions to produce carbonate minerals. This silicate weathering of continental minerals is more active in lower latitudes, during higher temperatures and therefore high CO_2 levels, and when topography is high. This interaction between chemical weathering and CO_2 levels produces a negative feedback loop: high atmospheric CO_2 levels warm the globe and increase weathering rates, which draws down atmospheric CO_2, lowering the temperature and thus decreasing weathering rates. If CO_2 levels fall far enough, glaciers begin to form. The decline in weathering rate, however, allows the buildup of CO_2, eventually warming the climate and inducing a further round of weathering.

tillites (fig 3.3) are immediately overlain by extremely unusual rocks known as cap carbonates, as noted earlier for Marinoan rocks. These layers of limestone or dolostone (a calcium magnesium carbonate) range from a few meters to more than 30 meters thick, are thinly laminated, can be found across entire basins, and may even correlate globally. Distinctive cap carbonates are associated with both Sturtian and Marinoan-age glacial deposits. The conjunction of glacial debris with carbonates, which generally reflect warm if not tropical conditions, is unusual enough, but cap carbonates, particularly those atop Marinoan glacial deposits, include such puzzling features as abundant seafloor cements, microbial structures with long vertical tubes, and unusual isotopic compositions (Corsetti and Grotzinger 2005; Fraiser and Corsetti 2003; Grotzinger and Knoll 1995; Hoffman et al. 1998; Hoffman and Schrag 2002; James, Narbonne, and Kyser 2001; Shields 2005; D. M. Williams, Kasting, and Frakes 1998). These features are unusual in carbonates of any age. For example, some cap carbonates have long crystals of what was originally aragonite. The formation and preservation of these fragile and unstable aragonite

Figure 3.3 Photograph of glacial tillites from the 635 million year old Marinoan glaciation, southern Namibia. Photograph by Doug Erwin.

crystals requires diffusion-limited growth and bottom waters saturated in carbonate. These deposits must have formed rapidly during flooding from the melting of the ice sheets. Equally perplexing is that glacial deposits in Australia were formed in low latitudes; indeed, paleomagnetic evidence suggests that many of the glacial deposits may have formed below 35° latitude (Evans 2000; Kilner, Niocaill, and Brasier 2005; Sohl, Christie-Blick, and Kent 1999).

To produce cap carbonates, the oceans must have been highly alkaline and supersaturated in carbonate. There are three leading hypotheses for such unusual geochemical conditions: (1) stratification of the oceans and sequestration of ^{12}C-enriched dissolved inorganic carbon in deep waters during glaciation followed by upwelling and overturn of high-alkalinity waters from the deep sea (Grotzinger and Knoll 1995); (2) destabilization of methane-enriched permafrost and submarine methane-rich gas hydrates (Jiang, Kennedy, and Christie-Blick 2003; Kennedy, Mrofka, and von der Borch 2008; Kennedy, Christie-Blick, and Sohl 2001); and (3) the deglaciation of continental ice postulated to be so widespread that it has been termed a Snowball Earth (Harland 2007; Hoffman et al. 1998; Hoffman and Schrag 2002; Schrag et al. 2002). Although hypothesis 1 helps explain the cap carbonates, it does not help us understand why they formed at low latitude. The methane scenario, hypothesis 2, accounts for the very light carbon isotope values in some cap carbonates but is contradicted by sedimentologic evidence from the cap carbonates. More significantly, the volume of some cap carbonates would seem to require a much larger volume of carbon than would be available from gas hydrates, although some methane release may have been involved (as noted to DHE by P. Hoffman in 2001).

Snowball Earth

Evidence that the glaciations may have been global, stretching to the equator and perhaps lasting 10 million years, spurred Hoffman, Schrag, and colleagues to invoke the Snowball Earth scenario (hypothesis 3), which was originally proposed to explain the Marinoan glaciation (Kirshvink 1992). In this hypothesis, an extreme drawdown of carbon dioxide cooled Earth and plunged it into a positive ice-albedo feedback loop where continental ice sheets eventually extended equatorward of 35°, rapidly causing glaciers to spread through tropical regions. The latitude of 35° appears to be a climatic tipping point. Climate models suggest that once continental ice sheets reach that point, the entire Earth rapidly becomes covered with ice. In this scenario, the concentration of continental masses in low latitudes increased the land-area feedback as growing ice sheets reduced the sink for carbon dioxide via chemical weathering and increased the planetary albedo. Thus, these low-latitude continents may have been a prerequisite for Snowball Earth initiation. In such a Snowball Earth, ice would eventually cover the oceans, to the obvious detriment of photosynthetic organisms. The thickness of such an ice cover, however, has been a major source of debate.

Although many alternative models have been proposed to account for the depth of these glaciations, we will mention only one here because it is relevant for subsequent discussions. Tziperman et al. (2011) have suggested that the trigger for glaciation was an increase in the export of organic material from surface to deep oceans by an expansion of marine eukaryotes, followed by anaerobic remineralization of the organic matter by sulfate- or iron-reducing bacteria. One of the intriguing aspects of this model is the invocation of biologically driven environmental change, a topic to which we will return in the context of other events.

Once established, release from Snowball Earth conditions depends on the buildup of atmospheric carbon dioxide generated by volcanic eruptions. Carbon dioxide would not be removed from the atmosphere because chemical weathering would be much reduced by the blanket of ice on the continents, so it would eventually increase to extreme greenhouse levels (Higgins and Schrag 2003; Pierrehumbert, Abbott, and Voigt et al. 2011). Release of another greenhouse gas, methane, from gas hydrate reservoirs has also been suggested as a means of freeing the planet from Snowball Earth conditions (Kennedy, Christie-Blick, and Sohl 2001).[2] Any such buildup of carbon dioxide would likely require millions of years, given estimated rates of volcanic activity, so a Snowball Earth could be a prolonged affair, even by geological standards. (A recent test for carbon dioxide levels during cap carbonate deposition suggests, in contrast to this model, that levels may not have been nearly as high as predicted: see Sansjofre et al. 2011). The intense greenhouse of the deglaciation phase is hypothesized to cause incredibly high silicate and carbonate weathering, generating the alkalinity required to form cap carbonates. The rapid release of sediments during the deglaciation phase could have increased the rate of burial of organic carbon, as well as the supply of nutrients to the oceans. Analysis of the ratio of phosphorus to iron in iron oxide–rich sediments indicates a high phosphate input to the oceans after the Sturtian glaciation, consistent with this model; similar data do not yet exist from rocks postdating the Marinoan glaciation (Planavsky et al. 2010).

The Snowball Earth hypothesis envisions an Earth very different from the experience of geologists and, not surprisingly, has proven to be highly controversial within the geological community. (The press, forever in search of the spectacular, seems to think it is great.) This book is not the venue for an extended discussion of the hypothesis or the many criticisms and alternatives, however. Given the highly unusual nature of the cap carbonates and the evidence of continental ice sheets near sea level in low latitudes, any explanation for them is almost necessarily unusual, if not downright weird, simply as a consequence of the very different, nonuniformitarian behavior of Earth's geochemical and climate systems at that time. The Snowball Earth hypothesis does serve, however, as a useful end member for the degree of environmental perturbation possible immediately before the diversification of animals.

In his first Snowball Earth paper, Hoffman raised the question of whether the Snowball Earth and its aftereffects were related to the metazoan diversification. All major eukaryotic clades, including a variety of multicellular algae and nonalgal unicellular eukaryotes, survived the Sturtian and Marinoan glaciations. Evidence that animals had evolved prior to the Marinoan glaciation is indicated by pre-Marinoan sponge body fossils (Maloof, Rose, et al. 2010) and sponge biomarkers (Love et al. 2009) that place boundary conditions on the extent of this glaciation. Thus, either the Snowball Earth hypothesis is too extreme a view of events or sufficient refugia existed to allow these various clades to persist during the widespread reductions in photosynthesis sea ice imposed (Runnegar 2000).

As the glacial phase ended, surviving lineages would have faced an unforgiving environment, including rapidly increasing temperature and changes in ocean water chemistry. One imagines that surviving multicellular lineages would be confined to small, isolated refugia where the environmental conditions would have been less harsh. When normal conditions returned, surviving lineages could diversify rapidly, as in many other post–mass extinction recoveries. An assemblage of stromatolites (finely layered sedimentary structures produced by the trapping and binding activities of microbes) and prokaryotic and eukaryotic microfossils found just above Sturtian-age glacial deposits in Death Valley reveals little loss of diversity or ecological complexity as a result of that glaciation (Corsetti, Awramik, and Pierce 2003).

The Marinoan event does, however, correspond to a fundamental shift in both the composition and the evolutionary dynamics of Proterozoic acritarchs (Butterfield 2004; Peterson and Butterfield 2005). Prior to the Marinoan glaciation, acritarchs exhibited low diversity and extremely long species durations. In the aftermath of the glaciation, an entirely new suite of acritarchs appeared, the Doushantuo-Pertatataka microbiota, discussed further in chapter 5. Here it is sufficient to note the novel appearance of these large, shallow-water and probably benthic single-celled organisms. The diversification of the new acritarch biota occurs after the Marinoan glaciation but is also roughly coincident with a carbon isotope anomaly (discussed below) and may not be directly related to deglaciation.

Ediacaran-Cambrian Climate

Independent of the validity of the Snowball Earth interpretation of the Neoproterozoic glaciations, the end of the Marinoan glaciation ushered in the Ediacaran period during a supergreenhouse interval, with high atmospheric CO_2 levels and increased chemical weathering (Kasemann et al. 2010), which, of course, eventually reduced the levels of CO_2. Glacial conditions returned at least one more time during the Ediacaran, with the relatively short-lived Gaskiers glaciation about 580 Ma, but this event appears to have been largely regional (Halverson et al. 2002, 2006). There is no evidence for glacial or cold climates after the Gaskiers glaciation and into the Cambrian, and geochemical proxies suggest very high atmospheric carbon dioxide levels may have persisted throughout much of the Cambrian (Halverson et al. 2009). High temperatures reduce the ability of the oceans to absorb oxygen, which may be relevant to the oxygenation of the oceans, as discussed later.

GEOCHEMICAL SIGNATURES OF ENVIRONMENTAL CHANGE

Studies of geochemical proxies for the redox state of the oceans and atmosphere during the Neoproterozoic and Ediacaran have demonstrated great changes in a variety of elemental cycles. Since the 1980s, carbon and sulfur isotopes, because they are sensitive to fluctuations in the oxidizing power of the oceans and atmosphere, have provided the bulk of the evidence of such redox changes. In the past few years, they have been joined by studies of other systems, including strontium, molybdenum, and iron. By chronicling these patterns through the Neoproterozoic and Cambrian, geochemists can identify not only the long-term history of the redox state, but shorter-term excursions as well. The difficulty, however, is that these geochemical proxies generally record relative changes on a global scale and do not provide evidence of the absolute amounts of oxygen available for animals in shallow marine waters.

Carbon Isotopes

The Cryogenian and Ediacaran experienced the largest fluctuations in carbon isotopes yet recorded on Earth (Grotzinger, Fike, and Fischer 2011). Carbon exists in two great reservoirs. One is a massive inorganic reservoir that includes everything from the deep carbon in the mantle and the limestone in a reef to the carbon dioxide in the atmosphere; the other is a smaller reservoir of organic carbon, including living organisms as well as their organic remains such as coal and oil. The carbon cycle transfers carbon between these two reservoirs through the oceans and atmosphere via the weathering of silicate minerals, which absorb carbon dioxide from the atmosphere; the release of carbon dioxide in volcanic eruptions and other plate tectonic processes; and the formation and burial of organic debris. Buried organic material is reintroduced into the cycle in the short term by the exposure of organic materials through weathering and erosion; over the longer term, it is reintroduced by the weathering of buried organic material and the release of methane buried on the continental shelves (fig. 3.4).

INORGANIC CARBON ISOTOPES Carbon isotopes are the primary tool used in reconstructing past changes in the carbon cycle. As with uranium and many other elements, carbon exists in multiple isotopes (atoms with the same number of protons but different numbers of neutrons). Carbon-14 is probably the best known carbon isotope because its rapid decay provides a means of dating recent archaeological and fossil material. In Neoproterozoic and Cambrian rocks, however, the stable carbon isotopes (i.e., those that do not experience radioactive decay), carbon-12 (^{12}C) and carbon-13 (^{13}C), are extremely helpful. Geochemists record shifts between the reservoirs of these isotopes as $\delta^{13}C$, the difference between the ratios of ^{13}C to ^{12}C in a sample relative to a known standard and reported in parts per thousand (‰). A zero value for the carbonate carbon or inorganic carbon reservoir (denoted $\delta^{13}C_{carbonate}$ and abbreviated $\delta^{13}C_{carb}$) means that the sample has the same ratio of ^{13}C to ^{12}C as the standard, which happens to be a fossil belemnite (cephalopod) from the Cretaceous Pee Dee Formation. Geochemists generally assume that $\delta^{13}C_{carb}$ values record the isotopic composition of dissolved inorganic carbon in surface waters (Hayes, Strauss, and Kaufman 1999; Kaufman and Knoll 1995). Assuming that the ocean and atmosphere are in equilibrium, variations in $\delta^{13}C_{carb}$ from shallow areas of the sea will reflect changes in the *global* carbon cycle because shallow-water carbonates should be relatively homogeneous with respect to carbon isotopes on a global scale.

The long-term average value of $\delta^{13}C$ is about 0‰, and deviations from this value usually reflect movements of carbon between the inorganic and organic carbon reservoirs. In general, long-term (millions of years) positive values of $\delta^{13}C_{carb}$ reflect increased burial of organic carbon (rich in carbon-12), or nutrient loading. Long-term negative values of $\delta^{13}C_{carb}$ are interpreted as

Figure 3.4 (right) The composite $\delta^{13}C$ isotopic record for the Ediacaran and Cambrian. The general outline of this curve is based on analyses of many different sections around the world as compiled by Condon et al. (2005) and Maloof et al. (2005). Colored boxes along the right margin identify some of the radiometric age constraints; see fig. 2.6.

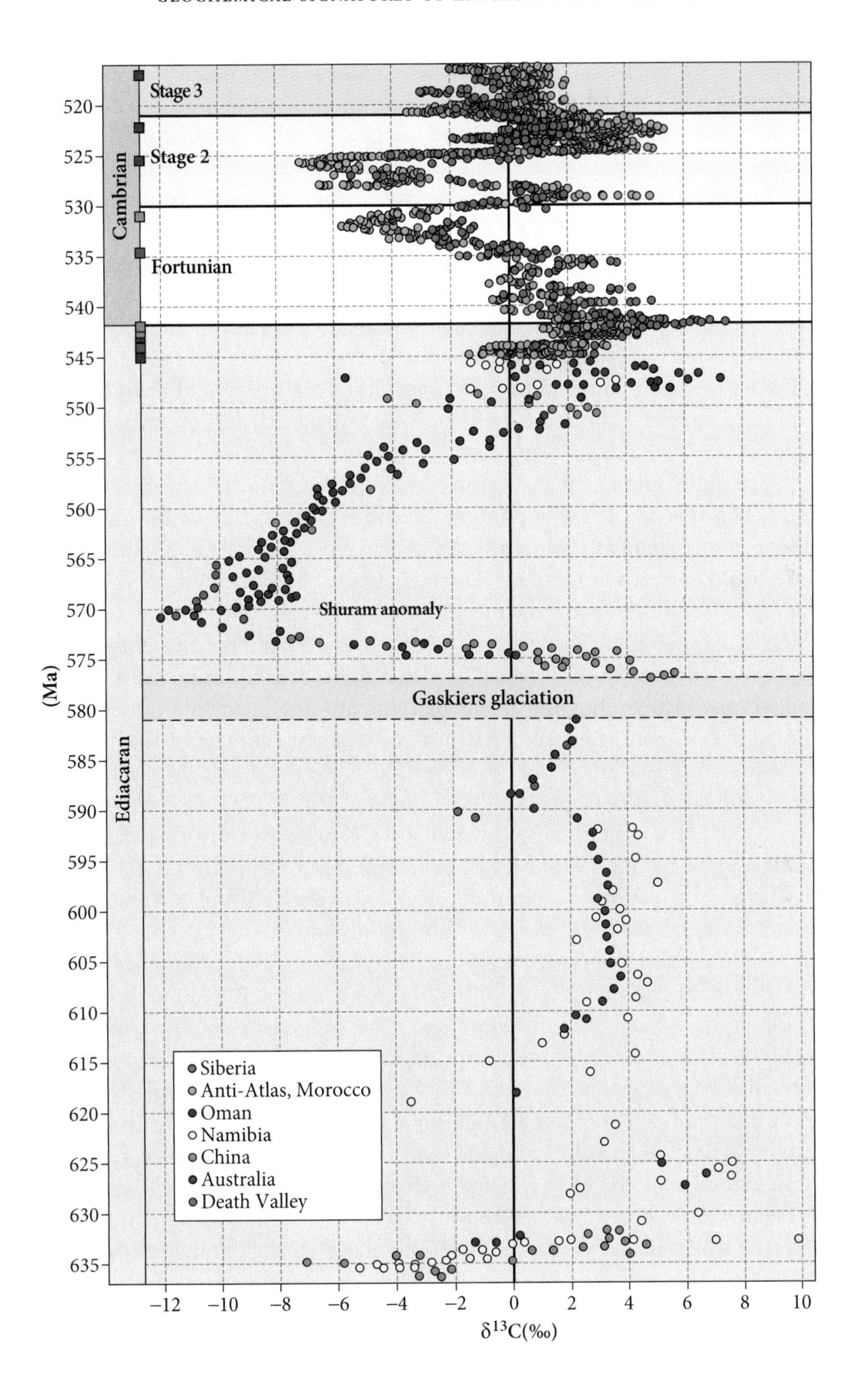
Stage 3
Stage 2
Fortunian
Cambrian
Ediacaran
Shuram anomaly
Gaskiers glaciation
(Ma)
520
525
530
535
540
545
550
555
560
565
570
575
580
585
590
595
600
605
610
615
620
625
630
635
−12 −10 −8 −6 −4 −2 0 2 4 6 8 10
δ13C(‰)
Siberia
Anti-Atlas, Morocco
Oman
Namibia
China
Australia
Death Valley

a decrease in the burial of organic carbon either because of a loss of primary productivity or a reduction in the rate of burial of organic-rich sediments (e.g., because of a reduction in the number or extent of shallow marine basins). The inorganic carbon isotope record is also replete with shorter-term shifts in carbon isotopes, most of which are negative. These $\delta^{13}C$ excursions indicate non-steady-state conditions of the carbon cycle and generally reflect the oxidation and input of large amounts of light carbon. The duration and magnitude of these $\delta^{13}C$ shifts can reveal both the volume of carbon added and its $\delta^{13}C$ value. For example, marine algae generally have a $\delta^{13}C$ value of about −25‰, while methane trapped in sediments after being generated by methanogenic microbes has values of −65 to −80‰ or more. So, a −2‰ shift in $\delta^{13}C_{carb}$ requires a smaller volume of methane than does oxidation of carbon from algae.

With these fundamentals, we can turn to the patterns in inorganic carbon isotopes during the late Neoproterozoic and Cambrian. Determining when these isotopic changes occurred is dependent on available radiometric dates and various correlation methods. During parts of the Ediacaran, there is continuing uncertainty about chemostratigraphic correlations, and, as we have noted, biostratigraphic correlations are often very broad. Consequently, there has been considerable controversy over the higher-resolution patterns of carbon isotope change, but the broad patterns are now generally understood. The pattern shown in figure 3.4 is our best current estimate of the shifting isotopic record, but it is certain to be revised as the quality of the radiometric dating and correlations improves (Condon et al. 2005; Halverson et al. 2002, 2006, 2009, 2010; Kaufman and Knoll 1995; Kaufman et al. 1996; Kaufman, Knoll, and Narbonne 1997; Maloof et al. 2005; Maloof, Porter, et al. 2010).

The general features evident in figure 3.4 are relatively high values of inorganic carbon, about 5‰, which are commonly interpreted as representing a sustained, net burial of organic carbon. Because the global amounts of carbon and oxygen are conserved, a net burial of organic carbon in sediments implies a buildup of oxygen in the atmosphere. This long-term pattern is punctuated by several pronounced, very large negative excursions, including two that immediately precede the Sturtian and Marinoan glaciations. Regional studies of Ediacaran-age rocks in south China suggest a steep gradient in $\delta^{13}C_{carb}$ between shallow-platform sediments and deeper slope and basinal sections. Thus, rather than the oceans being well mixed, with similar $\delta^{13}C_{carb}$ values in shallow and deep waters, there must have been some differentiation into different water masses (Ader et al. 2009; Jiang et al. 2007; McFadden et al. 2008). A recent study of rocks from the Death Valley area in California reveals three, small negative $\delta^{13}C_{carb}$ anomalies during the later Ediacaran, culminating in a negative shift at the Ediacaran-Cambrian boundary (Verdel, Wernicke, and Bowring 2011). This shift at the boundary is widespread and has been recorded from sections in China, the western United States, Siberia, Oman, and Morocco (Amthor et al. 2003; Brasier et al. 1994, 1996; D. Li et al. 2009; Maloof et al. 2005). Carbon isotope fluctuations continued through the first two stages of the Cambrian but were rapidly dampened as biological diversification accelerated in Stage 2; large carbon isotope fluctuations disappeared by Stage 3 (Maloof et al. 2005). Maloof suggested that these high-frequency excursions involved a combination of short-term releases of ^{12}C-enriched organic carbon (1–3‰ variability) and longer-term changes in the extent of organic carbon burial and nutrient cycling.

ORGANIC CARBON ISOTOPES Because less energy is required to break a $^{12}CO_2$ bond than a $^{13}CO_2$ bond, photosynthetic organisms have a slight bias toward the incorporation of ^{12}C rather than ^{13}C as they produce carbon compounds. Consequently, all photosynthetic organisms and everything that eats them, and their remains, are enriched in ^{12}C relative to the inorganic reservoir. This fractionation of organic material in ^{12}C also pushes the isotopic ratio of inorganic carbon in the opposite direction: as photosynthetic organisms preferentially incorporate ^{12}C, the isotopic ratio in the inorganic reservoir is shifted more toward higher ^{13}C. Burial of organic carbon has the same effect: as more organic material is buried, the amount of ^{12}C circulating through the oceans and atmosphere is reduced. Conversely, if organic material is released, perhaps through erosion of organic-rich shales, the volume of circulating ^{12}C increases.

As mentioned, the average value of the organic carbon ratio of most marine organisms whose carbon is fixed during photosynthesis is −25‰ to −28‰, whereas carbon fixed by methane-producing microbes can be −60 to −80‰. Because organic carbon is formed in shallow waters by photosynthetic activity, changes in organic carbon isotopes ($\delta^{13}C_{org}$) should parallel those in $\delta^{13}C_{carb}$ but be offset by the effect of fractionation. Through most of the past 542 million years, the values of $\delta^{13}C_{carb}$ and $\delta^{13}C_{org}$ (denoted $\Delta\delta^{13}C$) measured in rock samples have varied in tandem. This tandem variation is taken as evidence that the values have not been altered after the rocks were buried and that the record of $\delta^{13}C$ geologists recovered is a faithful record of the values of the ocean at the time the rocks were deposited.

This pattern of $\delta^{13}C_{org}$ tracking $\delta^{13}C_{carb}$ broke down during the most unusual and pronounced feature of the carbon isotope curve between the

Marinoan glaciation and the Ediacaran-Cambrian boundary: the Shuram-Wonoka anomaly. At that time, the carbon isotope record from the Huqf Supergroup in Oman exhibits a sharp drop from +5 to −12‰ followed by a slow recovery to positive values (Condon et al. 2005; Fike et al. 2006; Guerroue et al. 2006). The Shuram-Wonoka anomaly has also been documented in rocks in South Australia, in Namibia, in Death Valley in the United States, on the Yangtze Platform in China, in Siberia, and elsewhere (Grotzinger, Fike, and Fischer 2011). In addition, data from these rocks show a clear decoupling between $\delta^{13}C_{carb}$ and $\delta^{13}C_{org}$: there is considerable variation in $\delta^{13}C_{carb}$, whereas the $\delta^{13}C_{org}$ values are relatively invariant (fig. 3.5). At the end of the anomaly, about 550 Ma, the coupling between $\delta^{13}C_{carb}$ and $\delta^{13}C_{org}$ returns. The picture during the Cryogenian is more complex. One study found no evidence for coupling between $\delta^{13}C_{carb}$ and $\delta^{13}C_{org}$ before the Sturtian glaciation (746–663 Ma), but the development of covariation after the Sturtian glaciation and its persistence until the Marinoan (Swanson-Hysell et al. 2010). A second study found covariation between $\delta^{13}C_{carb}$ and $\delta^{13}C_{org}$ through much of the Neoproterozoic in sequences from Mongolia and northwest Canada, but not in northern Namibia (Johnston et al. 2012).

Uranium-lead radiometric dates from the Doushantuo Formation in south China suggest the Shuram-Wonoka anomaly ended by 550 Ma, but there is considerable dispute about when it began (Condon et al. 2005; Guerroue et al. 2006; S. H. Zhang et al. 2005). In Oman, the Shuram persists through 500 to 800 meters of rock; models of sedimentary subsidence suggest that it began about 600 Ma (Guerroue et al. 2006), whereas others suggest it may have lasted only 5 million to 11 million years (Bowring et al. 2007). Fossil evidence suggests that the onset of the anomaly likely occurred after about 575 Ma (Verdel, Wernicke, and Bowring 2011), providing a duration of just more than 20 million years. The Shuram carbon isotope excursion plays a critical role in ideas about the oxygenation of the deep sea, and we will return to it later in this chapter and again in chapter 7.

The Sulfur Cycle

The movement of sulfur through the oceans, atmosphere, and sediments is very sensitive to the amount of available oxygen. Sulfate enters the ocean mostly from rivers and is removed from the ocean through deposition and burial of evaporite sulfate minerals such as anhydrite and gypsum or through formation of the mineral pyrite in organic-rich, oxygen-poor muds. Pyrite is formed via a microbially mediated process known as bacterial sulfate reduction (BSR) that reduces sulfate to sulfide. The sulfide is converted to pyrite, and burial of the mud moves sulfide from the oceans into the sedimentary reservoir. This process is very sensitive to oxygen levels. In principle, then, information on the cycling of sulfur during the Neoproterozoic and Cambrian should provide important information on the oxygen history of the oceans. Unfortunately, there are many complexities and conflicts associated with interpreting sulfur isotope data, so despite the promise, the insights remain limited.

Changes in sulfur are recorded by measuring isotopic changes in seawater sulfate and pyrite in the same way that variations in the carbon cycle are recorded in shifts in carbon isotopes. Sulfur isotope values reflect shifts between the oxidation of pyrite during weathering to dissolved sulfate and its transport to the ocean and the formation of sedimentary pyrite (as sulfide) by BSR. Geochemists employ a ratio of two of the isotopes of sulfur, sulfur-32 (^{32}S), and sulfur-34 (^{34}S) to track the changes in the sulfur cycle, with the results reported as $\delta^{34}S$. Seawater sulfate is measured from evaporites, phosphorites, and carbonates (as carbonate-associated sulfur) and reported as $\delta^{34}S_{sulfate}$ (abbreviated $\delta^{34}S_{sulf}$). The pyrite sulfur is reported as $\delta^{34}S_{pyrite}$ (abbreviated $\delta^{34}S_{pyr}$). Just as photosynthesizing organisms and other carbon-fixing organisms prefer ^{12}C to ^{13}C, sulfate-reducing bacteria prefer the lighter isotope of ^{32}S to ^{34}S. Buried pyrite will be enriched in S^{32} relative to the sulfur in evaporites formed at the same time, a process that fractionates the sulfur reservoir, raising the value of $\delta^{34}S_{sulf}$. An increase in $\delta^{34}S_{sulf}$ can also reflect enhanced continental weathering, which can deliver additional nutrients to the ocean and thus increase bacterial sulfate reduction (Canfield and Teske 1996; Knoll et al. 1996).

The value of $\delta^{34}S_{sulf}$ was high in the immediate aftermath of the Marinoan glaciation and dropped through the middle of the Ediacaran to levels close to those of today, before climbing again to high levels across the Cambrian boundary (fig. 3.5). Sulfate $\delta^{34}S$ values then begin a long-term decline in the Cambrian that lasted through the Paleozoic, further emphasizing the unusual nature of the sulfate record of the Ediacaran (Fike et al. 2006; Halverson and Hurtgen 2007; Hurtgen, Arthur, and Halverson 2005; Hurtgen et al. 2006; Kah, Lyons, and Frank 2004; Mazumdar and Strauss 2006).

Sulfur isotope fractionation, the difference in the sulfur isotopes of sulfate and sulfide ($\Delta^{34}S = \delta^{34}S_{sulf} - \delta^{34}S_{pyr}$), has been widely used as an indicator of an Ediacaran increase in atmospheric oxygen. In theory, because both the sulfate reservoir and $\Delta^{34}S$ are linked to atmospheric oxygen, the Neoproterozoic increase in $\Delta^{34}S$ can be interpreted as reflecting the growth of the marine sulfate reservoir through weathering and oxidation of continental pyrite, which requires a sufficiency of oxygen in the atmosphere. Recent laboratory studies, however, indicate that some microbes are producing depleted ^{34}S that can

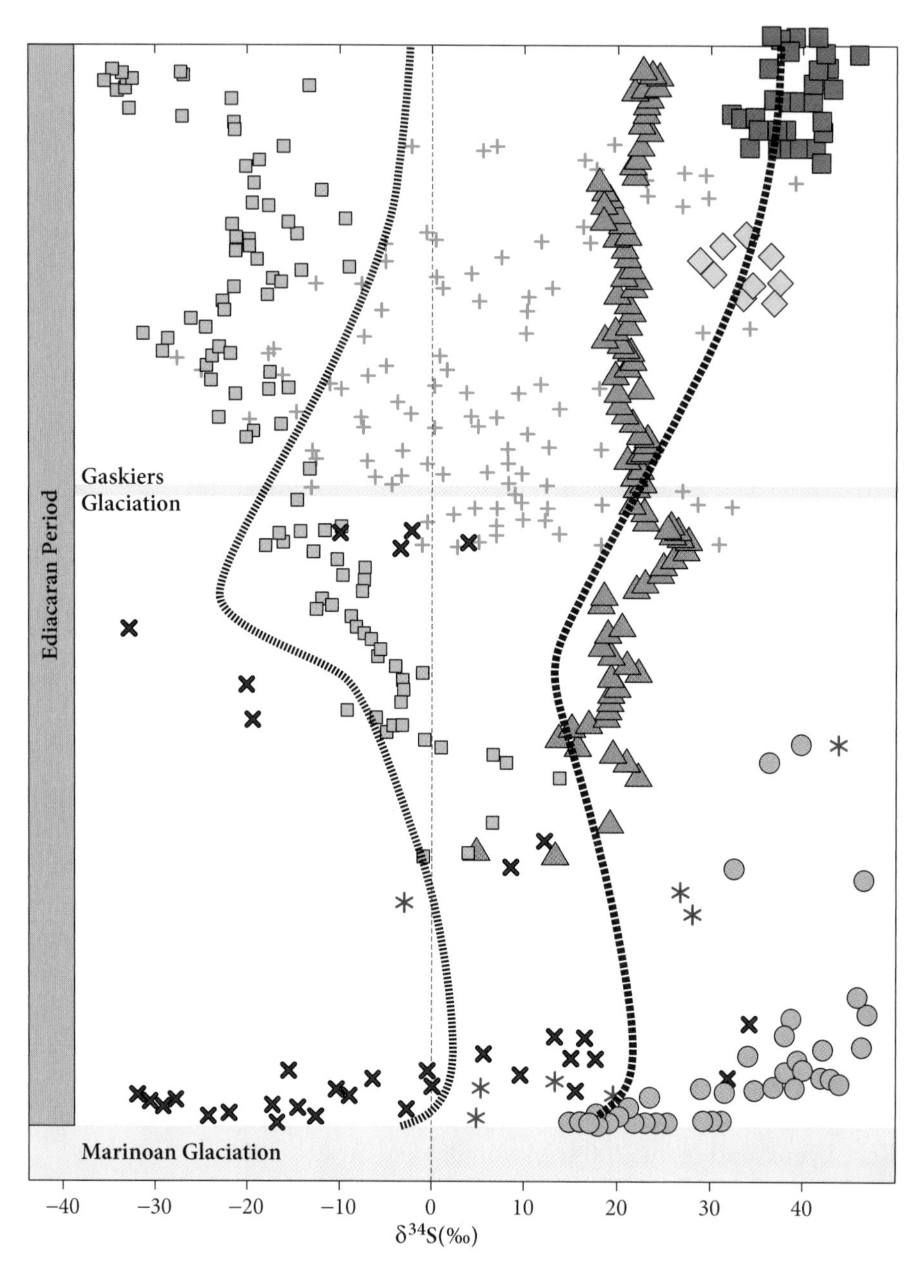

Figure 3.5 Sulfur isotopes from the Oman sections, with additional data from correlative sections in Namibia, Australia, Newfoundland, and Siberia. See text. From Fike et al. (2006).

mimic the values used to interpret the oxygen history of marine waters (Sim, Bosak, and Ono 2011).

Let us turn to the sulfur isotope data developed by Fike et al. (2006) for Oman, correlative with the carbon isotope data discussed earlier. This study reveals three phases in the sulfur isotope data, which can be interpreted with the help of the carbon isotopic record (fig. 3.5). The pattern of $\Delta^{34}S$ is low immediately after the Marinoan glaciation, which has been interpreted as indicating low marine sulfate and low levels of BSR. Then $\Delta^{34}S$ gradually increases, recording an increase in marine sulfate and BSR that is best explained as increased availability of atmospheric oxygen. During the Shuram carbon isotope anomaly, $\Delta^{34}S$ is relatively constant. After the end of the Shuram, at about 550 Ma, $\Delta^{34}S$ increases again, indicating another change in the sulfur cycle, plausibly interpreted as owing to increased oxidation. Fike et al. (2006) note that this increase may correlate to an expansion of microfossil diversity at about the same time. Some geologists now believe that the Shuram carbon isotope anomaly records ventilation of the deep sea with oxygen. The $\Delta^{34}S$ record is consistent with this interpretation (Fike et al. 2006; Halverson and Hurtgen 2007; Hurtgen, Arthur, and Halverson 2005; Hurtgen et al. 2006; Kah et al. 2004; Mazumdar and Strauss 2006), although without identifying the mechanism involved.

Today, gypsum is an important evaporite mineral, and its formation is a key means of removing sulfur, as sulfate, from the oceans. A study of sulfur isotopic behavior by Canfield and Farquhar (2009) indicates that gypsum deposition only became important in the Phanerozoic, replacing sulfur reduction and pyrite formation as the primary means of sulfur removal. Canfield and Farquhar implicate the burrowing activity of metazoans in this shift, suggesting that by stirring atmospheric oxygen into the sediments, bioturbation oxidized buried pyrite and increased marine sulfate levels, which enhanced the likelihood of gypsum deposition. Although the temporal resolution of their study is insufficient to identify when in the Neoproterozoic or early Cambrian this shift occurred, it is broadly congruent with the evidence for increasing metazoan bioturbation (chap. 5).

Other Redox Sensitive Elements

Carbon and sulfur have received the bulk of the attention by geochemists seeking to decipher the redox history of the oceans during the Neoproterozoic, but several other elements, such as molybdenum and iron, are also sensitive to oxygen levels and are playing increasingly important roles in the story. The data from these elements independently confirm, at least in broad outline, the oxygenation of the oceans during the Ediacaran.

Although anoxic conditions are usually viewed simply as involving low oxygen levels, they may also involve high levels of iron or sulfur, elements that are famously reactive with oxygen. Studies of iron speciation provide a sensitive indicator of anoxia. By measuring the ratio of highly reactive iron to the total amount of iron in sediments, geochemists can determine whether bottom waters were anoxic when the sediments were deposited. Moreover, they can discriminate between iron-rich (Fe^{2+}; ferruginous) and sulfur-rich (euxinic) conditions. Applied to rocks dating between 850 and 530 Ma from a variety of localities, such studies suggest a stratified ocean with sulfur-rich, or euxinic, waters above deep ocean, iron-rich waters, and with relatively low levels of oxygen in the surface mixing zone, at least toward the end of the interval (Planavsky et al. 2011). A study of the > 742 Ma Chuar Group in the Grand Canyon showed that even sediments deposited in shallow marine settings, and thus in active contact with the atmosphere, were anoxic. So, low atmospheric oxygen conditions prevailed at that time (Johnston et al. 2010). Later, sediments from the Drook Formation on the Avalon Peninsula in Newfoundland, where the oldest of the Ediacara macrofossils have been described, reveal development of oxygenated deep waters in the immediate aftermath of the Gaskiers glaciation, near 575 Ma. Moreover, the oxygenation of the deep ocean between 580 and 560 Ma was not universal, although anoxic conditions persisted in some settings and may have been stronger at some times than others. In contrast, shallow-water environments seem to have been oxygenated through much of the Ediacaran (Canfield et al. 2008).

Like iron, molybdenum (Mo) also provides an independent proxy for global redox conditions. Molybdenum weathers from continental crust, enters the ocean in the oxygenated state, and is removed from the ocean in largely sulfidic, anoxic environments. By analyzing Mo accumulation in black shales, geochemists can assess the level of oxygenation in deep-water, sulfidic environments. Analyses of Mo from the Chuar Group are consistent with widespread anoxic and euxinic waters and are consistent with the iron speciation data described above indicating low oxygen content in the atmosphere and oceans (Dahl et al. 2011). In contrast, rocks from the Miohe Member (551 Ma) of the fossiliferous Doushantuo Formation in south China exhibit Mo patterns typical of Phanerozoic black shales, conditions that required both limited sulfidic environments and oxygen (C. Scott et al. 2008; see also Dahl et al. 2010).

Additional Chemical Changes

The Late Neoproterozoic and Cambrian were also times of considerable changes in two other geochemical systems: the magnesium:calcium ratio, which may have affected the type of skeletons the earliest animals formed, and strontium isotopes, which reflect changes in weathering of the continents and in tectonic activity.

The seawater chemistry at the time skeletal formation began in various metazoans clades is reflected in the mineralogy of their skeletons. As seawater chemistry changed during the Ediacaran and Cambrian, so, too, did the mineralogy of the earliest skeletal organisms. The first animals with carbonate skeletons appear to have used aragonite, matching the mineralogy of the "aragonite seas" (intervals of time when the chemistry of the oceans favored the precipitation of aragonite) of the late Ediacaran through Cambrian Stage 2 as documented by fluid inclusions (Porter 2007, 2010). Studies of fluid inclusions within evaporitic rocks of this age from Oman reveal an increase in calcium ions (Ca^{2+}) in seawater (Brennan, Lowenstein, and Horita 2004). This enrichment in Ca^{2+} and a corresponding depletion in magnesium and sulfate concentrations evidently occurred *after* the onset of skeletonization but *before* the strong biotic diversification seen in Stage 2 of the Cambrian (Porter 2007; Zhuravlev and Wood 2008; see also chap. 6). Calcitic skeletons (with either high- or low-magnesium content) appeared during Cambrian Stage 3, corresponding to the evidence from fluid inclusions for "calcitic seas" (intervals during which precipitation of calcite was favored). During Stage 3, echinoderms and other groups with high-magnesium calcite skeletons appeared. This final shift may hold the key to the factors driving skeletal mineralogy, for it did not correspond to a wholesale change in oceanic chemistry, but rather to smaller oscillations of the magnesium:calcium ratio.

The magnesium:calcium ratio that evidently played such an important role in controlling the mineralogy of early skeletal organisms is in turn regulated by continental weathering and by ion exchange as fluids flow through mid-ocean ridge systems. Changes in strontium isotopes serve as a proxy for both continental weathering rates and hydrothermal alteration along midocean ridges. Of the four naturally occurring isotopes of strontium, two are informative for geological purposes: strontium-86 (^{86}Sr) and strontium-87 (^{87}Sr). The standard geological interpretation is that changes in the ratio of these two isotopes reflect changes in weathering rates and tectonic activity. During hydrothermal alteration of basaltic magma along the midocean ridges, fluids pick up relatively more ^{86}Sr and release it to the surrounding seawater, lowering $^{87}Sr/^{86}Sr$ ratios in seawater. Thus, lower values of the ratio of these two isotopes ($^{87}Sr/^{86}Sr$) are usually interpreted as indicating an increase in the release of mantle-derived ^{86}Sr associated with faster spreading on the midocean ridges. In contrast, radiogenic strontium, ^{87}Sr, is derived from weathering of continental crust, so higher values of $^{87}Sr/^{86}Sr$ denote an increase in continental weathering. (A complication is that not all crustal rocks have similar Sr compositions; today, there is much more strontium in Himalayan rocks, for example, than in other mountain ranges.)

The $^{87}Sr/^{86}Sr$ ratio rose steadily from early in the Neoproterozoic through the Marinoan glaciation and the Ediacaran period, reaching a peak at about 580 Ma (fig. 3.6) (Brasier et al. 2000; Halverson et al. 2007, 2009; Kaufman, Jacobsen, and Knoll 1993; Mazumdar and Strauss 2006; Nicholas 1986; Shields 2007). Difficulties in correlating between sections, and the low density of radiometric dates, have created problems in constructing a global curve later in the Ediacaran. Some workers identified a Sr plateau from about 570 Ma to the base of the Cambrian (Mazumdar and Strauss 2006). In contrast, another data set, which relies on fewer sections but with better correlations, indicates that the $^{87}Sr/^{86}Sr$ ratio decreased into the early Cambrian (Halverson et al. 2007).

Continental rocks are also rich in phosphorus, a critical nutrient for life. Increased weathering delivers phosphorus to the oceans, and understanding the cause of the increased strontium ratios is therefore critical to evaluating the extent of increased nutrient input during the Ediacaran and early Cam-

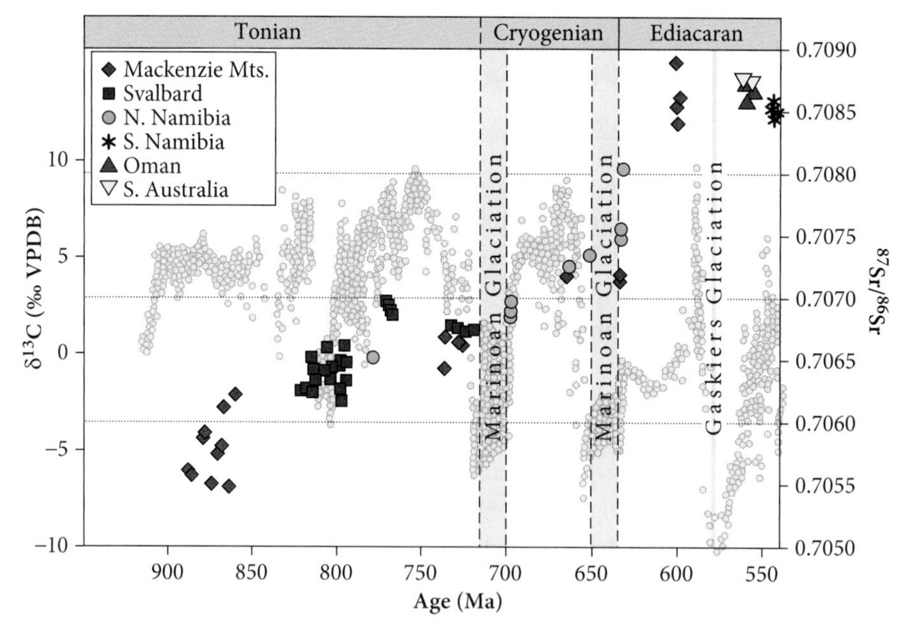

Figure 3.6 A composite Neoproterozoic-to-Cambrian record of $^{87}Sr/^{86}Sr$ overlain on the composite record of $\delta^{13}C$ (shaded circles). Notice the steady increase in $^{87}Sr/^{86}Sr$, with possible stabilization through the Ediacaran, and the continuing uncertainty over the values through the mid-Ediacaran because of difficulties correlating between sections. From Halverson et al. (2010).

brian. Although increased strontium ratios have generally been attributed to increased mountain-building activity, work by Halverson and colleagues (2007) shows that the increase actually began during the initial breakup of Rodinia and is incompatible with most models of tectonic assembly. Instead, Halverson and colleagues attribute the increase in strontium ratios to the breakup of an early Neoproterozoic supercontinent that increased moisture in what had been the core of the supercontinent. This increased moisture in turn increased weathering rates in the ^{87}Sr-rich continental interiors. The change in strontium isotopes is consistent with a strong input of material from continental weathering and probably nutrient-enhanced increases in primary productivity via an influx of phosphorus.[3]

Cardenas and Harries (2010) found a significant positive relationship between nutrient availability, as measured by the $^{87}Sr/^{86}Sr$ ratio and phosphorus recycling, and Phanerozoic global marine origination rates. The Ediacaran and Cambrian strontium ratios suggest that a similar pulse of nutrient input may have contributed to the biological events at that time.

EDIACARAN-CAMBRIAN OCEAN REDOX AND ENVIRONMENTAL DRIVERS

The largely geochemical studies reviewed above have clearly identified several episodes of nonanalog conditions. These include the extreme Sturtian and Marinoan glaciations, conditions that favored the deposition of the unique cap carbonates on glacial debris and the major, long-lasting Shuram-Wonoka carbon isotope anomaly associated with a decoupling of inorganic and organic carbon isotopes. In this section, we discuss the evidence and models that build on these patterns and, specifically, those that address the oxygenation of the oceans. Some of these models will use nonanalog environmental conditions to invoke nonuniformitarian explanations, particularly in the behavior of the carbon cycle, but other models have been proposed that are more strictly uniformitarian and that assume that the Ediacaran and Cambrian carbon and sulfur cycles behaved then much as they do today.

Three hypotheses have been presented as interpretations of the late Neoproterozoic carbon isotope record. The classic interpretation is that organic carbon is produced from a well-mixed reservoir of dissolved inorganic carbon. The lack of covariance between $\delta^{13}C_{carb}$ and $\delta^{13}C_{org}$, at least in some localities, suggests that a large pool of dissolved organic carbon was present. The third set of hypotheses proposes that the large isotopic anomalies, such as the Shuram, are the result of postburial alteration of the carbon isotopic record, and thus provide little information on geochemical cycles. We address each of these alternatives in this section. Although these differing hypotheses have yet to be definitively resolved, the issue is critical for determining the history of oxygen levels in the ocean. The magnitude of changes in carbon isotopes, however, means that an enormous volume of oxidizing power was required, and there is no consensus on its source, the timing of this transition, or of the processes responsible. Oxygen levels were clearly low early in Earth's history, and studies of the iron geochemistry of the Chuar Group (> 742 Ma) indicate that even sediments deposited in shallow water were formed under very low atmospheric oxygen levels (Johnston et al. 2010). Beyond the distinction between uniformitarian and nonuniformitarian approaches, explanations for geochemical proxy evidence also differ in whether they involve physical drivers, with the diversification of animals as a consequence, or biological drivers, with the increased marine oxygen as a consequence. The distinction between abiotic and biotic explanations is critical to understanding the triggers of the Cambrian diversification. Although intense research continues in this area, we favor the view that the primary causes of oceanic ventilation involved the activities of early organisms and thus that this critical environmental transition was biologically mediated.

A challenge in applying geochemical proxies to understand changes in redox is that they primarily apply to deep marine waters where the sampled rocks largely originated rather than to the shallow marine waters where the diversification of animals was likely concentrated. Some level of atmospheric oxygen developed as early as the Proterozoic, and wave activity would ensure mixing of some of that oxygen into the shallowest marine waters even in the absence of any other processes. So, before proceeding with examination of the geochemical evidence of marine redox, we must acknowledge an alternative view, most cogently advanced by Butterfield, that shallow reaches of the oceans were well oxygenated before the Ediacaran and that increased oxygen levels played little role in the diversifications of the Ediacaran and Cambrian (Butterfield 2009). Butterfield argues that the advent of suspension feeding animals fundamentally altered the structure and dynamics of the plankton, reengineering the water column from a stratified, cloudy, anoxic system dominated by cyanobacteria to a well-mixed, clear system dominated by large, eukaryotic algae that are easily removed from the water column by suspension feeders (fig. 3.7). Butterfield interprets the geochemical evidence for increasing marine oxygen as a consequence of biological activity rather than a cause. We agree with Butterfield on the importance of biological activity in driving the redox transformation of Ediacaran oceans, but we are less certain that other geochemical processes were not at work as well; we will return to his model below and in chapter 7.

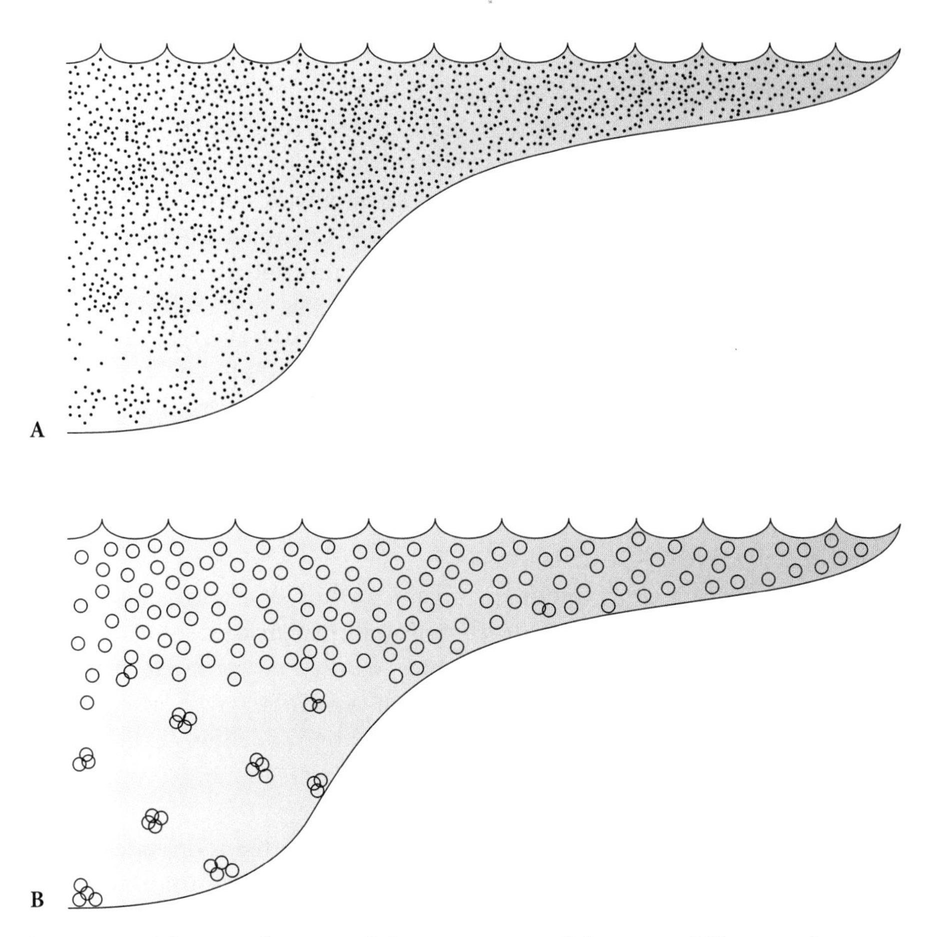

Figure 3.7 Schematic diagrams of alternative states of the ocean, differing in the activity of the biological pump. (A) A stratified, cloudy, and possibly anoxic to euxinic water column develops in the absence of predation on algae because the system is dominated by cyanobacterial picoplankton. (B) After the evolution of zooplankton, selection pressure for larger eukaryotic algae changes the structure of the phytoplankton, increasing the export of phytoplankton to the bottom and clearing the water column. After Butterfield (2009).

One of the most obvious uniformitarian, abiogenic explanations for the geochemical patterns involves changes in the seawater composition of carbon isotopes. These changes could occur either through increased carbon burial or through a significant addition of organic carbon to drive the extremely negative Shuram carbon isotope anomaly. Either of these mechanisms could cause a global signal in geochemical proxies. Increasing the burial of organic matter is the simplest way to increase oxygen levels. If we assume that the carbonate and organic carbon reservoirs tracked each other, buried organic carbon was no longer subject to oxidation, which freed oxygen to build up in the atmosphere and oceans. As noted earlier, the high inorganic carbon values through the Neoproterozoic have often been interpreted as reflecting increased carbon burial because the continental configurations after the rifting of Rodinia produced many smaller continents with deep, anoxic basins that would have allowed burial of organic carbon. Organic carbon burial also would have facilitated glaciations by reducing the atmospheric carbon dioxide levels. Studies in the Mackenzie Mountains of Canada, the Adelaidean Basin of Australia, the Otavi Platform in Namibia, the Chuar Group in the Grand Canyon, and sections in south China all record increased carbon burial in association with rifting and extension during the breakup of Rodinia (Derry, Kaufman, and Jacobsen 1992; Des Marais et al. 1992; Karlstrom et al. 2000; Kaufman, Knoll, and Narbonne 1997). This carbon burial model may explain positive $\delta^{13}C$ values during the Neoproterozoic, but does not offer an explanation for the negative $\delta^{13}C$ excursions.

The addition of a large volume of organic carbon is often treated as the major cause of short-lived negative carbon isotopic excursions, some of which are associated with various Phanerozoic mass extinction events. If, however, the Shuram event were 20 million to 50 million years in duration, as suggested by the thickness of the sequence in Oman and by modeling studies, there would simply be no reservoir of organic carbon large enough to produce a shift of that magnitude and duration.

In 1995, most geochemists assumed that changes in the carbonate carbon pool ($\delta^{13}C_{carb}$) moved in parallel with changes in the organic carbon reservoir ($\delta^{13}C_{org}$) as happens today and has happened through most of the Phanerozoic. As discussed earlier, however, covariation in $\delta^{13}C_{carb}$ and $\delta^{13}C_{org}$ was decoupled between the Sturtian and the Marinoan glaciations and then again during the Shuram carbon isotope anomaly. Rothman and colleagues proposed a highly nonuniformitarian explanation for this pattern: the presence of a very large reservoir of deep-water organic carbon (dissolved organic carbon, or DOC) in a stratified ocean, with oxidation of organic matter in surface waters, but no ventilation of deep oceans (Rothman, Hayes, and Summons 2003). If oxygen and sulfate levels in the water column were low enough through the Neoproterozoic, mineralization (oxidation) of the DOC pool would have been inhibited. The Shuram-Wonoka event would thus represent oxidation of the DOC pool.

Studies of carbon isotopes from south China imply just such a stratified water column and have been interpreted as supporting the existence of a DOC reservoir in the deep oceans (Ader et al. 2009; Jiang et al. 2007; C. Li et al. 2010; McFadden et al. 2008). Further support comes from carbon and sulfur isotopes from the Huqf Supergroup in Oman (Fike et al. 2006) and by studies in Australia (Swanson-Hysell et al. 2010). Sulfur isotopes have been interpreted as indicating increased oxidation of sulfide to sulfate, which in turn increased bacterial sulfate reduction. Together, these results imply that even the deep oceans had become largely ventilated by 551 Ma, the end of the Shuram anomaly (C. Scott et al. 2008; Shields-Zhou and Och 2011).

Subsequent dissection of Rothman's model has made clear that although there may have been an increase in the size of the DOC pool during the Neoproterozoic, as suggested by the results noted above, remineralization of the DOC pool is not a viable explanation for the Shuram-Wonoka event because the model requires an unrealistically enormous amount of oxidants. Rothman's model would require about 10,000 times current levels of DOC (Bristow and Kennedy 2008), but other models suggest that available oxidants would have been depleted in about 1.3 million years. To drive the oxidation, Rothman and colleagues invoked fecal pellets from pelagic bilaterians, increased abundance of algae with resistant cell walls, and biomineralization, yet there is little evidence of a sufficiently large fauna of pelagic bilaterians to supply fecal pellets before the Cambrian or of significant biomineralization. In addition, maintaining such a large pool of unreactive DOC is highly implausible if the atmosphere (and thus surface waters) had even a small amount of oxygen, unless the oceans were chemically stratified. Indeed, stratification of the whole ocean may not be possible for physical reasons, although the presence of local or regional unreactive pools of DOC may have occurred. If the DOC pool were limited to regional water bodies, the degree of the redox changes and the volume of oxidant required would be reduced. We will return to the Ediacaran DOC pool in chapter 5; it may explain much about Ediacara macrofossils even if it does not explain the Shuram-Wonoka event.

Finally, an abiogenic, nonuniformitarian mechanism proposes that the Shuram-Wonoka isotopic anomaly was *not* a signal of global oceanic conditions at all, but rather a postdepositional, diagenetic artifact. Derry notes that the δC^{13}_{carb} signal covaries with a δO^{18} signal, although the latter data are normally not presented (Derry 2010a, 2010b). The covariation between carbon and oxygen is often diagnostic of postburial alteration of carbonates by fluids containing high levels of carbon dioxide. When these geochemical studies began, the simplest assumption was that if rocks of the same age from different regions record similar patterns of isotopic change, they record a signal of the global ocean. If this diagenetic model is correct, however, the carbon isotopic evidence may provide little information on the oxygenation of the oceans at this time. A problem with the diagenetic model is that it is difficult to conceive of a realistic model that would produce such a globally similar carbon isotope pattern: why should five widely dispersed sections all exhibit such a similar pattern of isotopic shift? So, although diagenetic modification must always be considered in evaluating geochemical proxies, it is not clear that this model resolves our problems more convincingly than those discussed above (Grotzinger, Fike, and Fischer 2011). In addition, Johnston and colleagues (2012) have shown that for Sturtian carbon isotope anomalies covariation occurs in some sections but not others, and thus cannot be the result of postburial diagenetic alteration. Grotzinger and colleagues proposed a modification of the Derry model, suggesting a globally synchronous drop in sea level, associated with glaciation, during which sediments were "primed to more efficiently oxidize organic compounds that they encounter" (p. 6). A coincident increase in atmospheric oxygen would trigger oxidation of the sediments and produce a global diagenetic signal, also tied to the increased oxygen levels.

The limited radiometric dates available for long stretches of the early Ediacaran and no dates at all from 573 to 555 Ma greatly complicate correlation efforts and the ability of geologists to establish a global pattern. Moreover, the stratigraphic resolution of early geochemical studies was relatively coarse. These models assume that the geochemical proxies reflect a global record, even though the Shuram-Wonoka anomaly is only documented in five sections. Recent intensive studies have revealed more complexity and growing evidence for regional variations in the carbon isotope record, suggesting an attractive resolution to the difficulties with the models described above: a more complex history of regional variation in redox patterns.

Sulfur isotopic data and studies of iron speciation in south China provide evidence for a complex, stratified ocean during the Ediacaran (C. Li et al. 2010). In the Nanhua Basin in south China, outcrops of the Doushantuo Formation preserve rocks along an onshore-to-offshore gradient from the continental shelf into the deep basin. Although most Proterozoic basins are either ferruginous or euxinic, the Nanhua Basin appears to have been mixed. The iron speciation data are interpreted to reflect the presence of a complex, metastable stratified ocean, with an oxygenated surface water layer, ferruginous deep water, and a metastable zone of euxinic waters that impinged on the shelf. When sea level rose, these euxinic waters could spread across the shelf, with deleterious effect on any organisms there. With a drop in sea level, the euxinic layer would retreat off the shelves. Sedimentary studies confirm that euxinic conditions were more common during rising sea levels. Importantly, Li and colleagues conclude that this setting could persist for such a long time because of the continued low levels of sulfate in deep waters, and the sulfur isotope data are consistent with this conclusion. Together with the iron speciation data from the Drook Formation in Newfoundland described earlier, it seems increasingly likely that our interpretations of geochemical proxy data may have been off the mark because it was assumed that the geochemical proxies reflected the state of the whole ocean. Although some of the geochemical proxies may reflect a global signal, it seems likely that the oxygenation of the oceans from about 600 Ma into the Cambrian was a prolonged and complex affair, with different regions becoming ventilated at different times. There may also have been shifts between anoxic and oxic conditions that varied over time in the same regions. Such patterns of variation would have created great challenges for metazoans during that time, but, on the other hand, their activities may well have been at least partly responsible for the pattern of oxygenation. Up to this point, we have largely ignored the activity of organisms, but the active involvement of animals in constructing their own environment may have played a significant role in the rate and pattern of oxygenation of the oceans during the Ediacaran.

There are several different models in which the late Neoproterozoic increases in oxygen were primarily mediated by biological activity, particularly during a change from a carbon cycle with extensive microbial reworking of phytoplankton in the water column (thus depleting dissolved oxygen) to a situation in which most organic materials are rapidly transferred to the seafloor and oxidized in the sediments there. Logan and colleagues proposed that such a change was driven by the evolution of pelagic bilaterians (Logan et al. 1995). Once bilaterians moved into the water column as marine zooplankton, they packaged their waste into fecal pellets, which would settle to the bottom, thus removing suspended organic matter from surface waters and delivering it to the seafloor. This process would shift the oxygen-sulfide boundary from high in the water column into the bottom sediments, freeing oxygen in the water column and permitting additional exploitation of the water column, and the seafloor, by animals. Despite the attractions of the model, there is little evidence that planktonic bilaterians were ecologically abundant (or even existed) prior to 551 Ma. Indeed, fecal pellets probably did not become significant ecological factors until Stage 2 of the Cambrian, near 530 Ma, well after the ventilation of the ocean.[4]

Butterfield has proposed an ingenious alternative to Logan's model, also involving the evolution of zooplankton (Butterfield 1997, 2009). In the absence of selection by zooplankton predators, phytoplankton today are smaller (0.2 to 2.0 microns, rather than 2 to 200 microns) and consequently have a much lower settling rate. Where phytoplankton are abundant today, as in lakes with high nutrient inputs from industrial pollution or farm runoff, the waters are turbid, have high levels of suspended organic carbon, and high nutrient level, and the deeper waters are anoxic. Large size is often an effective refuge from predators, so the evolution of zooplankton forces phytoplankton to evolve toward larger sizes with a subsequent greater likelihood of settling, producing a clearer, better-oxygenated water column with low suspended carbon and lower nutrients, all conditions that favor eukaryotic algae (fig. 3.7). Thus, the evolution of zooplankton would have triggered a state change in a number of physical and chemical features of shallow-water marine ecosystems, although it probably would not have been stable until the appearance of mesozooplankton in Stage 2 of the Cambrian (see chap. 7). Although this scenario provides a biological driver for the changes in ocean structure, it does not necessarily require an increase in shallow-water oxygen levels.

An attractive additional possibility is that the ventilation of the deep sea was affected by sponges, progressively migrating from shallow waters over the many tens of millions of years of the Ediacaran (Sperling, Peterson, and Laflamme 2011; Sperling, Pisani, and Peterson 2007; Sperling et al. 2009). Because of their unique method of suspension feeding, sponges pump large volumes of water, removing particulate organic matter and sequestering carbon. Essential to this process would be the burial of some proportion of the sponge populations after death, sequestering carbon in the sediments. Over time, sponges may have sequestered large volumes of carbon simply upon their burial and thus have achieved the state change Butterfield described.

How do we make sense of all this apparently conflicting information on the timing and pattern of oxygenation of the oceans through the Ediacaran? Of course, it is possible that the data are simply misleading, perhaps more heavily influenced by diagenesis than most geochemists have yet realized. Assuming, however, that most of the carbon and sulfur isotope data reflect a primary signal, resolving the quandary seems to require abandoning the hope that there is a single, global pattern or that the change was unidirectional. At present the data suggest that in the aftermath of the Marinoan glaciations at 635 Ma the very shallow oceans were oxygenated but the deep oceans were largely anoxic, sulfur rich in some places, and iron rich in other places, and that the deep ocean's oxygenation was a stuttering and episodic process (C. Li et al. 2010). At least some oxygenation of the deep oceans may have occurred after the Gaskiers glaciation, but the more pervasive oxygenation of waters on the continental shelf appears to be associated with the mid-Ediacaran Shuram carbon isotopic excursion. Given the piecemeal record we have and the general absence of good radiometric dates, it is quite possible that this transition was more complex than it now appears, with persistence of some areas of anoxia and some shifting back and forth between anoxic and oxygenated environments through the mid-Ediacaran, perhaps regionally. Although there may have been some involvement of abiogenic processes in the changes in ocean chemistry documented by various geochemical proxies, it appears that the primary drivers of this process were likely biological, involving the advent of sponges and other suspension feeders and changes in the structure and dynamics of the plankton communities. There is some evidence for a rapid shift in the redox state of the oceans near the Ediacaran-Cambrian boundary, including a sharp negative carbon isotope excursion near the boundary apparently unassociated with any glaciation and the transient shift in molybdenum isotopes sampled from black shales (Wille et al. 2008). Molybdenum is sensitive to redox changes, and these data suggest that deep ocean euxinia persisted, at least locally, into the early Cambrian, before the transition to a better oxygenated early Cambrian deep sea. So it may be that ventilation of the ocean was not complete until near the Ediacaran-Cambrian boundary, raising profound questions about the relationship between this change in redox state and the diversification of animal ecosystems. Fortunately, many geochemists are actively pursuing these problems, acquiring more detailed data sets, and testing alternative hypotheses. Our understanding has advanced quite rapidly over the past decade, and current work holds the potential of resolving this issue in the near future.

NOTES

1. The number and correlation between glacial episodes remain subjects of active research. For the Sturtian glaciation, see Halverson et al. (2005); the number of glaciations, and whether they are globally synchronous, remains unresolved (see Kendall, Creaser, and Selby 2006; Condon et al. 2005 on the Marinoan; and Thompson and Bowring 2000 on the Gaskiers glaciation). There have been claims for as many as four distinct glacial epochs based on carbon isotope excursions (Brasier et al. 2000; Kaufman, Knoll, and Narbonne 1997), although not all are necessarily associated with global, or near-global, glaciations. For example, three tillites in the Quruqtagh of northwest China appear to represent the Sturtian and Marinoan glaciations, and the youngest, from the Hankalclough Formation, may represent either the Gaskiers glaciation or, less probably, evidence of a glaciation associated with the Shurum carbon isotope excursion near 550 Ma (Xiao, Bao, et al. 2004).

2. Very negative $\delta^{13}C$ values have been recorded from the Marinoan cap carbonates of south China. These values of 48‰ are difficult to explain other than by widespread methane release. Kennedy, Mrofka, and von der Borch (2008) propose that low-latitude permafrost contained sufficient methane clathrates and that their release induced a positive feedback loop and increased warming. See also Bao, Lyons, and Zhou (2008).

3. That the $^{87}Sr/^{86}Sr$ ratio almost never falls below 0.707 after about 600 Ma may suggest a fundamental shift in the style of continental weathering, enhanced by the development of the soil biota (Derry 2006; Kennedy et al. 2006), but this rather speculative hypothesis requires considerable confirmation.

4. Studies of modern marine systems reveal considerable export of organic carbon to sediments independent of animals. Such export can occur via physical agglutination, formation of marine snow (Turner 2002), and even picoplankton [evidently in rates proportional to their primary productivity (Richardson and Jackson 2007)], suggesting that animals were not required to lower the oxygen-sulfide boundary. Logan et al. (1995) also note that although Proterozoic oceans were low in phosphorus, the extensive Cambrian phosphorite deposits may mark the increased ventilation of the bottom.

PART TWO

ɞ

The Record of Early Metazoan Evolution

CHAPTER FOUR

The Metazoan Tree of Life

The world around us teems with a marvelous diversity of animals, seemingly of all shapes and sizes and exhibiting clever designs that equal or exceed the functional efficiencies of our engineered products. Imaginative combinations of design elements in early civilizations produced wings on horses, snakes in hairdos, serpents tongues on too many subjects, and, of course, fauns, sphinxes, dragons, and countless other members of this mythical menagerie. That there are coherent morphological themes among actual animals was surely understood long before they were formally studied. These themes were first established in the fossil record during the Ediacaran and Cambrian periods, following a shadowy interval of diversification in the Cryogenian. Our methods for identifying the patterns underlying the distribution of characters and establishing the evolutionary relationships between different animal groups has become increasingly sophisticated and powerful, providing us a far better understanding of these evolutionary patterns and, equally importantly, greater insights into where our ignorance is deepest.

ANIMAL CLASSIFICATION

Philosophers began trying to order and classify the natural world in antiquity, but this effort was abandoned for centuries before being taken up again by serious scholar-naturalists during the Renaissance. Although some of those early naturalists recognized the significance of fossils, the classifications that eventually arose were based on living organisms. Carolus Linnaeus, a Swedish naturalist, produced the most successful early animal classification scheme, and the tenth edition of his *Systema Naturae* (Linnaeus 1758) became a model for the field. Linnaeus's system was hierarchical (fig. 4.1), with the most morphologically similar animals grouped into species, similar species grouped into genera (singular, genus), the more similar genera grouped into orders, and similar orders grouped into classes, all embraced within the Kingdom Animalia

A Linnaeus's Usage		B Modern Usage	
Regnum	1	Kingdom	1
Classis	6	Superphylum	4–5
Ordo	64	Phylum	30–35
Genus	312	Class	Uncertain
Species	4,400	Order	Uncertain
		Family	Uncertain
		Genus	600,000–2,000,000
		Species	2,000,000–5,000,000?

Figure 4.1 (A) The five levels in Linnaeus's hierarchical classification of animals, with the number of taxa he included in each category at the time of writing *Systema Naturae* (Linnaeus 1758). (B) The principal levels in use today, with the approximate numbers of described taxa where estimates are available. Linnaean systematists added new levels to contain the large numbers of described species. In diverse groups, the number of levels is increased further by adding such subdivisions as superfamilies and subfamilies to the hierarchy.

(Regnum Animale of Linnaeus); the animal kingdom is now formally termed the Metazoa ("higher life"). Additional levels within the taxonomic hierarchy have been erected for certain highly diverse groups, and, as relations among the phyla have become clearer, the level of superphylum has been increasingly used between the kingdom and phylum levels. For more than 250 years, the names that most scientists used for taxa were based on the Linnaean hierarchical system, and they remain the names most in use for animal groups today.[1]

In 1858, a century after Linnaeus's work, Charles Darwin and Alfred Wallace published their hypothesis of evolution by natural selection. At the heart of their proposal was the idea of descent with modification: when organisms resembled one another, it was because they were historically related and had inherited their shared traits from common ancestors. If that were the case, the Linnaean hierarchy had been formed through evolution along branches in a tree of life. Such a tree would display the genealogical relationship among organisms, exactly as a family tree does for related humans. Shared morphology does not necessarily reflect a shared evolutionary history, however; many organisms resemble one another because they have evolved similar morphological solutions to similar problems of adaptation. These morphologies appear to be homologous—to have descended from a common ancestor—when in fact they are only analogous; that is, the similarity is due to common function, but they have independent origins. Thus, general morphological resemblances may fool us into assuming a closer kinship than is actually the case. To cope with this and other problems, Hennig (1950, 1966) proposed an improved system for identifying branches in the tree of life. This system, called cladistics or phylogenetic analysis, uses a tree rather than a hierarchy as a basis for classification and has become widely adopted. In cladistics, branching order inferred from the distribution of shared characters is of primary importance, and a variety of sophisticated statistical approaches to establishing branching order have been developed.[2] Neither the Linnaean hierarchy nor the Hennigian tree structures cope very well with continuous evolution along the branches of the tree of life, however, and in identifying units along branches as taxa, we must break the continuum in some way. Thus, we must impose our own definitions of discrete units within what are actually historical continua. Every living animal has an ancestry tracing back along a continuous pathway that leads down along the branching pattern of the tree of life, through node after node, to the first animals and, of course, even further back into the world of unicellular organisms.

The Growth (and Decline) of Morphological Complexity

Multicellular groups have evolved many times from various unicellular ancestors, but only a few of them have gotten beyond the stage of colonies, that is, of associations of essentially identical cell types (Buss 1987; Knoll 2011). Metazoans, by contrast, have many different cell types, from four or five kinds among the simplest free-living forms at present to several hundred kinds in humans. The vast majority of metazoans use cells as basic building blocks in a hierarchical system of organization—forming tissues from cells, organs from tissues, and organ systems from organs—leading to highly complicated body plans. The most basal metazoan clade, a sponge-like form, had only a weakly hierarchical morphological architecture, however.

Cell types, like clades, have family trees and in many cases appear to have evolved from multifunctional ancestral cells to types that became specialized for a narrower range of functions. This sort of specialization is likely to have been involved in the evolution of metazoan multicellularity from unicellular ancestors because eukaryotic unicells must each perform the whole range of metabolic and behavioral functions necessary to their lives, whereas cells in metazoans can be relieved a some of these tasks by specialized cell types and can become simpler (McShea 2002). Even though the cells may become simpler, however, the morphology of the organism as a whole tends to become more complicated as cell types are multiplied, for in the most part they are being used in tissues and, especially in eumetazoans, in organs that are associated with increased morphological complexities. In addition, judging from the relatively low number of cell types in lineages suspected of having undergone morphological reductions, when lineages are simplified, the number of cell types is evidently reduced.

A model of the functional segregation of cell types as they branch and diversify during the rise of morphological complexity has been presented by Arendt, Hausen, and Purschke (2009) based on the evolution of photoreceptors. The earliest "eye" may have been a cell that sensed only light intensity but required shading of light-sensitive parts of the cell, likely achieved by intracellular granules. Later, cilia were incorporated into the primitive eye to respond to light cues by cell movements (phototaxis). Arendt and colleagues consider the possibility that all three of these functions (photoreception itself, light-shading via pigment granules, and steering via locomotory cilia) were once served by a single cell type. As eyes evolved further, however, three distinct functional cell types eventually appeared, separating each of these functions. Of course, many cell lineages have also evolved functions that were outside the range of their ancestral cell types. A likely example in eye evolution is the appearance of rhabdomeres, microvilli of cells that are arrayed at the receptive surface of many protostome eyes.

The differentiation of specialized cell types thus accompanies the evolution of tissues and organs as they become more specialized and, commonly, more complex and as they presumably contribute to improved adaptation. Distinctive evolutionary trajectories have underlain the integration of the morphological, physiological, and behavioral systems of each metazoan body plan. We cannot attempt to trace the myriad and disparate contributions to the rise of cell-type diversity within all metazoan body plans, but we can attempt a brief look at some of the distributional patterns of major morphological features that are revealed by the branching pattern in the tree of life, to which we now turn.

BUILDING EVOLUTIONARY TREES

Understanding animal diversification before and during the Cambrian explosion requires knowledge of the relationships among the various major groups of animals as well as between animals and their closest relatives. With these relationships as a foundation, we can plot the timing of divergences between different groups with fossil and molecular evidence, calculate evolutionary rates, and begin to puzzle out the patterns of morphological transformation and eventually the ecological and developmental processes that must have been involved. The introduction of molecular DNA sequence data to investigate the relationships among the metazoan phyla began in 1988,[3] and with the development of new analytical techniques and reinvestigation of morphological data, a revolutionary new topology has been established for the metazoan tree. In reconstructing the tree of life, however, neither morphological nor molecular data are foolproof, for many analytical techniques introduce their own errors and challenge our ability to discern relationships. In addition, as with attempts to reconstruct the history of environmental change during the Ediacaran-Cambrian explosion from geochemical data, attempts to reconstruct from molecular data the topology of the tree of life during those remote times are beset with special difficulties, with relevant evidence still lacking for some important branches. Our understanding of the tree has nevertheless been advancing rapidly so that we can now infer the role of the Cambrian events in the origins of major branches of the tree of life with some confidence.

Phylogenetic Evaluation of Morphological Patterns

An important objective of cladistics is to identify the sequence of branching within a phylogenetic tree. When an evolving lineage branches, the last common ancestor of the daughter branches lies at the node (fig. 4.2, species a), and each branch has a founding species, the daughter species of that common ancestor (fig. 4.2, species b and b′). These daughters are by definition sister species; that is, they arise at the same node. Each of the branches, called a clade, becomes a separate taxon, with one of the sisters as the founding species. In cases in which we do not know the branching order among related clades, they can be treated as arising from a single diversification, creating a polytomy (producing more than two branches), although with higher taxa, polytomies are simply covers for our ignorance of the actual branching order. When two clades emanate from the same node, they are sister clades (fig. 4.2, clades 1 and 2 and clades 3 and 4). By definition, each clade must include all the

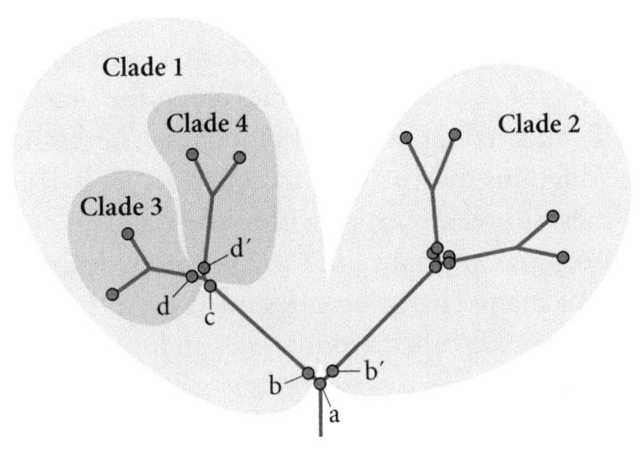

Figure 4.2 Some concepts in cladistics: a, the last common ancestor of clades 1 and 2; b, b′, the founding taxon of clades 1 and 2, respectively; and d, d′, the founding taxon of clades 3 and 4, respectively.

subsequent descendants of its founding species. Each branching within a clade produces two new clades, however, each with a founding species (as species d and d′ give rise to clades 3 and 4, respectively, in fig. 4.2). Later members of a clade may be quite unlike the founding species, even in basic architecture, but owe their clade membership to their position as descendants of that founding species. Thus, in cladistics, the occurrence of a node rather than the appearance of a morphological innovation as in Linnean taxonomy is used to break the evolutionary continuity. Clades are often given names, which are based on a branching in the tree of life rather than on the origin of a morphologically defined taxon. There are no ranks in cladistic classifications, which are not hierarchical, although later clades are nested within earlier ones.

Sister species will share most of their features because they are inherited from their common ancestor, but usually they will have some morphological difference(s) as well. The inherited characters are called ancestral or plesiomorphic, and the novel feature(s) are called derived or apomorphic. These relations can be illustrated in a cladogram, a tree-like figure that is not a phylogenetic tree but rather is one that shows the distribution of characters among species, the sequence of which can be used to infer their branching order. Figure 4.3 is a cladogram showing basic types of character distributions among three species. An apomorphy is a derived character, found in a given species and not in others. When a species with an apomorphic feature speciates, both daughter species will usually inherit the apomorphy and thus will have a derived feature shared with their parent, now called a synapomorphy. The synapomorphy can be used to identify members of the clade founded by their parent, provided that all the descendants share the same derived feature (i.e., that it is not lost somewhere along the line). As each of the descendants becomes a distinct species, however, it will usually evolve at least one morphological apomorphy of its own, which may be passed on to its descendants in turn. Thus, there will be accumulations of nested apomorphies along the branches that become synapomorphies in succeeding daughter branches. The branching sequences worked out in cladograms may be used to construct a phylogenetic tree, representing a hypothesis of relationships within the clade.

True synapomorphies must be due to common descent—homologous, in other words—but many characters are similar even though independently evolved. Such characters are called homoplasies and can reflect convergence of form due to similar evolutionary pressures but reflecting different genetic and developmental mechanisms or due to parallel evolution. Characters that are present in forms ancestral to a clade and that are thus not only inherited by all clade members but also are shared with the ancestors are called plesiomorphies. Although plesiomorphies should be homologous characters, they can-

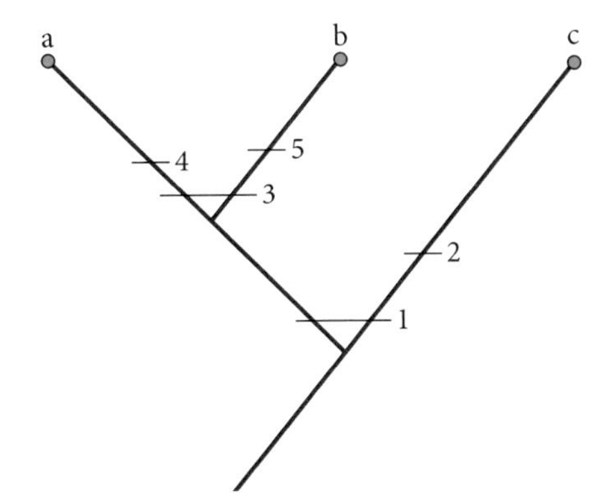

Figure 4.3 Some terms in cladistics illustrated in a cladogram of three species a, b, and c that show characters 1 through 5. Character 1 is synapomorphic for the clade of species a, b, and c, but is plesiomorphic (ancestral) for the clade of species a plus b. Character 2 is apomorphic for species c. Character 3 is synapomorphic for species a plus b. Character 4 is apomorphic for species a, and character 5 is apomorphic for species b.

not be informative of relations within a clade because they are common to all its members, unless they are lost in some later branch. When characters that form important criteria for positioning a lineage are subsequently lost, the branching order may be obscured.

Until the rigorous character definition that was required by cladistic methodology came into general use in the 1980s, systematists did not always realize the frequency of character loss and homoplasy. If a taxon is indeed a true clade, founded by one last common ancestor, it is monophyletic. By contrast, when homoplasies are misinterpreted as shared derived features—as synapomorphies—lineages with distant ancestors can be mistakenly placed together in the same clade. Such a taxon, which does not contain the last common ancestor of all the contained species, is polyphyletic. When systematists identify polyphyletic clades, they emend them to produce monophyletic ones. Computer programs now allow large data matrices of many species and characters to be analyzed using a variety of different algorithms. Although parsimony analyses, relying on the fewest number of character changes, were initially favored, most phylogenetic analyses today rely on maximum likelihood or Baysian approaches that implement sophisticated modeling of evolutionary processes. The resulting phylogenetic trees are hypotheses based on the available character information and the best analytical methods, but the trees are subject to revision. When we use Linnaean categories in this book, we treat them as if they are monophyletic clades; this assumption is sometimes uncertain, however, and we will mention it if cogent.

Throughout this book, we will focus on a number of critical nodes: the last common ancestors (LCA) of various clades. The bilaterian LCA, for example, is the last common ancestor of all bilateria and defines a clade that includes all the descendant bilaterian groups (not all of which necessarily look bilaterally symmetrical, such as many echinoderms). Divergences that occurred before an LCA of living groups are called stem branches (Jefferies 1979), and they form stem clades. By definition, they cannot be ancestral to any living group. Living clades are called crown clades (because their descendants are at the crown of the tree of life), and their LCAs can be partly reconstructed from features held in common among their descendants, including molecular sequences and their functional implications.

Phylogenetic Analysis of Molecular Data

Nucleotide sequences provide another critical source of phylogenetic information. Although we cannot recover molecular data from long-extinct taxa, those sequences are an invaluable source of information for resolving the relationships among living clades. With a framework for living clades in hand, phylogenetic systematists can then resolve the phylogenetic position of fossil groups. Fossils often provide critical information in resolving phylogenies because they incorporate combinations of characters that are not present in living taxa. The relationships between onycophorans and arthropods and among the basal lophotrochozoan clades (annelids, mollusks, and their allies), as we will see, can only be resolved through a combination of molecular and morphologic data. The evolution of molecules is hardly more straightforward than that of morphology, however, and there are many complexities to deep phylogenetic reconstruction. For example, the rate of molecular sequence change can differ along different branches of a tree; consequently, a species with more rapidly evolving sequences may appear to be less closely related to its last common ancestor than a species with more slowly evolving genes that will retain more of the ancestral sequence. Two species that are only very distantly related may share a highly conserved gene, one that has experienced little change even over hundreds of millions of years, suggesting that the two species are closely related. And, oddly enough, if two species share a gene that evolves rapidly in both, they may appear more closely related than is the case[4] because evidence of their relation to their actual common ancestor has been degraded by all the changes. Because even two random sequences of nucleotides—three in sixteen, on average—will be similar by chance (as there are only four different nucleotides), independently and rapidly evolving genes may come to resemble one another more closely than they resemble those of their more slowly evolving relatives. Therefore, the sequences in rapidly evolving genes may suggest a closer relationship than is the case. This phenomenon is called long-branch attraction, so named because if branches on the tree are scaled to the distances (i.e., the differences) between sequences, the branches based on rapidly evolving genes are relatively long.

Different genes commonly evolve at different rates, even within the same lineage, and change rates at different times according to selective values arising from the physical environment, from interactions with other species, or even from interactions between the gene's products and the products of other genes in the same genome. Therefore, trees based on different genes commonly differ. Gene duplication presents an additional challenge. Many genes have been duplicated within a genome, giving rise to more than one copy. Genes that are descended from a gene in an ancestor are said to be orthologs of that gene, whereas genes descended by duplication from a gene in the same genome are paralogs. Although the presence of paralogs in the genomes of different groups indicates that there was a common ancestor in which the paralogs were present, it does not help much to clarify the relations between the groups. Finally, genes may also be transferred between the genomes of different species, which confounds a tree-like interpretation of their evolutionary descent. Although such lateral gene transfer is common among unicellular forms, it seems much rarer among metazoans and is not likely to influence the metazoan tree of life significantly.

MicroRNA (miRNA) molecules, which are processed from RNA transcripts and serve as important regulatory molecules in metazoan development (see chap. 8), provide another line of molecular evidence for phylogenetic position that is important in assessing major branching patterns. The number of miRNA molecules present in metazoan genomes increases more or less in concert with increases in body-plan complexity, and many of the miRNAs are conserved within the lineages in which they originate. Thus, the presence of a given suite of miRNAs in a metazoan genome provides evidence of the clade to which it belongs and—if it is assumed that miRNAs have not commonly been lost, which appears often to be the case—can provide somewhat weaker evidence of clades to which it does not belong.

Despite this catalog of difficulties, phylogenies based on molecular evidence are invaluable, providing new data, independent of morphology per se, with which to place living clades in the tree of life. False phylogenies created by morphological homoplasies can often be uncovered by molecular sequence and miRNA data. On the other hand, when divergences suggested by molecular data are affected by such artifacts as noted above, morphological (including embryological) evidence may indicate the more correct tree topology. Because large volumes of sequence data can now be obtained relatively easily, and because entire genomes have been sequenced for a growing number of animals, many of these problems are becoming resolved for living clades.

Phylogenetic Evaluation of Fossils

Many of the clades found in the Cambrian have living representatives, and molecular data from these extant taxa provide a phylogenetic framework for resolving the relationships among their more ancient ancestors, but one of the great problems in assigning ancient fossils to places in the tree arises from the incompleteness of the fossil record. There are huge gaps in our knowledge of the early branches of Metazoa. We have hardly any record of soft-bodied forms, and even among those groups with durable skeletons that are most likely to have been fossilized, far more taxa are missing than are preserved so that the gaps among preserved lineages effectively erase the morphological pathways that once connected them. This problem of missing fossils is most acute for the phases of Metazoan evolution between the invention of the sponge body plan (i.e., the appearance of the first metazoan) and the Cambrian explosion, an interval when nearly all the fauna lacked hard parts. During that time, there were many important divergences in the tree of life, and characters accumulated along separate evolutionary pathways to produce distinctive body architectures recognized today as separate phyla. If, however, the ancestor is separated from two (or more) descendant clades by a significant morphological gap, it becomes difficult to decide whether common features are truly ancestral or are homoplasies. This situation is common when dealing with the early fossil record, where the gaps are long and encompass the key divergences that established major living groups of animals. By the time of the Cambrian explosion, there were more than thirty lineages with distinctive body architectures. Alas, the founding ancestors of most of those phyla were soft-bodied, whereas unique body plans appear suddenly as fossils and then disappear from the record, with no clear connection to any living clade.

Molecular Clocks

Soon after publication of the structure of DNA in 1953, it was suggested that comparison of DNA sequences in separate lineages could provide a measure of the time since their divergence, and in 1965 the first attempts were published using such a molecular clock (Zuckerkandl and Pauling 1965). Molecular clocks assume that molecular sequences evolve at a relatively constant rate and that the rate of change can be calibrated from the age of first appearances of living clades with particularly good fossil records.

Turning these notions into a reliable tool has proven difficult, especially for nodes originating as deep as the Cryogenian to Cambrian. Different genes, and even different portions of a single gene, evolve at different rates. Clearly, faster-evolving genes are appropriate to date more recent divergences, as among Cenozoic radiations, whereas slowly evolving ones are required to date more ancient divergences, such as those between the LCAs of phyla. Furthermore, because different lineages can show significantly different rates of molecular evolution for the same molecules, homologous sequences that have evolved at closely similar rates must be found.

To date the LCAs of lineages for which appropriate molecular sequence differences have been identified, we must have some measure of the rate of change of the measured sequences, which requires using the fossil record to calibrate the results, converting changes in sequences to absolute age dates. The reliability of the calibration depends on the distribution and quality of calibrated ages, however; numerous, well-constrained calibrations across the tree provide the most reliable estimates of divergence times. Because different clades can show significantly different rates of molecular evolution, rates for a calibrated clade may give erroneous results if applied to another clade. Vertebrates, for example, display slower rates of molecular evolution than invertebrates; thus, vertebrate calibrations overestimate invertebrate divergence times (Peterson et al. 2004).

Fossil calibrations are most reliable if available calibration points are both older and younger than the interval of interest so that we are interpolating the ages of LCAs between known divergence times. This goal is clearly extremely difficult to achieve using the Cambrian forms themselves. During the later Phanerozoic, however, when the fossil record is much better (especially for crown groups), the dates of first appearances of crown clades can commonly be estimated and used to calibrate the evolutionary rate of molecular sequences. For early metazoan divergences, though, most if not all the calibration points must be younger than the divergences we are interested in dating, so our molecular clock dates are extrapolated beyond the calibration dates. Such extrapolated clocks can be particularly challenging because a substantial change to the rate of molecular evolution would be hard to detect.

There are still other problems with molecular clocks, but approaches that minimize most of these problems have led to dates that tend to show internal consistency (i.e., seem to be in the correct order) and seem reasonable when tested by the fossil record. In general, however, the nodes dated by molecular clock techniques are significantly older than nodes judged from the fossil record, commonly as much as one-third older, and some, surely erroneous, have been literally 1 billion years older.

At least part of the differences in molecular clock and fossil-based dates for ancestral events is that the two methods are dating different things and will differ even if both methods happen to be accurate. Molecular clocks date LCAs, which are essentially identical to the founding ancestors of the lineages that began at that time. By contrast, fossil ages are based on morphology, and for major clades they date the appearance of new body plans. As noted, the body plans of sister phyla may not appear for many millions of years after their LCA (see also chap. 9). The higher the taxon, the longer this temporal gap between the founding ancestor and the appearance of synapomorphies that characterize its morphology, and for major taxa such as phyla it is no surprise that LCAs date earlier than the first appearances of the phylum-level morphologies in the fossil record; indeed, it is inevitable if all dates are accurate. Molecular clocks, however, should not yield dates for the origins of clades that are younger than fossil calibration dates.

A HYPOTHESIS OF ANIMAL RELATIONSHIPS

The latest metazoan phylogeny, based on molecular data from living species but with fossil clades included, is shown in figure 4.4. The tree is based chiefly on phylogenetic analyses[5] of molecular sequence data, but where it seems likely that an artifact is creating a false branching pattern, either embryological or adult morphological information is used to find a more reasonable pattern. Polytomies, as in the lophotrochozoans, reflect unresolved phylogenetic relationships, a convention to convey our ignorance. Although this tree is based on the latest evidence, it will not be the last word. The tree has been continually revised and continues to be changed as new findings are made through laboratory and field work. Figure 4.5 shows the results of a combined molecular phylogeny and a molecular clock analysis of the divergence points for a large suite of extant taxa.

Three critical points are evident from the molecular clock results. First, the origin and divergence of sponges and cnidarians occurred in the Cryogenian. Second, most bilaterian clades diverged during the Cryogenian and Ediacaran, although fossil evidence for them is not found until the late Ediacaran or Cambrian. Third, bilaterian crown groups diverged in the late Ediacaran and Cambrian, congruent with their first fossil appearances; many of the fossils found during the early Cambrian represent stem clades. Not surprisingly, divergences from common ancestors precede the appearance of the descendant body plans, which makes perfect sense because the stem ancestors were likely to have been soft-bodied and commonly morphologically non-

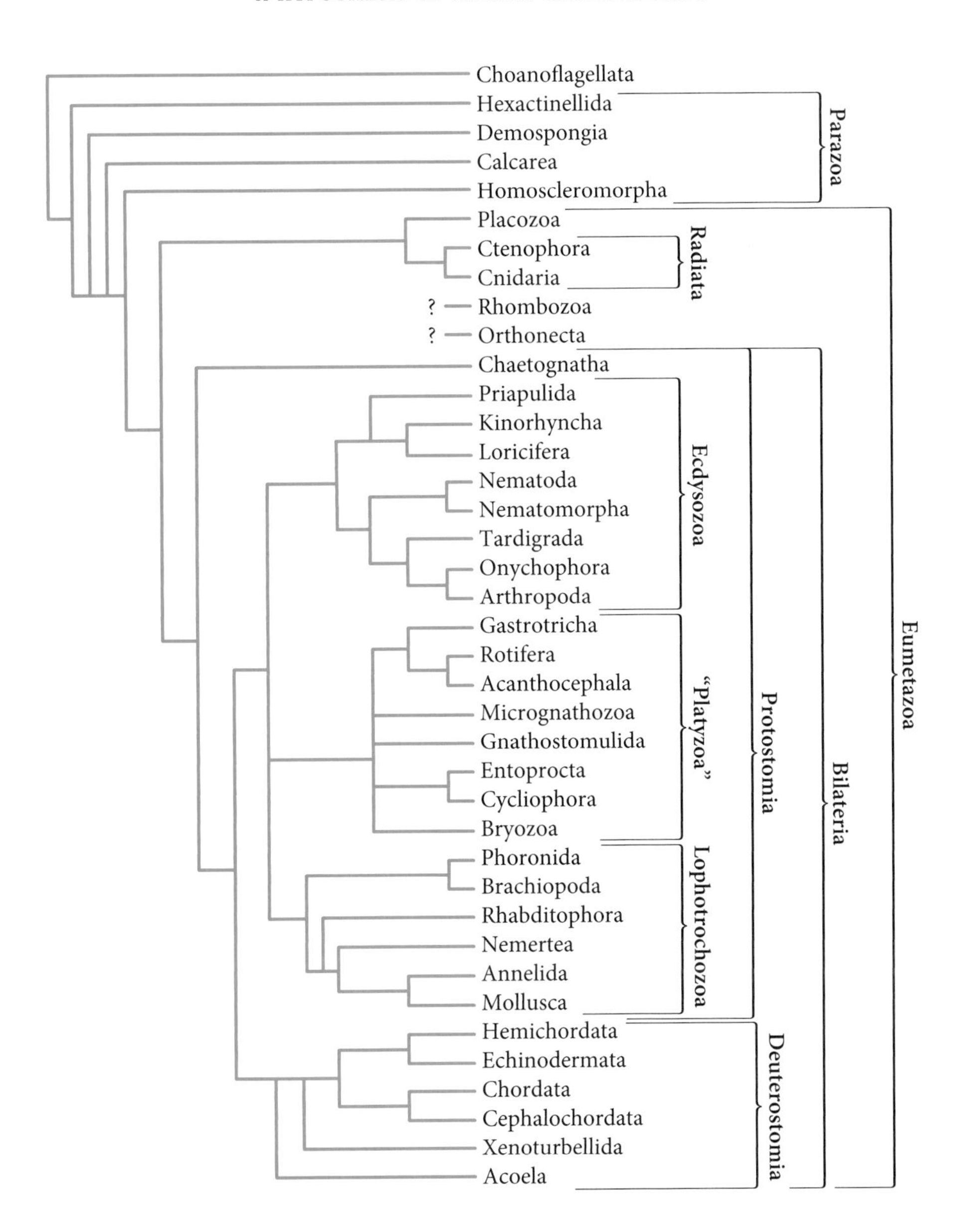

Figure 4.4 A phylogenetic tree representing a hypothesis of relationships among major metazoan clades. The tree is based on the more common relations found in molecular phylogenies, supplemented or modified by morphological and developmental data when the available molecular sequences are few or are ambiguous.

diagnosable (chap. 9). On the other hand, the gaps do not appear so large as to be considered patently misleading given the nature of the fossil record. For example, the protostome/deuterstome LCA is dated at 700 Ma, but the first

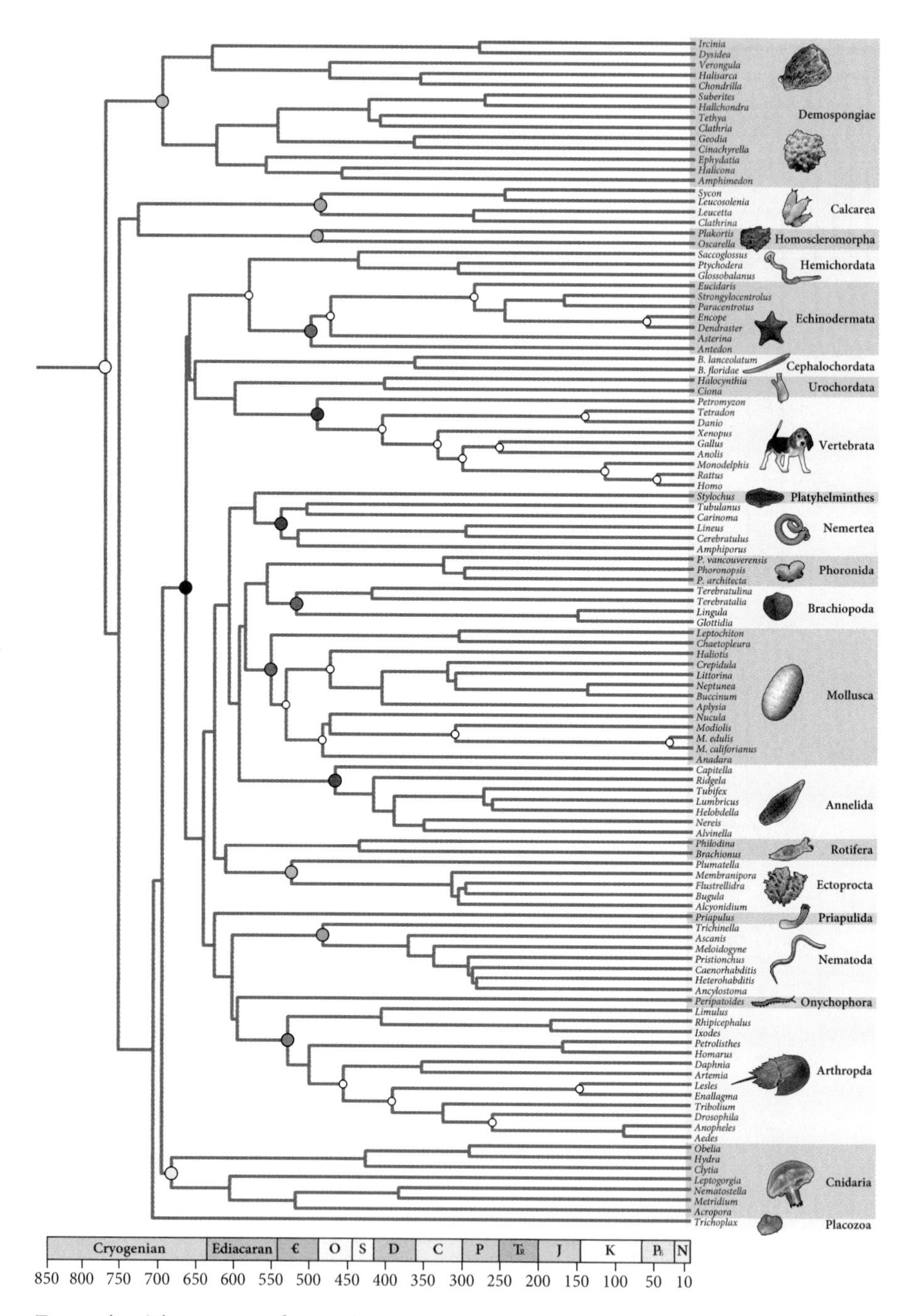

Figure 4.5 The pattern of animal taxa as inferred from the molecular clock. Seven different housekeeping genes from 118 taxa were used to generate this chronogram. Twenty-four calibrations (open circles) were used and treated as soft bounds. (*continued*)

body fossils of crown phyla descended from that LCA appear at about 530 Ma. Most of the "missing" phyla are small-bodied and, with one exception (most Bryozoa), lack mineralized hard parts. We will return to the nature of those missing early bilaterian body plans in chapter 9. That is not to say that this tree is very misleading; the major groupings are almost certainly correct, corroborated by many lines of evidence, and some of the unknown branching sequences do not really matter for our purposes.

In the remainder of this chapter, we provide a *précis* of the major elements of this tree, beginning with the sister clade to Metazoa and then progressing through the major clades. This basic commentary—including relevant details of their anatomy, biomechanics, and physiology—is based on living representatives. Many elements of metazoan architecture—for example, whether a fluid skeleton is present, much less whether it is pseudocoel, coelom, or hemocoel—cannot usually be answered by evidence from the fossils themselves, even the wonderful critters of Burgess Shale–type faunas. When discussing the fossils, we shall lean heavily on the basic elements found in their living descendants, which also serve to determine their places in the phylogenetic tree.

Choanoflagellata

Both molecular evidence (Carr et al. 2008; N. King et al. 2008) and morphological evidence indicate that the closest living relative of sponges is the unicellular group Choanoflagellata, which unfortunately lacks a fossil record. Choanoflagellate cells have a collar of minute "tentacles" (microvilli) (fig. 4.6A, 1) surrounding a flagellum. The flagellum forces water up and away from the collar, thus drawing water in through the microvilli where bacteria and bacteria-sized algae are trapped and ingested. Choanoflagellates are both solitary and colonial, and they include both benthic and pelagic species (fig. 4.6A–C).

Figure 4.5 (*cont.*) There is general concordance of bilaterian phylum-level crown groups (colored circles; the color of each circle is the same as the corresponding taxonomic bar and label on the far right), with the first appearance of most animal groups at the Ediacaran-Cambrian boundary. In contrast, the origins of the demosponge (dark blue) and cnidarian (yellow) as well as the bilaterian (black) and metazoan (gray) crown groups are deep in the Cryogenian. Geological period abbreviations: Є, Cambrian; O, Ordovician; S, Silurian; D, Devonian; C, Carboniferous; P, Permian; Tr, Triassic; J, Jurassic; K, Cretaceous; Pe, Paleogene; N, Neogene. Adapted from Erwin et al. (2011).

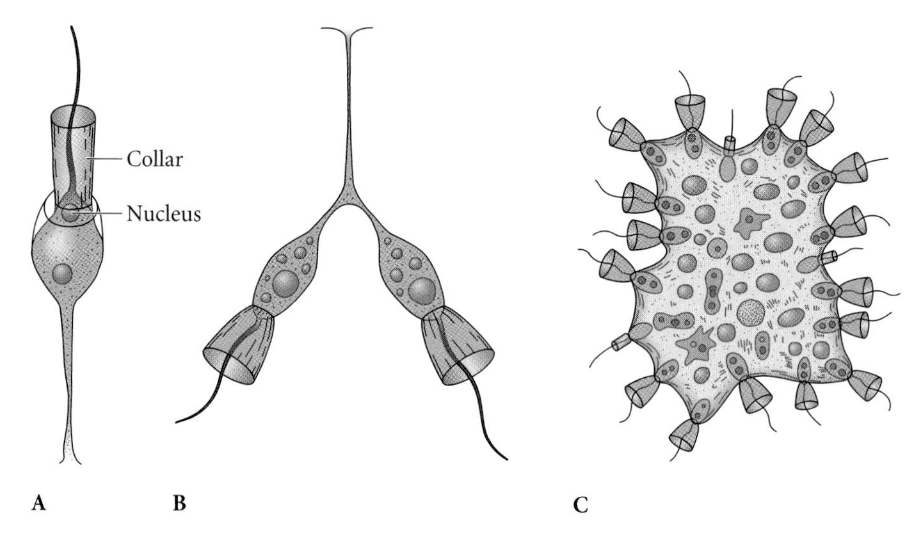

Figure 4.6 Living choanoflagellates. (A) The benthic stalked choanoflagellate, *Codosiga*; 1, collar, 2, nucleus. The collar comprises numerous closely packed microvilli, tentacle-like extensions of the body wall. (B) Paired cells of a species of *Codosiga* that lives in tide pools by attaching to the surface film. (C) Colonial choanoflagellate, *Codonosiga*; cells are embedded in a gelatinous matrix, and those in more central positions have absorbed their flagellae and are dividing. The spherical objects in the cells of these three species are vacuoles containing food or metabolites. Cells, with collars, are on the order of 5 to 15 microns tall. (A) and (C) adapted from Hyman (1940); (B) adapted from R. E. Norris (1965).

Sponges

The critical innovation associated with the origin of Metazoa was the ability to form multicellular, differentiated organisms. Among the six to ten cell types found in most sponges, some are arranged in layers (fig. 4.7). The remaining Metazoa all possess epithelial tissues in which cells are bound into sheet-like tissues by various types of adhesion molecules. Most workers do not regard the sponge cell layers as true tissues because they are not usually underlain by basal membranes (although some are), nor are the cells usually connected by molecular junctions (although they are in some larvae), nor are organs differentiated from the cell layers. Sponges are the only animals that can be disassociated into single cells and still reassemble. Thus, sponges are commonly characterized as being at the cellular stage of construction; in other words, sponge architecture does not necessarily rely on epithelial tissues. We will revisit this issue in chapter 9 when we consider developmental information based on genomics; it turns out that demosponges have many of, but not all,

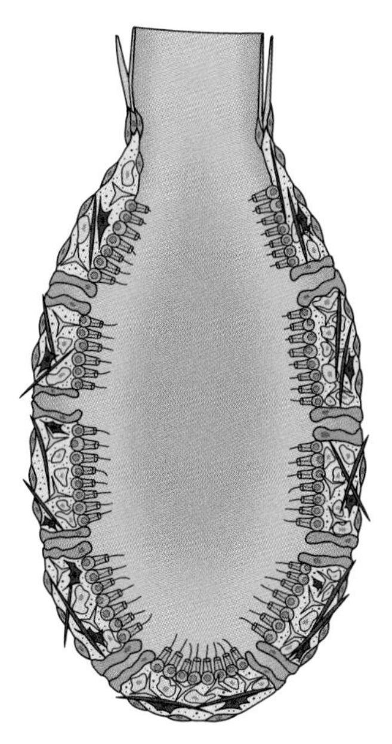

Figure 4.7 Sponges. Schematic of the simplest sponge type with choanocytes lining the interior, which is essentially a feeding chamber, and with a variety of other cell types and mineralized spicules forming the body wall. Adapted from Hyman (1940).

the genetic instructions for components of eumetazoan epithelia (Fahey and Degnan 2010). Sponges usually have two cell layers separated by an extracellular gelatinous matrix, and during early development, certain sponges show growth patterns and cell movements that give rise to this double-layered form. Sponges have no nerve or muscle tissues. Most sponge body walls contain spicules of silica or calcite that stiffen the wall and are sometimes fused, thus preserving the body form. Feeding in sponges is carried out by flagellated cells called choanocytes, cells that resemble choanoflagellates, which are arrayed in feeding chambers (fig. 4.7).

Recent molecular evidence suggests that living sponges belong to at least three monophyletic crown groups (Sperling, Peterson, and Pisani 2009; Sperling, Pisani, and Peterson 2007): Hexactinellida and Demospongia (with siliceous spicules when spiculate and sometimes grouped in the clade Silicea), Calcarea (with calcareous spicules), and the Homoscleromorpha (Gazave et al. 2012). Hexactinellids are the most basal of these groups based on molecular evidence, and many of their important tissues are syncytial (containing numerous nuclei but lacking cell walls), possibly evolving from earlier cellular forms that are now extinct and were descended from choanoflagellates.

The Homoscleromorpha bridge the gap to other metazoans, with basal membranes beneath some cell layers and other features uncommon or lacking in sponges but known among epithelial metazoans (fig. 4.4; Nichols et al. 2006). Basal membranes are part of the extracellular matrix of metazoans and contain collagen; the homoscleromorph sponges have type IV collagen, a form that is characteristic of epithelial basal membranes in eumetazoans (Boute et al. 1996). Because homoscleromorph sponges are more closely related to eumetazoans than to the other sponge groups (Sperling, Pisani, and Peterson 2007), sponges are paraphyletic, which also means that the LCA of "sponges" must have had a sponge-like morphology.

The Simplest Epithelial Metazoans

The simplest free-living modern metazoan phylum that exhibits a tissue-based architecture is the Placozoa, represented by a single living genus, *Trichoplax*. Placozoans are minute, flattened, irregularly discoidal, and although differentiated dorsoventrally, they are not bilaterally symmetrical, suggesting an affinity with radiates (cnidarians and ctenophores). There are only four somatic (nonreproductive) cell morphotypes, the smallest number known in free-living metazoans. The defining characters of the clade include upper and lower epithelial-like tissue layers, both ciliated, which are separated by a fluid-filled cavity containing fiber cells. The genome includes genes that code for extracellular matrix constituents, including collagen IV, although extracellular matrix has not been detected in placozoan bodies by conventional staining methods (Srivastava et al. 2008). Although the two tissue layers may be candidates for ectoderm and endoderm homologues, there is no gut and the tissues do not give rise to organs; thus, placozoans are not usually classed among the Eumetazoa (see below). There are no muscle cells and no neurons. Locomotion is ciliary or "amoeboid" (based on body-shape changes mediated by contractions of the fiber cells). In aquaria, placozoans creep over algal films on glass walls and digest the algae extracellularly, and when they move on they leave an area somewhat cleared of algae. Reproduction is achieved by fission and by budding off of "swarmers," cells derived from the upper epithelium, but because possible embyros have been observed, sexual reproduction has been suspected and is supported by possible signs of recombination found in a small sample of genes (Signorovitch, Dellaporta, and Buss 2005). The recent sequencing of the complete genome of *Trichoplax* strongly supports their placement between sponges and the radiate clades (Schierwater et al. 2009; Srivastava et al. 2008). Placozoans have no fossil record, but if their phylogenetic placement is correct, they diverged during the Cryogenian.

Classic Eumetazoa

Eumetazoans exhibit more complex development in which fertilized eggs undergo cleavage—that is, they split into smaller and smaller cells—to produce a ball of cells, the blastula, which commonly has more than sixty cells and may be either solid or hollow (fig. 4.8). The cells of the blastula then proliferate to produce a gastrula, a double-layered embryo with two distinct epithelial tissue layers, the ectoderm (the outer layer) and the endoderm (the inner layer) (fig. 4.9). Organs arise from these two "germ layers" as development proceeds. For example, the integument and nervous system arise from ectoderm, and

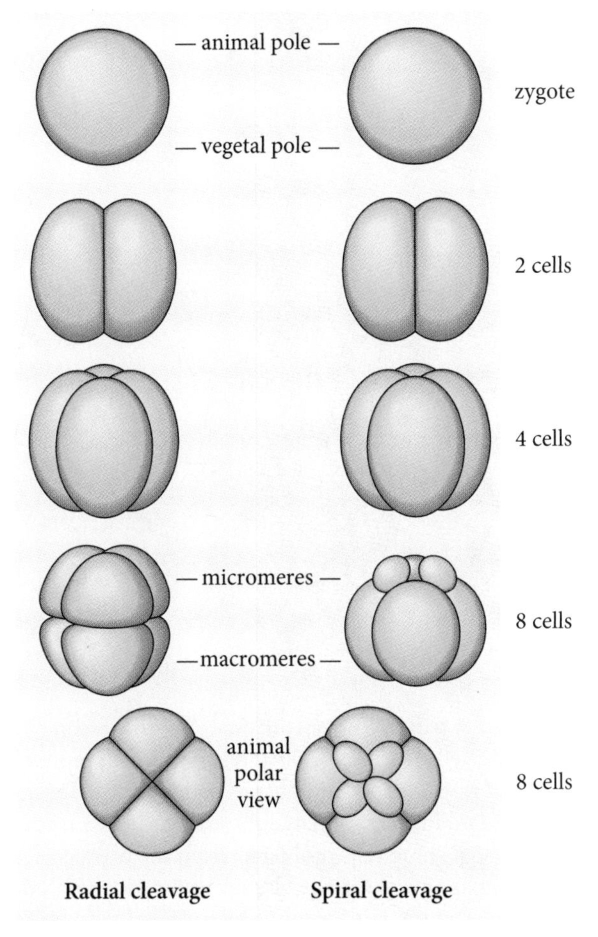

Figure 4.8 Two important metazoan cleavage patterns, radial and spiral, shown only to the eight-cell stage; most cleavages continue to a sixty-four-cell stage or thereabouts, forming a blastula, before cell proliferation through mitoses of cell lines produces tissues. Adapted from Brusca and Brusca (1990).

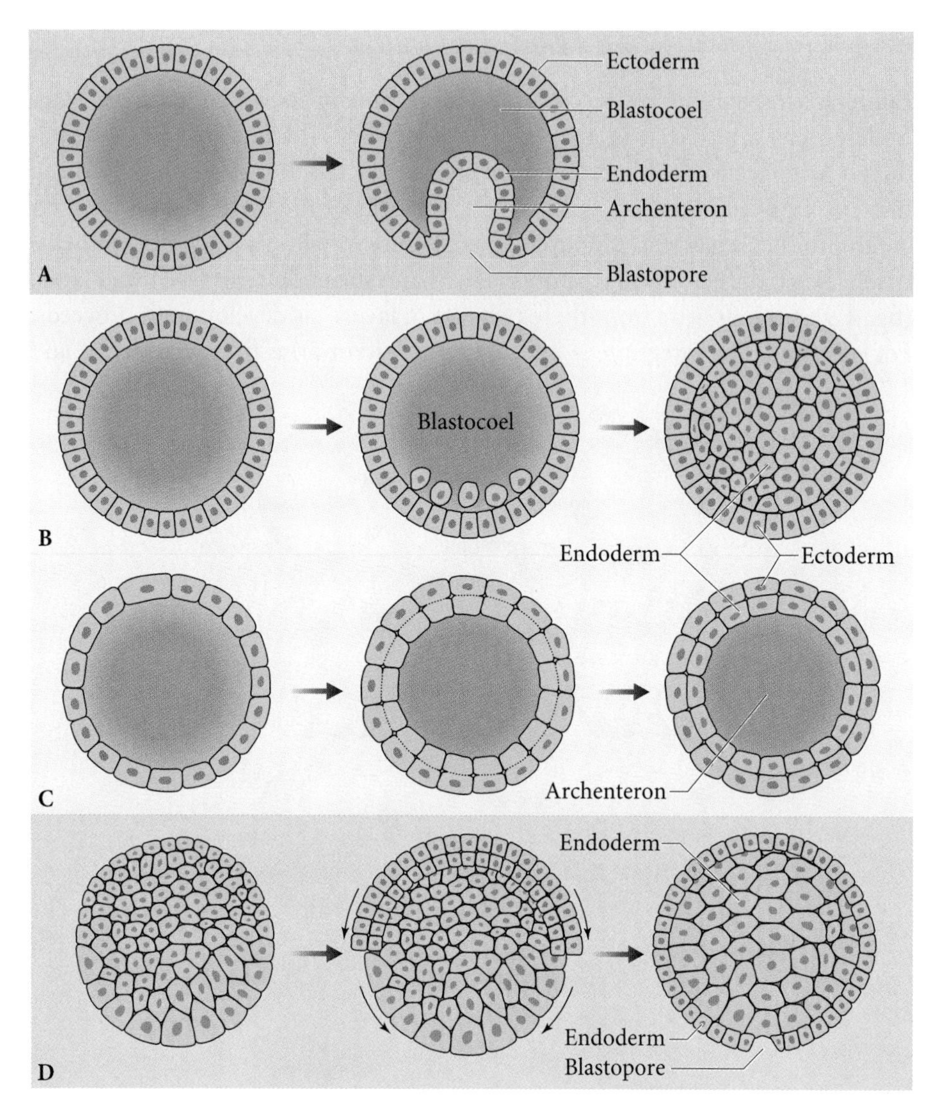

Figure 4.9 Four common patterns of gastrulation. As eumetazoan cleavage terminates at the blastula stage and cells begin to proliferate, gastrulation produces a two-layered embryo from the blastula via one of a number of pathways. (A) Invagination of the wall of a hollow blastula, found in some cnidarians; the space between the tissue layers is the blastocoel. (B) Introgression; cells proliferate from the blastocoel wall, common in cnidarians. (C) Delamination; cells in the blastula wall divide to produce an inner layer. (D) Epiboly; cells grow over the outside of the blastula and become ectoderm, usual in Ctenophora. Adapted from Brusca and Brusca (1990).

the gut and digestive glands arise from endoderm. In ctenophores and cnidarians, these layers are separated by a layer of extracellular matrix, the mesoglea, and the adult retains this two-layered or diploblastic structure. In bilaterians, a third germ layer, the mesoderm, arises between ectoderm and endoderm to produce a triploblastic architecture. Mesoderm is recognized as a germ layer because it also produces organs, most notably muscles. Eumetazoans include two major groups, the radiate phyla and the bilaterians (fig. 4.4).

The Radiate Phyla

The two radiate phyla, Ctenophora and Cnidaria, have muscle cells that form discrete muscular tissues in some groups, but they are not developed into an extensive third germ layer that gives rise to an array of organs as does mesoderm in Bilateria (see below). Both clades share a generally radial symmetry, although not a perfect one; for example, some organs or structures occur singly rather than being repeated in a radial pattern. Neither radiate group has a vascular system or a body cavity (digestive cavities, which connect directly to the exterior at least through the mouth, are not considered to be "body cavities"). The nervous systems are generally net-like, although nerves are condensed into tracts in some cases; in cnidarians, there are both ectodermally derived nerves, as in eumetazoans in general, and endodermally derived nerves, a feature that is evidently unique among metazoans and that includes the nerves connecting the nerve net (Marlow et al. 2009). Radiates are nevertheless relatively simple stucturally; cnidarians, for example, attain dimensionality largely by draping and folding their tissue layers, causing Shick (1991, 3) to comment that they "are at the origami level of construction." The phylogenetic relationship among cnidarians, ctenophores, and other eumetazoans remains cloudy. Cnidarians are clearly a sister group to the remaining eumetazoans, but the phylogenetic position of ctenophores has not been stable in molecular phylogenies, some of which have them diverging before cnidarians and, in one case, somewhat implausibly, even before sponges (Dunn et al. 2008). We have followed traditional arguments and linked them with cnidarians.

CNIDARIA The phylum Cnidaria (fig. 4.10A–C) of living sea anemones, corals, and jellyfish is chiefly carnivorous. Cnidarians have tentacles that bear specialized stinging cells (cnidocytes) that aid in prey capture and defense and are usually concentrated on tentacles and around the mouth. The digestive cavity (coelenteron) is endodermal and blind. The basal living subphylum Anthozoa (sea anemones and corals) is benthic. Benthic cnidarians tend to

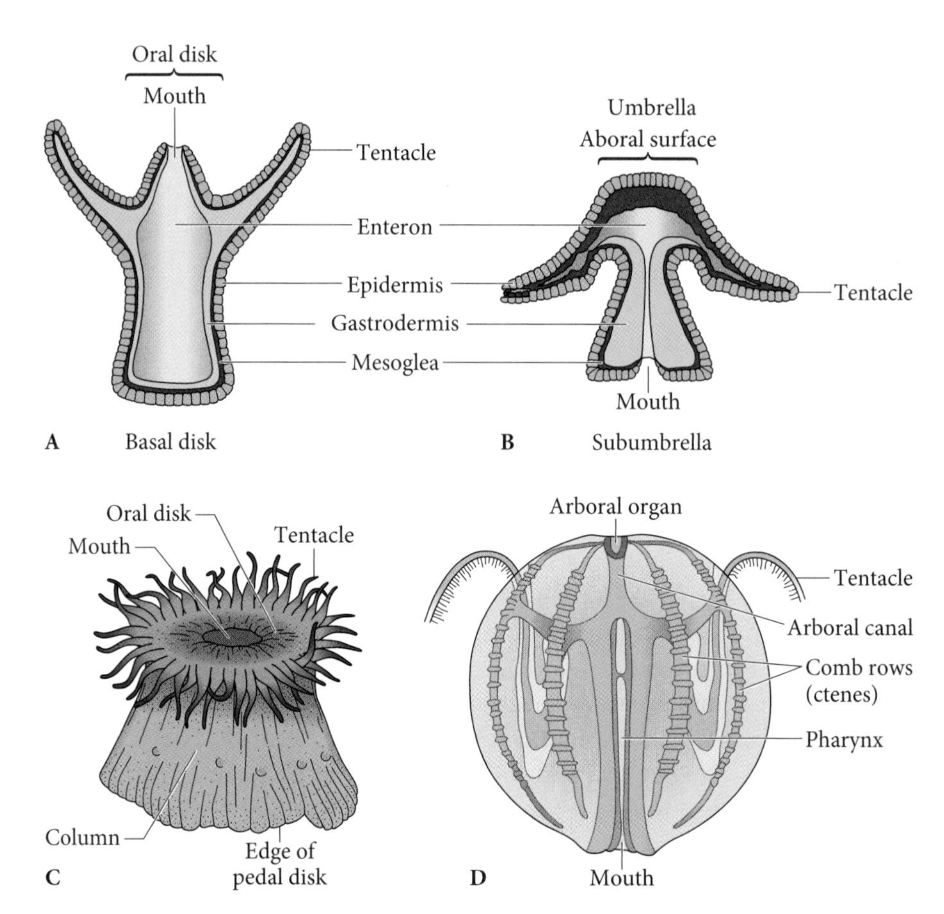

Figure 4.10 Radiate phyla. (A)–(C): Cnidaria. (A) Diagrammatic body plan of a sea anemone. (B) Diagrammatic body plan of a medusozoan jellyfish. (C) Sketch of a modern sea anemone, *Calliactis.* (D) Generalized ctenophore. (A) and (B) after Oliver and Coates (1987); (C) adapted from Bayer and Owre (1968).

capture small prey such as larvae and small members of the plankton, and pelagic forms are important members of oceanic food webs.

CTENOPHORA Ctenophores, also known as comb-jellies, are entirely soft-bodied and chiefly pelagic (fig. 4.10D). They are characterized by multiple rows of so-called combs, ciliary tracts that are embedded in the body wall, running from the apex of the body down toward the mouth. In all living ctenophores, there are eight such rows, each consisting of several thousand fused cilia, the combs beating in a coordinated manner to propel the organism; a swimming ctenophore is one of the more beautiful creatures in the sea. Ctenophores tend to be voracious carnivores preying on zooplankton, crustaceans, jellyfish, and one another; all but one group have two tentacles connected in a trailing, net-like structure used in fishing for prey. Like cnidarians, ctenophores have specialized cells to aid in prey capture, but instead of stinging prey with cnidae, they use an adhesive material packed as granules into a cell, the collophore, that also contains a filament that helps entangle and tether the prey. Other feeding methods such as prey engulfment are used in some groups. Ctenophores have well-developed muscle tissues (of mesodermal origin in bilaterians; see below) and seem morphologically derived when compared with cnidarians, although the branching order is uncertain.

The Bilaterians

Bilaterians (fig. 4.4) are triploblastic, having three germ layers: ectoderm and endoderm as in radiates, plus their third layer, mesoderm, which gives rise to muscle tissues (a large part of the bilaterian body mass) and to gonads, kidneys, blood cells, and so on. Mesoderm proliferates on the site of the blastocoel compartment; if a blastocoel space is retained, it usually lies between the mesoderm and endoderm. As the name implies, these forms are basically bilaterally symmetrical, although they are not perfectly so; some organs or structures are offset, like the human heart, or unique to one side, like the human spleen.

The presence and nature of body cavities—fluid-filled spaces enclosed within tissues—provide a way of characterizing bilaterian architectures (fig. 4.11). Body cavities are important in bilaterian biomechanics and physiology. Minute bilaterian phyla with solid bodies, without internal fluid-filled spaces, are called acoelomate (without cavities) (fig. 4.11A). Other chiefly small-bodied phyla have a fluid-filled body cavity that lies between mesodermal tis-

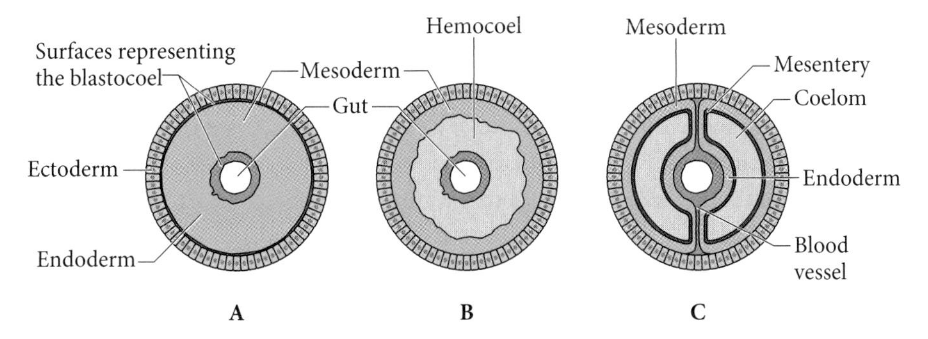

Figure 4.11 Relations of acoelomate bodies (A), primary body cavities (pseudocoels) (B), and secondary body cavities (coeloms) (C), to mesodermal tissues. Adapted from Wilmer (1990).

sues and endoderm—the site of the blastocoel. The blastocoel is somewhat inappropriately called a pseudocoel (false cavity) in adults (fig. 4.11B). The pseudocoel commonly acts as a fluid skeleton, transmitting forces generated by mesodermal muscles that lie along the body wall and helping provide body shape and turgor, with the fluid also furnishing physiological services. None of the acoelomate and pseudocoelomate groups have a circulatory system; their bodies are small enough that oxygen can be supplied to interior tissues by diffusion. Although such small forms share a distinctive grade of body-plan construction, they are found in each superphylum and thus are not necessarily closely related.

Bilaterian groups with more complex body plans attain greater body size than acoelomates and pseudocoelomates. In larger bilaterians, we find new architectural elements, and some are spectacularly preserved in Cambrian deposits (chap. 6). An important design element in larger body plans is a blood vascular system (BVS), a branching space in which hemal fluid (blood) is circulated by a heart or in some cases by more generalized muscular contractions. Like the pseudocoel, the BVS develops on the site of the embryonic blastocoel, so these two spaces are topologically identical. Unlike the pseudocoel, however, the BVS includes vessels that penetrate into tissues that may be derived from all germ layers, but the lumens of the vessels remain part of a branching blastocoel compartment topologically. Thus, the BVS can carry oxygen to tissues remote from the body wall and provides many other physiological services as well. In some vascularized groups, there are also open hemal channels and sinuses, not contained in discrete vessels, which usually lie along the gut. Just as pseudocoels commonly function as fluid skeletons, the BVS is sometimes enlarged to form a voluminous blood-filled space, the hemocoel (blood cavity), which serves as a fluid skeleton to create body turgor or to antagonize muscular contractions. The hemocoel is the main body cavity in a number of phyla.

Another architectural element found in larger metazoans is the coelom, a fluid-filled cavity that is contained entirely within mesoderm and is thus walled off from pseudocoels and the BVS (fig. 4.11C). The histology of the tissue lining the coelomic cavity may vary from place to place even within the same organism, but in many cases it is a muscular tissue. Sheathed in muscle and closed to the outside world (except for specialized ducts), coeloms can function as fluid skeletons in powerful hydrostatic systems and form both the main body cavities and the main axial skeletal elements in many larger-bodied groups.

For more than a century, early development provided the chief criteria for establishing phylogenetic relationships among the bilaterian phyla. Development in larger bilaterians that had blood vascular systems, hemocoels, or coeloms—and, in some cases, durable skeletons—seemed to imply that there were two bilaterian superphyla with contrasting suites of early characters. In one group, Deuterostomia (fig. 4.4), cleavage was usually radial (fig. 4.8), and after gastrulation, the mouth arose at a site remote from the blastopore (deuterostomy), whereas mesoderm arose from cells recruited from the walls of the larval gut or enteron (enterocoely); as these cells proliferated, they incorporated space captured from the gut lumen to serve as eucoeloms, and mineralized skeletons were largely internal (endoskeletons). In the other group, Protostomia, cleavage was commonly spiral (fig. 4.8) or at least not radial; the mouth formed at or near the site of the blastopore (protostomy), mesoderm commonly arose from a mesodermal or endomesodermal precursor cell so that coelomic space had to be derived from splitting within the proliferating tissue (schizocoely), and mineralized skeletons were largely external (exoskeletons).

Not all taxa assigned to the protostomes or the deuterostomes show these characters, and some clearly show striking combinations of characters of both groups. With the advent of molecular phylogenies and continued morphological and developmental work, however, it has become clear that aspects of early ontogenies have often been remodeled so that allied clades differ significantly, whereas some similarities among clades turn out to be homoplasies. The result of this molecular evidence has been that some taxa within the Deuterostomia have been transferred to Protostomia, which itself has been broken into two large clades (figs. 4.4, 4.5): Ecdysozoa (Aguinaldo et al. 1997) and Lophotrochozoa (Halanych et al. 1995). There are, however, still a number of phylum-level taxa, chiefly small-bodied, that have long branches on molecular phylogenies and whose phylogenetic position has not yet been clearly established.

No living bilaterians have been definitively identified as belonging to a clade that branched off before the split between protostomes and deuterostomes and thus preceding the P/D ancestor (fig. 4.4 and chap. 9). One group, the Acoelomorpha, a flatworm-like group, has been thought to be the sister clade to all the other bilaterians (Ruiz-Trillo et al. 1999; see also Baguña and Riutort 2004; Baguña et al. 2008; Ruiz-Trillo et al. 2002). More recent work has suggested that the phylogenetic position of acoels may be at or near the base of crown deuterostomes (see below). There are more than 250 living species of acoels, and they have no known fossil record, although they have been linked to some Ediacara macrofossils (see chap. 5).

The Bilaterian Superphyla

Each of the well-established bilaterian superphyla Deuterostomia, Ecdysozoa, and Lophotrochozoa (fig. 4.4) contains clades that have a range of body-plan architectures, from small, relatively simple forms with worm- or slug-like bodies to larger, complex organisms. Evolution has thus traversed a whole range of grades within each superphylum, although the direction of change, whether from simple to more complex or the reverse, is not always clear. Molecular evidence suggests that some simpler worm-like forms are found near or at the base, or on basal polytomies, in each superphylum.

DEUTEROSTOMIA The superphylum Deuterostomia includes echinoderms and hemichordates on one branch (called Ambulacraria) and chordates on another (fig. 4.4). It was long recognized partly on the basis of features of the early embryo—radial cleavage, for example—and of a coelom that was interpreted as being divided, at least originally, into three regional compartments. It has turned out that these characters are not restricted to deuterostomes (some lophotrochozoans have three compartmental regions and radial cleavage) and thus are not sufficient for assignment to this clade. Crown deuterostomes have radial cleavage, have blood vascular systems, and are eucoelomic (with one important exception, the chordate subphylum Urochordata), with mesoderm typically arising from cells along the archenteron and coelomic space captured from the gut lumen (enterocoely). Coelomic spaces in some derived groups do originate by splitting of mesoderm (schizocoely), but because their mesoderm arises from the archenteron, this pattern is considered to be a modification of an ancestral enterocoelic condition. Swalla and Smith (2008) review the phylogeny of major deuterostome taxa, including fossil evidence.

Xenoturbellida. The simplest animal that is likely to be a deuterostome is *Xenoturbella* (fig. 4.12), a small-bodied, vermiform, acoelomate bilaterian with a simple nerve net and a blind gut, resembling the Acoelomorpha in structural grade. *Xenoturbella* feeds on bivalves, perhaps their eggs and/or larvae, which contaminated early DNA sequencing efforts, suggesting that it was a mollusk itself. Based on fine-structure studies, some authors have suggested that *Xenoturbella* may be sister to Bilateria (Ehlers and Sopott-Ehlers 1997), a position usually accorded to the Acoelomorpha. However, subsequent studies indicated that *Xenoturbella* is allied to deuterostomes (Bourlat et al. 2003), and a sister to the hemichordate-echinoderm branch (Bourlat et al. 2006). Accordingly, *Xenoturbella* is now a phylum of it own, Xenoturbellida.[6] Needless to say, *Xenoturbella* lacks a fossil record; it may represent a deuterostome branch that arose before definitive deuterostome features were evolved, or it may be phylogenetically derived but morphologically simplified. A unique deuterostome feature is the perforation of the pharynx to produce gill slits or sieve-like feeding structures, but these are absent and, indeed, unneeded in the minute, morphologically simple *Xenoturbella.*

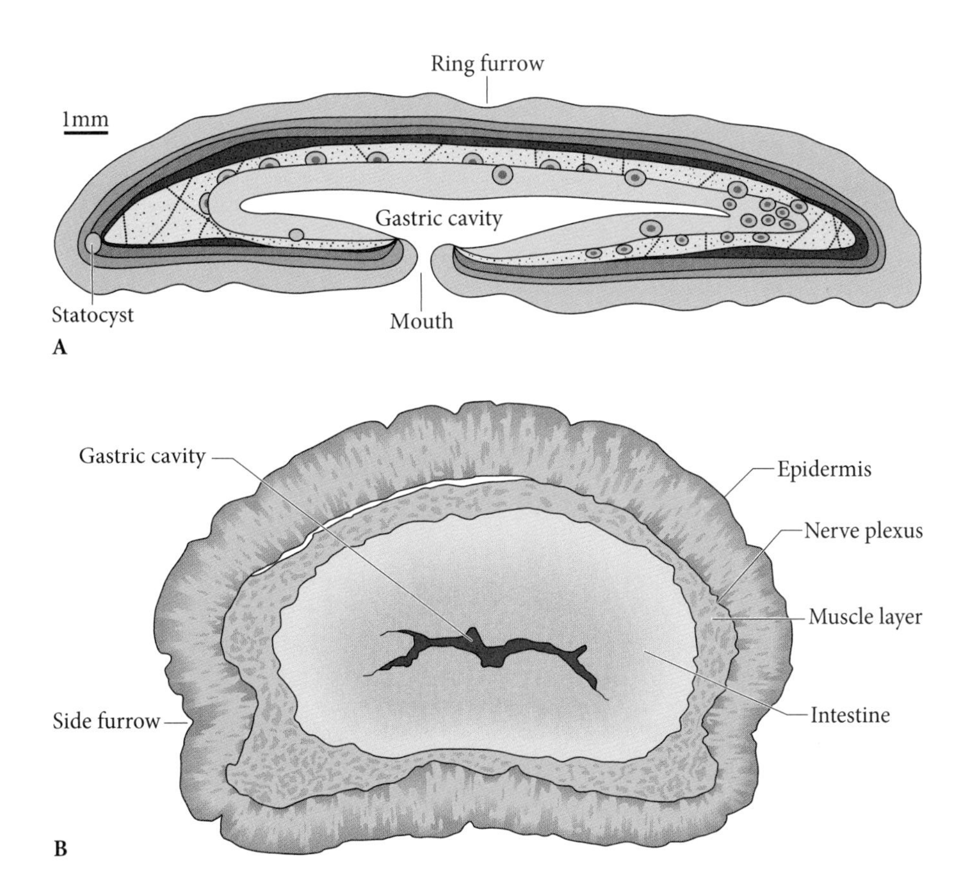

Figure 4.12 The xenoturbellid *Xenoturbella bocki,* likely a basal deuterostome. Xenoturbellids share a simple structural grade with acoels. (A) Longitudinal cross section. (B) Sketch of transverse cross section anterior to the mouth. Scale bars 0.1 mm. Adapted from Telford (2008) after Westblad (1949).

Acoela. Like Xenoturbellids, the acoels lack a known body fossil record. Although they are similar in structural grade, the relationship between these two groups is uncertain, and acoels also lack gill slits or other features that would ally them with deuterostomes on morphological grounds (fig. 4.13). Acoels have an unusual cleavage pattern in that it is spiral but involves two cells at a time (duets), rather than the four-cell spiral cleavages (quartets) found in lophotrochozoans (see below); it has been pointed out that this pattern could evolve from a radial one such as that which characterizes invertebrate deuterostomes (Henry, Martindale, and Boyer 2000). In both acoels and *Xenoturbella,* the nerve net is principally involved in coordination. The phylogenetic placement of acoels remains a subject of considerable debate. Recent molecular evidence indicates that they are deuterostomes, related to *Xenoturbella* (Philippe

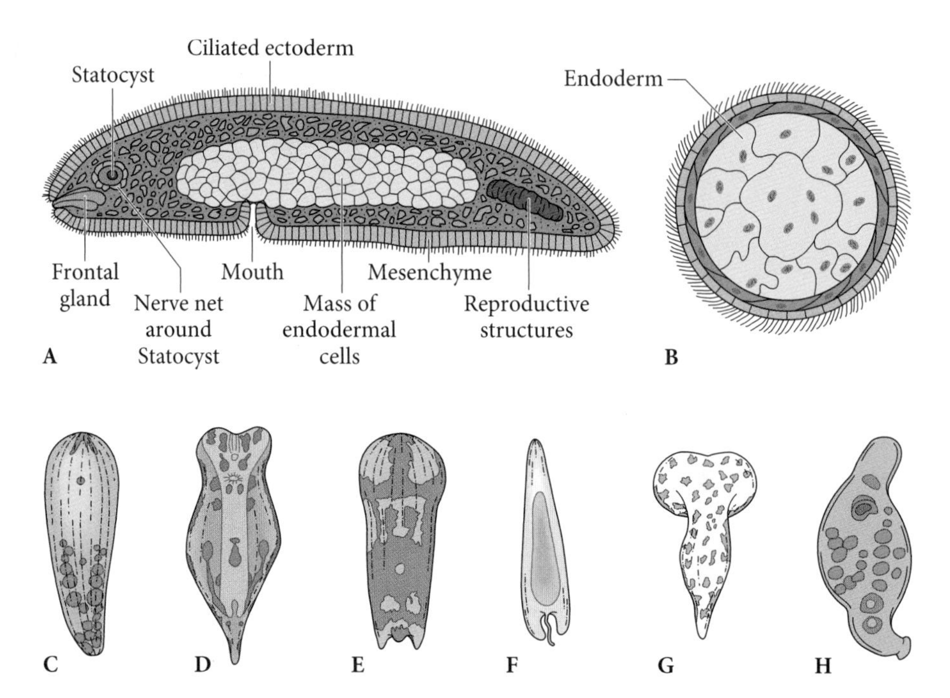

Figure 4.13 Gross form and structure of modern acoels, a group unknown as fossils but one that may represent an early deuterostome branch or possibly the earliest known bilaterian branch. (A) Saggital section of a generalized primitive acoel that has a cellular endoderm. (B) Diagrammatic cross section of acoel digestive tract showing another common acoel digestive tissue pattern, with a central syncytium. (C)–(H) Representative living acoels, which range to 5 mm in length. (A) adapted from Brusca and Brusca (1990); (B) adapted from J. P. Smith, Dellaporta, and Buss (1985); (C) adapted from Barnes, Calow, and Olive (1993) after Hyman (1951).

et al. 2011), but earlier studies suggested that they were the sister group to the bilateria (Baguña and Riutort 2004; Ruiz-Trillo et al. 1999).

Hemichordata. Hemichordates do not have mineralized skeletons (fig. 4.14A–C), but two of the three major groups, the extinct Graptolithina and the living Pterobranchia, secrete tough organic skeletons that fossilize under appropriate conditions (fig. 4.14B, C). With one possible exception, these two groups are colonial, with individuals commonly being on the order of 1 mm, and they are probably derived relative to the hemichordate stem ancestor. It is the third group, however, Enteropneusta (fig. 4.14A), that may contain the hemichordate crown ancestor. However, enteropneusts lack durable skeletons and generally have a poor fossil record. Living enteropneusts are solitary, vermiform sediment feeders of the benthos; large species can reach well over 1 m in length today. The coordinating functions of the nervous system are chiefly

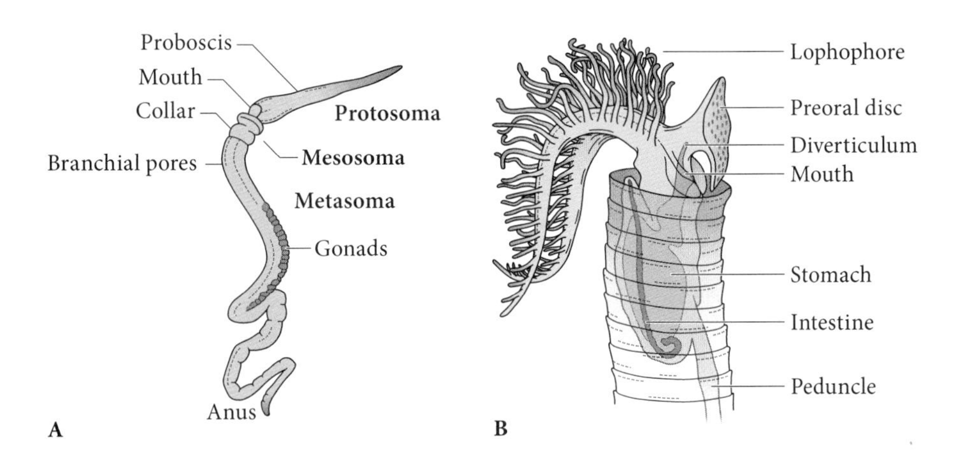

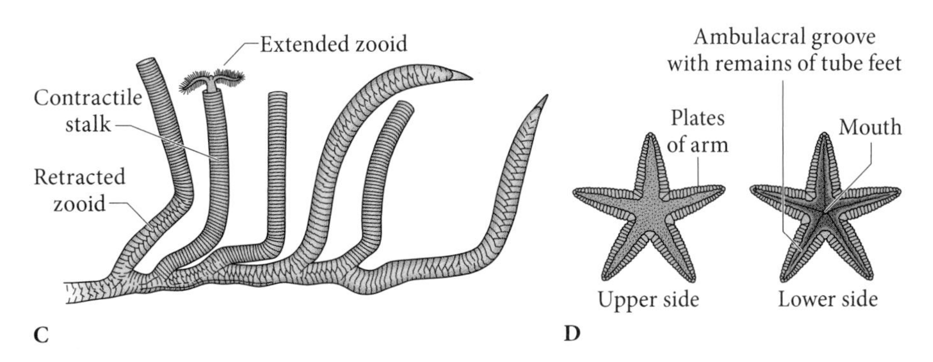

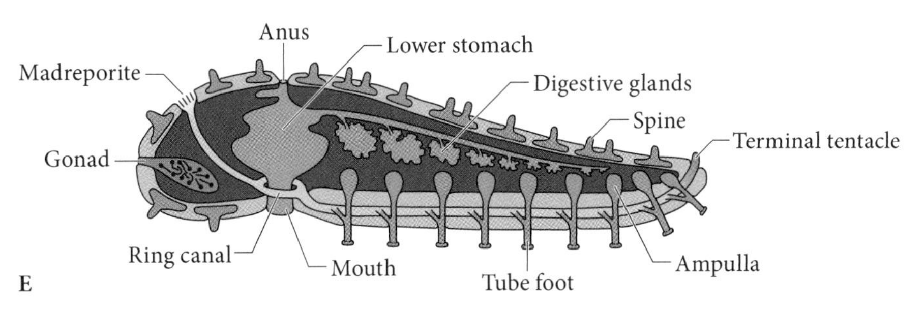

Figure 4.14 Ambulacrarian deuterostomes. (A)–(C) Hemichordata. (A) Enteropneusta, exterior of *Saccoglossus,* representing this solitary, soft-bodied vermiform group that may be similar to the last common ancestor of the hemichordates (on molecular evidence) but that has no explosion fossil record; the group is first known from the late Paleozoic. (B) Pterobranchia. Gross anatomy of an individual zooid of *Rhabdopleura*, tube about 0.5 mm wide; notice the tentacular feeding crown, analogous but not homologous with the lophophorate feeding system. (C) Colony of *Rhabdopleura.* (D) and (E) Echinodermata. (D) The classic pentameral symmetry of a modern starfish, class Asteroidea. (E) Cross section of a starfish arm to show gross internal anatomy. The black space is chiefly coelomic. (A)–(C) adapted from Bulman (1970); (D) adapted from Boardman, Cheetham, and Rowell (1987); (E) adapted from Pearse et al. (1987).

invested in a nerve net, recalling acoels and xenoturbellids. Indeed, juvenile enteropneusts have some structural similarities with acoels (Tyler 2001).

Echinodermata. The other ambulacrarian phylum, Echinodermata, has a body plan that is highly derived in living groups (fig. 4.14D, E). In early developmental stages, echinoderms show radial cleavage and enterocoely, are bilaterally symmetrical, and usually are triregionated, the classic features of coelomate deuterostomes. Crown group echinoderms undergo a unique metamorphosis, however, when they undergo torsion and transform to pentagonal symmetry, as epitomized by starfish. Echinoderm skeletons are chiefly composed of calcite plates secreted within mesoderm. The plates have a distinctive internal texture, called stereom, that is unique among living metazoans. Another unique echinoderm feature is a system of coelomic canals, with a central ring canal surrounding the anterior digestive tract from which radial canals, commonly five, extend laterally (e.g., one in each starfish arm); these radii are called ambulacra, whereas areas between them are called interambulacra. The radial canals give off series of structures (tube feet) that are used variously in food capture, respiration, and locomotion. In echinoderms enclosed by skeletons, radiating ambulacral plates can be easily distinguished and are separated by radiating interambulacral plates.

Chordata. Living chordates include three major groups, two of which, Cephalochordata and Urochordata, are invertebrates; the third is Vertebrata itself (fig. 4.15). All three groups appear to be represented in explosion faunas (chap. 6). Chordates are particularly characterized at some stage by the presence of a notochord, a stiff but springy dorsal rod that may have evolved for swimming because it antagonizes muscles whose contractions bend the body into lateral waves. Cephalochordates (acraniates), the earliest branching among living chordate groups, have well-developed notochords used in swim-

Figure 4.15 (right) Chordate deuterostomes. (A) Cephalochordata, acraniate invertebrate deuterostomes; the amphioxus *Branchiostoma floridae.* This group has a notochord and gill slits, but lacks a vertebral column. (B) and (C) Urochordata. (B) The tadpole larvae of a tunicate; notice the notochord. (C) An adult tunicate; the larval notochord and tail are lost, and the pharynx has expanded to become a branchial feeding basket. (D) A contemporary craniate, the lamprey *Lampetra,* an agnthan (lacking jaws); there are "teeth" within the oral funnel. Although living agnathans are eel-like, fossil groups have a diversity of body shapes. (A) adapted from J. Z. Young (1981); (B) adapted from Brusca and Brusca (1990); (C) adapted from Brusca and Brusca (1990); (D) adapted from Hardisty (1979).

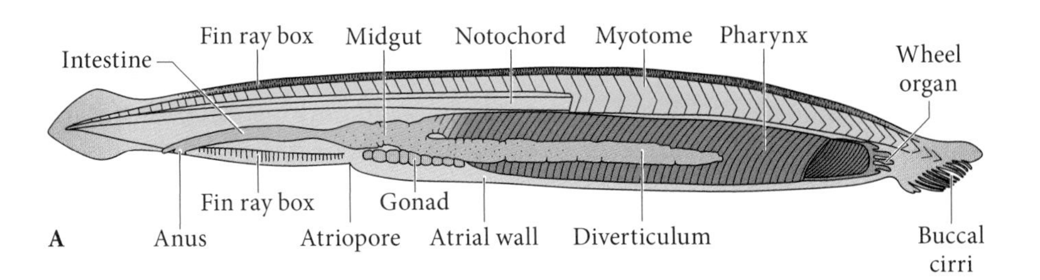
Intestine
Fin ray box
Midgut
Notochord
Myotome
Pharynx
Wheel organ
Fin ray box
Gonad
A
Anus
Atriopore
Atrial wall
Diverticulum
Buccal cirri

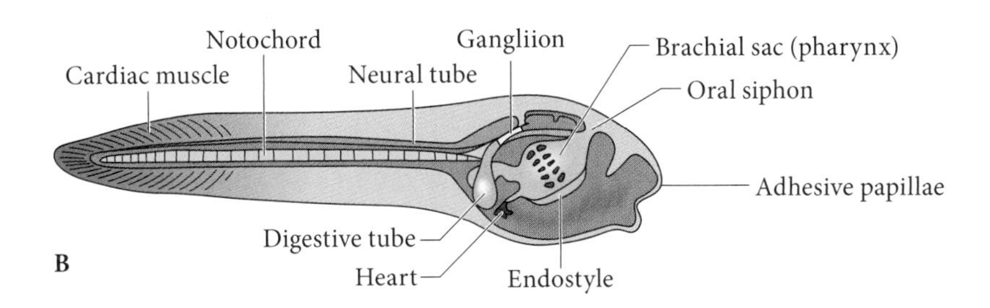
Notochord
Gangliion
Brachial sac (pharynx)
Cardiac muscle
Neural tube
Oral siphon
Adhesive papillae
Digestive tube
B
Heart
Endostyle

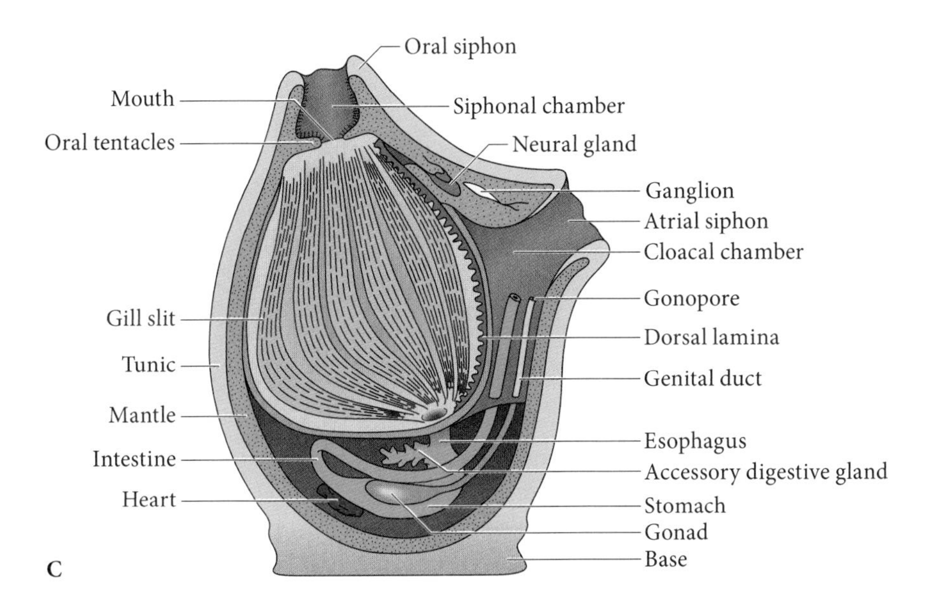
Oral siphon
Mouth
Siphonal chamber
Oral tentacles
Neural gland
Ganglion
Atrial siphon
Cloacal chamber
Gonopore
Gill slit
Dorsal lamina
Tunic
Genital duct
Mantle
Esophagus
Intestine
Accessory digestive gland
Heart
Stomach
Gonad
C
Base

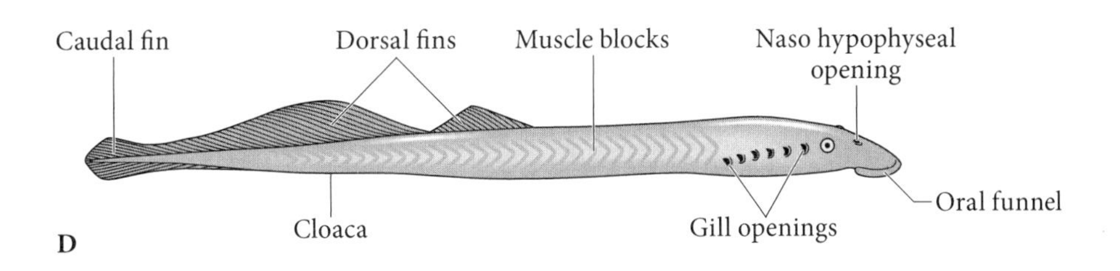
Caudal fin
Dorsal fins
Muscle blocks
Naso hypophyseal opening
Oral funnel
Cloaca
Gill openings
D

ming or wriggling through sand; they are filter feeders (fig. 4.15A). Urochordates, also filter feeders, have a notochord during a larval "tadpole" stage (fig. 4.15B) but lose it at metamorphosis; they include both benthic and pelagic groups. The benthic urochordate group Tunicata has a vastly expanded pharynx to form a feeding basket (fig. 4.15C). Molecular evidence indicates that the relative simple urochordates are sisters to Vertebrata (Delsuc et al. 2006).

Vertebrates (craniates), which have vertebral columns in addition to notochords, are represented in explosion faunas by very primitive jawless fish (agnathans)[7] that are cartilaginous rather than bony and that have vertebrae represented by small irregular elements, called arcualia, that flank the dorsal nerve cord. An important vertebrate feature is an embryonic tissue called the neural crest that lies between the neural tube and dorsal ectoderm. This tissue gives rise to cell types that are incorporated in a variety of organs.

ECDYSOZOA The protostome superphylum Ecdysozoa includes arthropods, onychophorans, priapulids, and a number of small-bodied phyla, all of which share a molting habit. A variety of cleavage types, including radial ones and a sort of irregular, spiral-like type, are exhibited by various phyla. Although larger members of this superphylum have coelomic spaces, none has an important fluid skeleton that is coelomic, instead using hemocoels for that function. Coelomic spaces arise by schizocoely in ecdysozoans. Smaller members of the ecdysozoa are pseudocoelomic (priapulids, nematodes, etc.), whereas larger hemocoelic forms have a BVS. Ecdysozoans are either direct developers (i.e., hatch as juveniles rather than as larvae) or have larvae that represent early stages of a direct-developing ancestor.

Ecdysozoans secrete an integument that ranges from thin and pliable, as in the terrestrial onychophorans (velvet worms), to complexly layered, tough, and therefore quite durable, even mineralized in some cases. The more durable integuments serve for muscle attachment and armor. Once secreted, the ecdysozoan integument cannot be enlarged, so, to accommodate growth, it must be molted periodically and a new integument secreted; a molting system is found in all ecdysozoans, even those, like nematodes, that are small-bodied.

Priapulida. The marine worms Priapulida have bodies divided into an anterior proboscis (a retractile introvert) and a trunk (fig. 4.16A, B). The unsegmented trunk is encircled by prominent rings, which are probably used for purchase in burrowing, as are the cuticular ribs, spines, and other structures on the proboscis. The cutical is molted with growth. The body cavity is fairly capacious but is not lined by a peritoneum, and because there is no BVS, it is interpreted as a pseudocoel. Cleavage is radial (Lang 1963; Wennberg, Janssen, and Budd 2008), unlike most ecdysozoans.

Loricifera. A group of exceptionally small-bodied forms, loriciferans have a cuticular skeleton of six plates, called a lorica, that surrounds their short trunk and that is molted. They are likely to be pseudocoelomic and have been grouped with priapulids and Kinorhyncha, another minute phylum with a molted cuticle, in the clade Scalidophora (Telford et al. 2008) or alternatively as sister to the Nematomorpha (Sørensen et al. 2008).

Onychophora. Living onychophorans are the sister group to the arthropods. All living species are terrestrial, and the fourteen or so genera in two families are similar in gross morphology (fig. 4.16C–E). The cylindrical body has a BVS and a large hemocoelic body cavity. The body is segmented and has a correlated series of muscles, organs, and appendages that include paired anterior antennae and conical to cylindrical walking legs called lobopods. Ridges that contain hemal channels ring both the body and legs. The body wall is surrounded by a thin, pliable cuticle that permits changes in body shape: living onychophorans can squeeze through narrow passageways.

Arthropoda. Arthropods are the most diverse living animal phylum and in modern seas are rivaled in species richness only by mollusks. The arthropod body (fig. 4.17) is hemocoelic and segmented, and most marine forms have tough organic integuments that are mineralized in some groups. Unlike onychophorans, marine arthropod limb integuments are usually hardened (sclerotized) and are jointed to permit complicated movements in locomotion and feeding. The body is usually differentiated morphologically into functional regions (tagmata), and the limbs commonly show a variety of morphologies, often correlated with the tagmata. Differences in tagmosis and in limb morphologies among taxa provide an extensive source of morphological variability, underlying much of the storied richness of arthropod biodiversity. Arthropods have left us the most diverse Cambrian fossil record of any phylum.

Relations among the main crown arthropod groups have proven difficult to establish, but mounting evidence indicates that there are two major clades: Pancrustacea, which includes crustaceans and insects; and Chelicerata, which includes horseshoe crabs, spiders, scorpions, and the pycnogonids. The other important crown group, the Myriapoda, which includes among others millipedes and centipedes, has been allied with each of those clades in different classifications that still have their supporters, although evidence seems stronger at present in support of their alliance with Pancrustacea in a clade called Mandibulata (Regier et al. 2010) [see the concise review by Edgecombe (2010)].

LOPHOTROCHOZOA Lophotrochozoa, the other major protostome superphylum, encompasses two and possibly three distinctive subclades, the

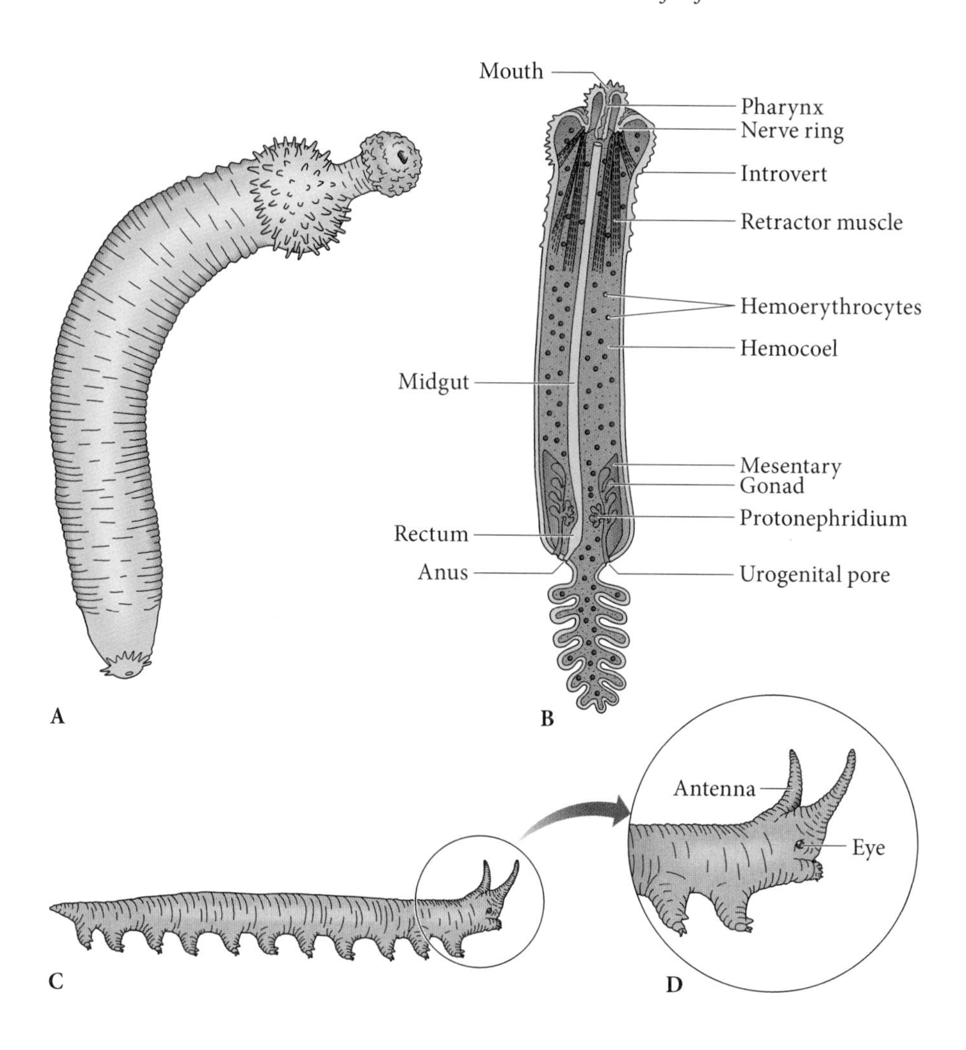
Mouth
Pharynx
Nerve ring
Introvert
Retractor muscle
Hemoerythrocytes
Hemocoel
Midgut
Mesentary
Gonad
Protonephridium
Rectum
Anus
Urogenital pore
A
B
Antenna
Eye
C
D

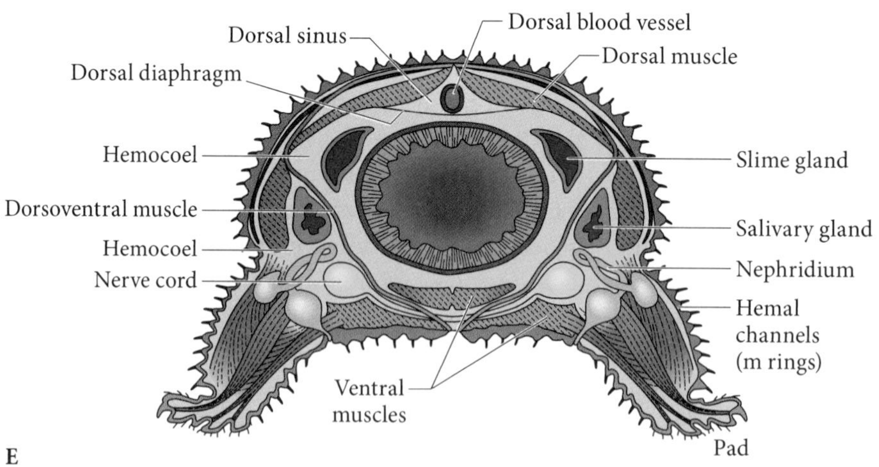
Dorsal sinus
Dorsal blood vessel
Dorsal muscle
Dorsal diaphragm
Hemocoel
Slime gland
Dorsoventral muscle
Salivary gland
Hemocoel
Nephridium
Nerve cord
Hemal channels (m rings)
Ventral muscles
Pad
E

Figure 4.16 (left) Two ecdysozoan phyla. (A) and (B) Priapulida. (A) lateral view of *Meiopriapulus.* (B) Generalized cross section based chiefly on *Meiopriapulus* to show gross anatomy. (C)–(E) Onychophora. (C) Lateral view of *Peripatoides.* (D) Head region; notice that the antennae arise in front of the eyes. (E) Cross section to show gross anatomy; the main body cavity is a hemocoel. Hemal channels of the blood vascular system run inside the ridges, encircling body and limbs. (A) adapted from Calloway (1988), (B) adapted from Ruppert and Barnes (1996), (C)–(E) after Snodgrass (1938).

interrelations among which are not entirely clear. These groups are Spiralia, Lophophorata, and possibly Platyzoa (fig. 4.4). Although they are clearly allied by molecular evidence, their most obvious morphological commonality is that they are a nonecdysozoan (i.e., nonmolting) group.

Spiralia. The Spiralia are united partly on the basis of their shared cleavage pattern, which is a classic spiral one in quartets, and their mesoderm formation, which arises chiefly from a single blastomere (numbered 4d) that usually divides to give rise to endoderm and to most mesoderm. This characteristic cleavage pattern and cell fate is assumed to have characterized the spiralian LCA. Because this pattern is shared by some acoelomate flatworm groups such as Rhabidtophora, they are commonly included in Spiralia, but molecular support for this relationship has been equivocal. Within Spiralia, both eucoeloms and organ coeloms arise by schizocoely.

Mollusca. The general molluscan body plan (fig. 4.18) can be described as comprising a head, foot, and visceral regions encompassed by a mantle, but evolution of molluscan architecture has proceeded through changes in the importance and arrangement of such body parts. For example, a head is present in gastropods and chitons, is associated with the foot in cephalopods, but is absent in bivalves and scaphopods, whereas organs and other structures are serially repeated in some groups (e.g., monoplacophorans) and not in others. In addition, different molluscan clades can show differential enlargement, reduction, displacement, folding, or rotation of organs (Lindberg and Ponder 1996). Such differences are often not reflected in skeletal morphology, so interpreting the anatomical arrangements within fossil molluscan groups is challenging (chap. 6). Although some mollusks have coelomic spaces, living mollusks do not have eucoeloms that form hydrostatic skeletons but instead rely on blood spaces or tissues for hydrostatic skeletal functions. Many mollusks do have an intramesodermal space developed around the heart that can be considered an organ coelom, and some have such spaces around gonads and excretory organs as well.[8]

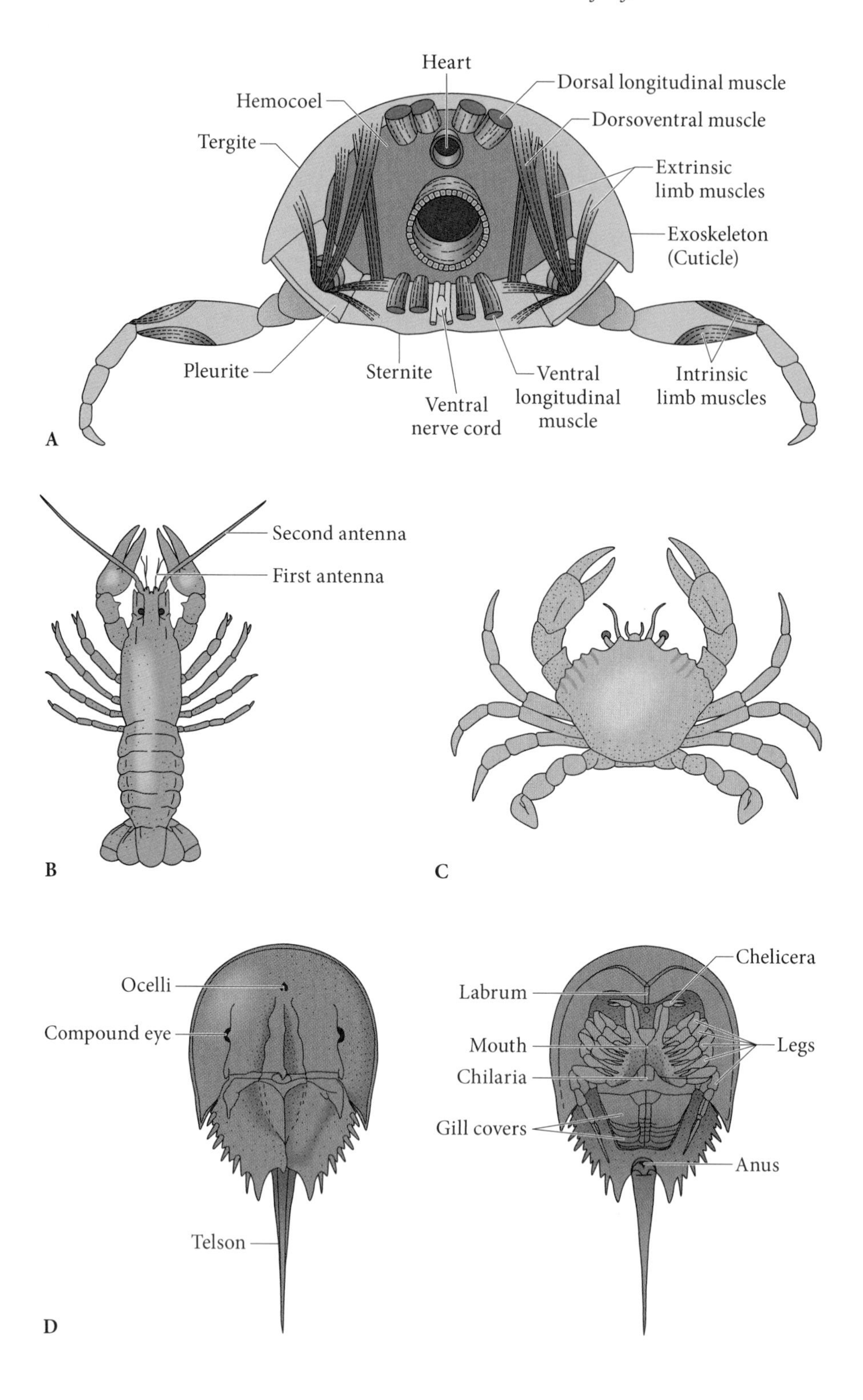
Heart
Hemocoel
Dorsal longitudinal muscle
Dorsoventral muscle
Tergite
Extrinsic limb muscles
Exoskeleton (Cuticle)
Pleurite
Sternite
Ventral longitudinal muscle
Intrinsic limb muscles
Ventral nerve cord
A
Second antenna
First antenna
B
C
Ocelli
Compound eye
Telson
Chelicera
Labrum
Mouth
Legs
Chilaria
Gill covers
Anus
D

Figure 4.17 (left) Arthropoda. (A) Transverse section of a generalized arthropod showing major anatomical features. (B) and (C) Two living crustaceans, a crayfish and a crab; anteriorly, they have two pair of antennae, both arising posterior to the eyes (beneath the carapace). (D) Dorsal and ventral views of a chelicerate, the horseshoe crab, which has an anterior pair of claw-like appendages, called chelicerae, also arising from behind the eyes. (A) adapted from Brusca and Brusca (1990), (B)–(D) adapted from Robison and Kaesler (1987).

Annelida. Most marine annelids are segmented and eucoelomic, with the coelom subdivided into compartments by septa that correspond to intrasegmental boundaries (fig. 4.19). Thus, different sections of their coelom may have their compartments deformed in different patterns by muscular activity so that, for example, locomotory waves may pass down the body to provide for efficient burrowing, crawling, or swimming. The blood vascular systems are sometimes closed, although hemal sinuses may be developed along the gut. Many of these annelids have paired appendages that include coelomic extensions (parapodia, fig. 4.19A). The appendages on trunk segments contain chitinous setae and are used chiefly in creeping or swimming. In addition, many marine annelids are suspension-feeding tube dwellers, forming the tubes either from sand particles that they glue together (e.g., Sabellidae) or by secreting a tube that is a mixture of calcium carbonate and a mucopolysaccaride (e.g., Serpulidae).

A new phylogenetic analysis of this previously puzzling group has greatly clarified their phylogenetic relationships and reveals that their life history features appear in general to be correlated with their phylogenies (Struck et al. 2011), although very small annelids that live interstitially (some of which lose their coeloms and become quite solid-bodied) are derived from a variety of ancestral lineages. The two large clades of annelids are the Class Errantia, which comprises the mobile forms, many of which are predators or algal specialists; and the Class Sedentaria, which comprises more sessile burrowers and tube dwellers, many of which are suspension feeders or feed off organic material in the sediment. Despite the view of annelids as highly segmented, many lineages of partially or entirely unsegmented forms are nested within them. The former phyla Echiura and Pogonophora turn out to belong within Polychaeta, and the former phylum Sipuncula, which is unsegmented and lacks setae, also turns out to belong among the Annelida (Schulze, Cutler, and Giribet 2007; Struck et al. 2007). The analysis of Struck and colleagues (2011) now confirms the nesting of these three groups within Anellida and indicates that the more basal crown annelids were segmented.

Lophophorata. Two phyla within the lophotrochozoan superphylum—the brachiopods and phoronids—have radial cleavage with, in some cases, meso-

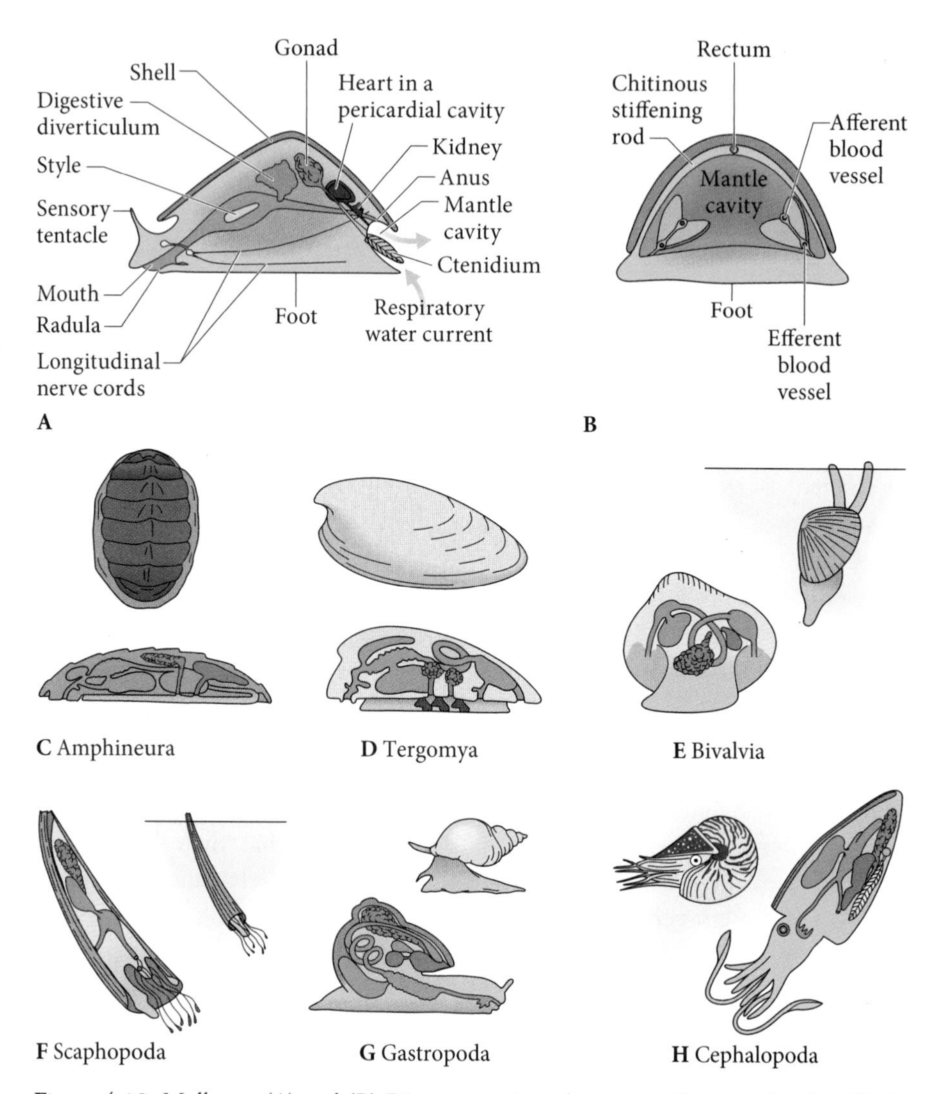

Figure 4.18 Mollusca. (A) and (B) Diagrammatic architecture of a generalized mollusk. (D)–(H) Variations on the basic molluscan theme shown in six of the classes. Dark areas are mantle cavities and ctenidia (gills); egg-like regions are reproductive organs. (A) and (B) adapted from Barnes, Calow, and Olive (1993), (C)–(H) adapted from Salvini-Plawen and Steiner (1996).

derm originating from the larval gut wall (and are thus enterocoelic) and with bi- or triregionated coeloms, developmental characters that are also associated with deuterostomes (fig. 4.20). These two phyla bear a coiled, tentacular feeding apparatus with a coelomic lumen, a lophophore; some deuterostomes (pterobranch hemicordates) have a somewhat similar tentacular feeding system.

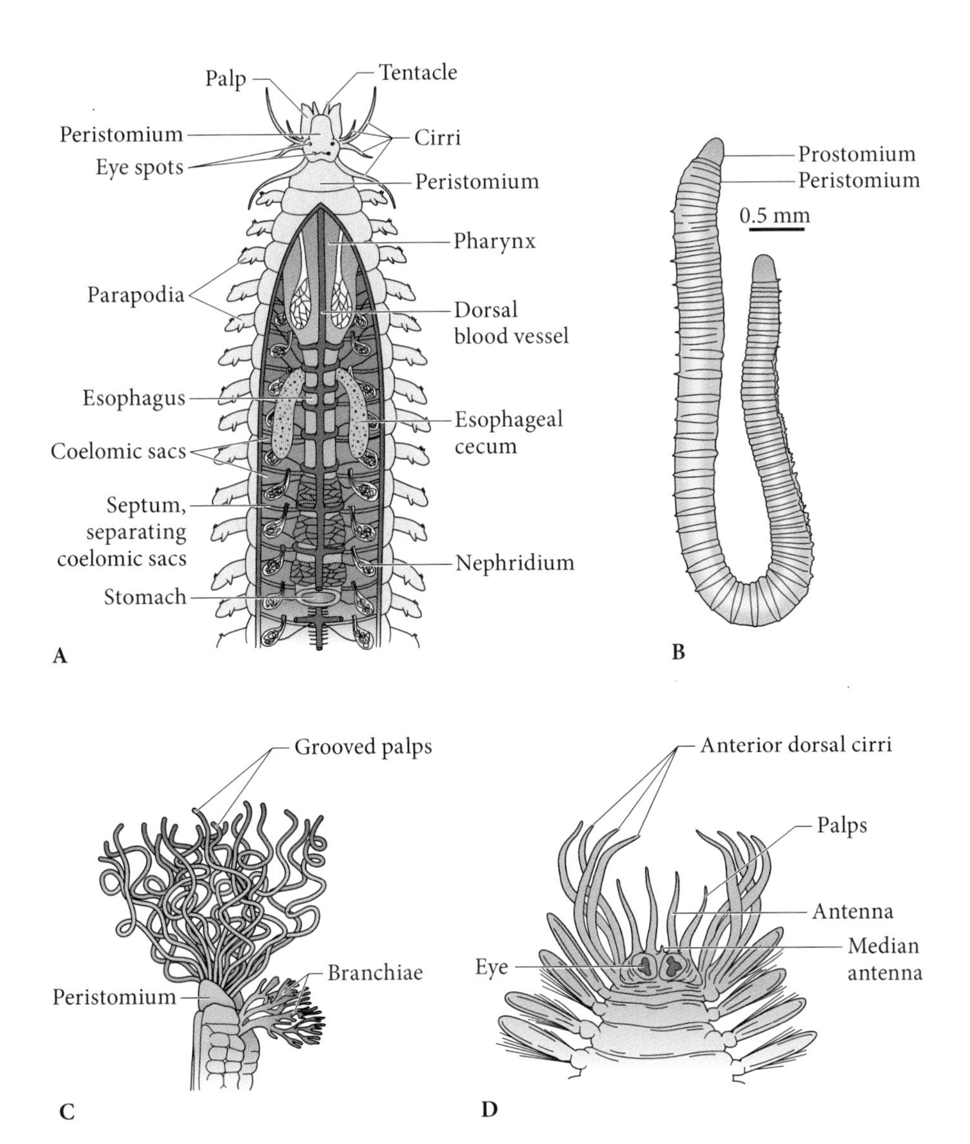

Figure 4.19 Annelida. (A) *Nereis,* gross anatomy of anterior in dorsal view; parapodial muscles are antagonized by coelomic lumenae. Many Nereids are active predators and scavengers, whereas others are relatively sessile. (B) External view of Arenicola, a burrowing sediment feeder with a very simple body type. (C) Anterior portion of *Terebella,* a tube dweller that feeds on organic materials chiefly collected from the sediment surface; the grooves on the palps are feeding grooves. (D) Anterior portion of *Amphiduros.* Species in this family are chiefly active predators, although some are commensal or parasitic. (A) after F. A. Brown (1950), (B) adapted from Rouse and Pleijel (2001) after Ashworth (1912), (C) adapted from Rouse and Pleijel (2001) after McIntosh (1885), (D) adapted from Rouse and Pleijel (2001) after Hartman (1961).

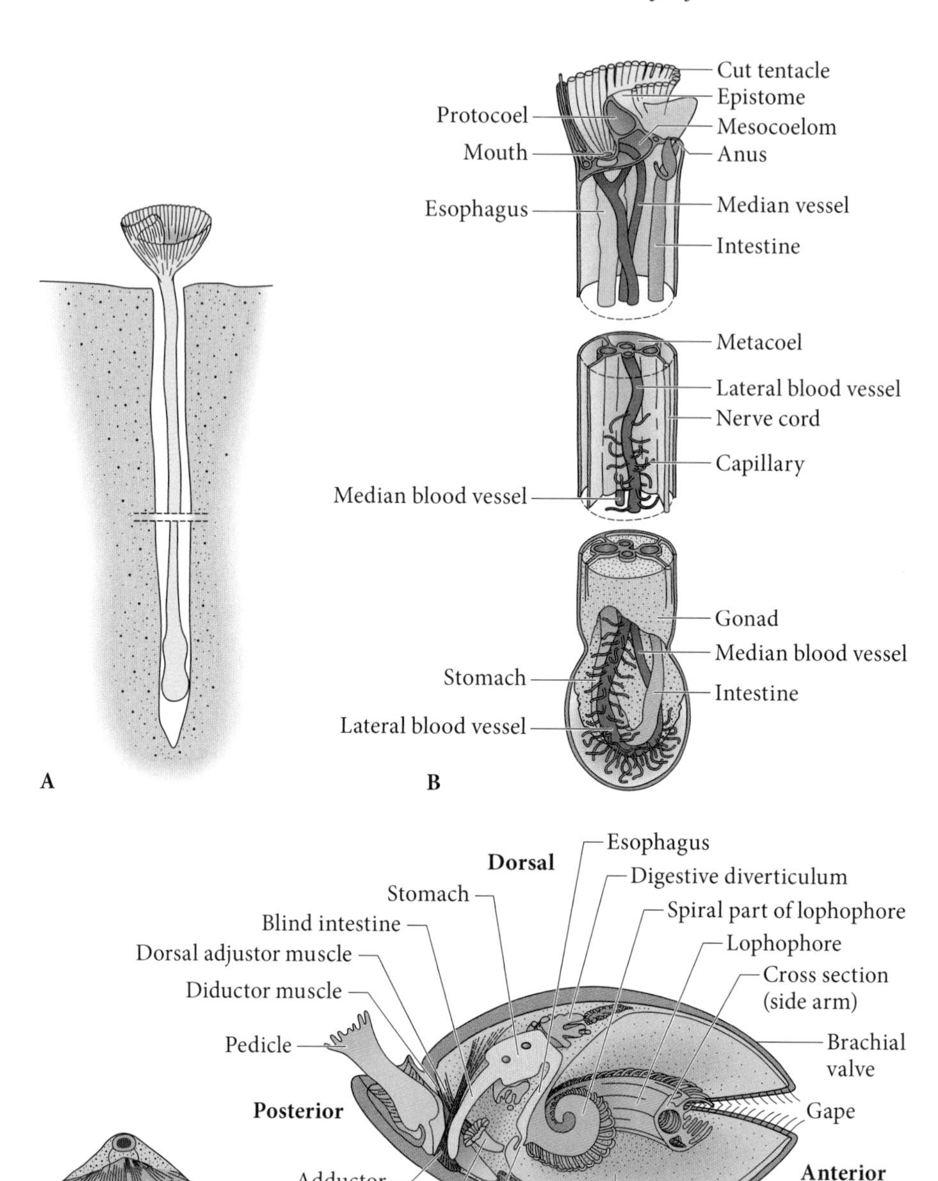

Cut tentacle
Epistome
Protocoel
Mesocoelom
Mouth
Anus
Esophagus
Median vessel
Intestine
Metacoel
Lateral blood vessel
Nerve cord
Capillary
Median blood vessel
Gonad
Median blood vessel
Stomach
Intestine
Lateral blood vessel
A
B
Esophagus
Dorsal
Digestive diverticulum
Stomach
Spiral part of lophophore
Blind intestine
Lophophore
Dorsal adjustor muscle
Cross section (side arm)
Diductor muscle
Brachial valve
Pedicle
Gape
Posterior
Anterior
Adductor muscle
Pedicle valve
Nephridium
Mantle cavity
Ventral mantle
Mouth
Gonad
C
D
Ventral

Figure 4.20 (left) Lophophorates. (A) and (B) Phoronids. (A) Individual in feeding position within a burrow; (B) gross anatomy. (C) and (D) Brachiopods. (C) External view of rhynchonellate brachiopod (*Terebratalia*). (D) Gross anatomy of a generalized rhynchonellate, which can be likened to a phoronid architecture that has accommodated to enclosure within a shell, although the order of origin of these two groups is not known. Brachiopods are commonly attached to objects on the seafloor by their pedicles. (A) and (B) adapted from Emig (1979), (D) adapted from Hyman (1951).

Strong molecular evidence has allied lophophorates with the spiralians rather than the deuterostomes, however. It seems that the deuterostome-like features of lophophoran development have been independently derived.

Other Nonecdysozoan Protostomes. Chaetognatha, a phylum of small-bodied, chiefly pelagic predators, has been allied with protostomes on molecular evidence. One of two possible phylogenetic positions seems most likely: either as basal protostomes (Helfenbein et al. 2004; Marletaz et al. 2006) or as basal lophotrochozoans (Matus, Copley, et al. 2006). Chaetognaths have radial cleavage and derive their mesoderm by enterocoely, although they form their body cavities in an unusual manner, called heterocoely (Kapp 2000). Thus, they share deuterostome-like developmental features with some lophotrochozoans. A group of small-bodied, chiefly pseudocoelomate phyla is sometimes recognized as a fourth bilaterian superphylum, called Platyzoa (fig. 4.4; Cavalier-Smith 1998; Giribet et al. 2000), but most of the groups originally placed here have been more confidently assigned elsewhere.

SUMMARY OF METAZOAN ARCHITECTURES AND ANIMAL RELATIONSHIPS

This brief introduction illustrates that there is little relation between the structural architecture or degree of morphological complexity of crown phyla and their phylogenetic relationships. On present evidence, each of the bilaterian superphyla contain relatively less morphologically complex forms that may be sister clades to the remaining groups: the acoelomate Xenoturbellida and perhaps the Acoela for Deuterostomia; the pseudocoelomate Priapulida for Ecdysozoa (although they are much more complex than xenoturbellids or acoels), and perhaps an acoelomate flatworm group for the Lophotrochozoa. Evidence from microRNAs, discussed in chapter 8, indicates that many of these clades may have lost these important regulators of cell types and thus have become simplified from morphologically more complex ancestors. At one time, similarities in architectural complexities were given significant weight in estimating relationships. For example, many of the acoelomate and pseudocoelomate groups now placed in different superphyla were placed in a single phylum (Aschelminthes) by many authorities, whereas the more complex arthropods and annelids were thought to be closely allied (and many authors believed that annelids were ancestral to arthropods) chiefly because they are both segmented. Molecular data, as summarized in figure 4.4, have uncovered new alliances—but therefore also new patterns of morphological disparity—within each of the superphyla, which has required evolutionary biologists to seek new scenarios to explain the morphological transitions and to seek new hypotheses to account for the rough similarities in the unfolding of complexity in each major group.

In hindsight, it is easy to see why systematists have sometimes been misled by morphological criteria when attempting to establish a tree of phyla. First, there are many parallels in design features among the superphyla (Valentine 2004). Thus, attempts to establish kinships among the phyla on morphological criteria alone have tended to unite forms that share a common morphological grade, resulting in erroneous trees that have been, and continue to be, emended by the findings of molecular-based phylogenetic studies.

Second, it is clear that the living fauna does not constitute a very good sample of the history of animal diversity. The tree in figure 4.4 is only a minimal sketch of the numerous divergences that occurred during metazoan evolution. At the level of phyla, the gaps in the tree are numerous and commonly so broad that we often have to extrapolate between body plans that were connected in life's history by ancestral forms that had little in common with the descendant body plans. Thus, the features held in common by sister phyla are often very basic ones—for example, mesodermal tissues in bilaterians—and are not much help in determining kinships in any detail, whereas the features that characterize and indeed define phylum-level clades were evolved independently within each clade. The requirements of the more successful morphological trees, which rely on the sequential appearance of nested synapomorphies, are generally not met by the disparities in metazoan morphologies that arose before the Cambrian explosion.

The Cambrian explosion encompasses the whole range of major morphological design features that we catalog among the crown phyla. Therefore, all the founders of the major metazoan groups, and possibly of all the living phyla, must have arisen before the Cambrian explosion, although, as mentioned above, the lineages present in explosion faunas are largely extinct stems. What we get, then, is a snapshot of those major groups of the time that had reasonable preservation potentials. We find them to represent quite a good cross section of body plans among the living metazoans, but body plans at an earlier stage of evolution, before many of the synapomorphies that characterize crown groups had risen to prominence.

Our task, then, is to describe those fossil faunas to account for the evolution of their more important synapomorphies. First, though, we must examine the unusual and intriguing fossils found before the explosion (chap. 5) and try to understand how they relate to the marvelous animals we review from the explosion itself (chap. 6). We will then attempt to reconstruct the ecological and evolutionary processes that together created the great ecosystems within which the living biosphere was fashioned (chaps. 7 through 9).

NOTES

1. By international agreement, the tenth edition of Linnaeus's *Systema Naturae,* published in 1758, is the basis for priority in the naming of animal species and genera; later publications provide the priority in naming families. For taxa on still higher levels, however, there are no formal rules of priority.

2. Cladistic methodology is nicely explained in Page and Holmes (1998) and Hall (2007).

3. The earliest large-scale attempts at a metazoan tree based on molecular sequences were by Field et al. (1988) and Lake (1989, 1990). Even the most up-to-date metazoan phylogenies using the largest available data sets vary somewhat, indicating that artifacts are still present. Recent examples of metazoan phylogenetic tree topologies include Dunn et al. (2008); Minelli (2009); Nielsen (2008); Paps, Baguña, and Ruitort (2009); and especially Peterson et al. (2008) and Philippe et al. (2009), who incorporate new phylogenomic data (and see note 5).

4. Felsenstein (1978) first pointed out that in long-branch comparisons, the longer the sequences that are being compared, the greater this error becomes, illustrating a case in which adding more data only produces more erroneous results.

5. This tree is based on the more recent topologies reported in note 3, supplemented by findings from whole-genome analyses such as King et al. (2008), Putnam et al. (2007), and Srivastava et al. (2008) and secondarily by developmental and morphological features.

6. Xenoturbellids have some fine-structural features that are similar to acoels (Ax 1996) and others that recall epidermal and subepidermal structure in the Enteropneusta (Hemichordata) (Pederson and Pederson 1988), although these similarities are not convincing evidence of affinities (see Ehlers and Sopott-Ehlers 1997) and may derive convergently from similar functional requirements. It is also possible, however, that those groups represent an alliance of small worms that gave rise to Bilateria and to its crown superphyla.

7. Agnathans are represented today by the essentially freshwater lampreys and the marine hagfishes, both with elongate bodies, but fossil agnathans had a broad variety of body shapes.

8. For an excellent discussion and cladistic analysis of morphological character distributions in spiralians that includes a discussion of coelomic spaces, see Haszprunar (1996).

CHAPTER FIVE

Dawn of Animals: The Ediacara Biota

To Charles Lyell, Charles Darwin, and other nineteenth-century geologists, the fossil record began with the first trilobites of the Cambrian. Indeed, in *The Origin of Species*, Darwin puzzled over the abrupt appearance of complex fossils in the rock record. Although Darwin probably never knew it, the veil on the early history of animals was being pulled back even as he edited later editions of *The Origin*. In 1868, Alexander Murray, the first director of the Geological Survey of Newfoundland, discovered some concave disks that he believed were fossils and that he described as *Aspidella terranovica*. For many years, other paleontologists doubted that the disks were evidence of life (see the review in Gehling, Narbonne, and Anderson 2000). Seventy-eight years later, in 1946, Australian geologist Reg Sprigg discovered a suite of impressions in what are now known to be Neoproterozoic rocks in the Ediacaran Hills north of Adelaide, Australia (Sprigg 1947), but they too remained controversial until 1959, when Martin Glaessner realized that they were similar to frond-like fossils from undoubted pre-Cambrian rocks at Charnwood Forest in England. Thus, Glaessner established the existence of a geographically widespread, pre-Cambrian suite of fossils that he called the Ediacaran fauna in which he included similar forms found in Namibia in 1933 by Georg Gürich (Glaessner 1958, 1984). The term *Ediacara* is probably derived from the Australian aboriginal name for the region, Idiyakra or Ideyaker, with the suffix at least descended from *jakara,* for spring or water (Jenkins 2007).

Today the Ediacaran fauna of soft-bodied impressions is known from diverse assemblages of fossils far beyond Newfoundland, Namibia, and Australia and now includes more than thirty localities on five continents (Narbonne 2005; Shen et al. 2008). There are surprising similarities between the structure of the Ediacaran radiation and the later Cambrian diversification. Both begin with the early establishment of a broad range of novel morphologies occupied by distinctive clades followed by diversifications within those clades; each event involves the construction of ecological relationships, and each ends with a winnowing and eventual extinction of multiple clades. In this chapter, we discuss the nature of the Ediacara macrofossils,* what they reveal about their general biology and ecology, and what they imply about the earliest steps in animal evolution. We begin, however, with the earliest records of animal life, including chemical fossils, or biomarkers, of early sponges, and the exquisitely preserved putative fossil embryos of the Doushantuo Formation in south China. We then discuss the Ediacara organisms and end with a look at the diverse assemblage of trace fossils and the first skeletonized fossils from the latest Ediacaran rocks in Namibia.

NEOPROTEROZOIC SPONGES

The earliest evidence for animal life comes not from the traditional sorts of fossil, but from a biochemical fossil, the remains of a lipid—believed to be produced today only by sponges—before about 635 Ma (Love et al. 2009). Such biochemical fossils, or biomarkers, play an increasingly important role in our understanding of the earliest history of life. Once the original molecules break down into a stable chemical configuration, they are relatively resistant to further degradation, and they are often more easily preserved than the remains of whole organisms (Peterson, Summons, and Donoghue 2007). Thus, the earliest signs of animal life are biomarkers from the Huqf Supergroup in southern Oman, which encompasses rocks deposited from the middle Neoproterozoic to the early Cambrian, including both the Sturtian and later Marinoan glacial episodes. Along with biomarkers of marine algae and of dinoflagellates, many samples from throughout the section contain a particular biomarker, 24-isopropylcholestane, which is the preserved form of a complex sterol produced today only in the cell wall of demosponges. This biomarker is relatively abundant in these rocks from Oman, indicating that the organisms from which it was derived were fairly abundant in these shallow-water environments, but no reliable evidence of this biomarker has been found below the Sturtian glacials. The biomarker is also missing from contemporaneous deeper-water rocks in Australia, suggesting that those environments had not yet been colonized. Although the biomarker data from Oman are quite robust, there are caveats (Brocks and Butterfield 2009). For example, although

*To distinguish the soft-bodied macrofossils from other fossils of Ediacaran age, we will follow recent practice and refer to them as Ediacara macrofossils or the Ediacara biota.

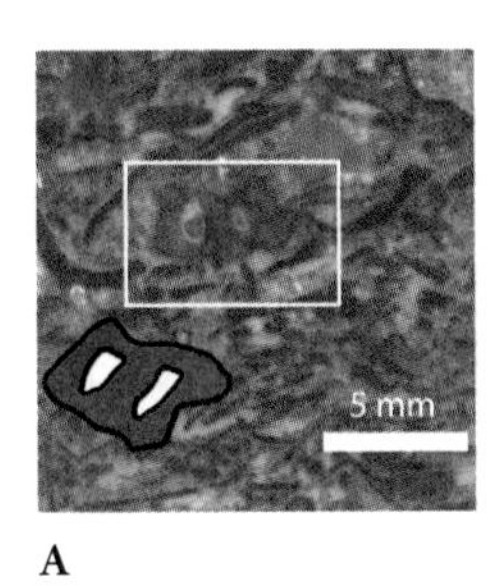

A

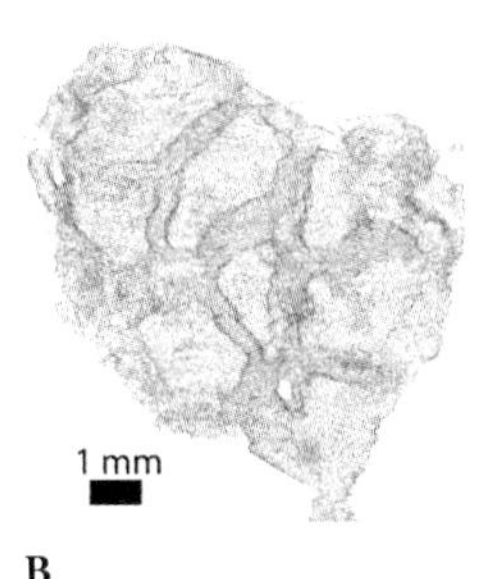

B

Figure 5.1 *Trezona,* a possible Cryogenian-age sponge. (A) Photograph of a thin-section showing a number of individuals. (B) Reconstruction based on serial sectioning and computer reconstruction of specimens. From Maloof, Rose, et al. (2010).

24-isopropylcholestane is produced only by demosponges today and has not been found in living choanoflagellates (Kodner et al. 2008), it is possible that during the Ediacaran other organisms produced it, perhaps by stem choanoflagellates or some now extinct clade between choanoflagellates and living demosponges. Possible sponge body fossils, also dating between the Sturtian and Marinoan glaciations, have been reported from the Trezona Formation in Australia, however (Maloof, Rose, et al. 2010). Reconstructions of specimens from this assemblage of weakly calcified structures show both asymmetric, irregularly shaped organisms with circular external apertures and a network of internal canals (fig. 5.1). Nonmetazoan explanations for these fossils can be excluded, and they are most parsimoniously described as very early sponges. Although both biomarker evidence and molecular clock evidence demonstrate the presence of sponges during the Cryogenian, recall from chapter 3 that iron speciation data from > 742 Ma rocks of the Chuar Group indicate very limited atmospheric oxygen at this time and thus little free oxygen in shallow marine waters (Johnston et al. 2010).

The earliest possible sponge body fossil associated with the Ediacara soft-bodied macrofossils is *Thectardis,* a simple cone-shaped fossil found with the Mistaken Point biota in Newfoundland (fig. 5.2) (Sperling, Peterson, and Laflamme 2011). *Thectardis* occurs low in the sections at Mistaken Point, where it sometimes dominates particular beds (Clapham et al. 2004). An enigmatic form from the Doushantuo Formation in south China, *Cucullus,* has recently been assigned to the demosponges in light of possible complex fibers (Y. Wang and Wang 2011). Other possible evidence of sponges has been recovered only much later in the Ediacaran, including bundles of possible hexactinellid sponge spicules from the Rawnsley Quarzite in South Australia (Gehling and Rigby 1996), in rocks along the White Sea in Russia (Serezhnikova and Ivantsov 2007), and in southwestern Mongolia (Brasier, Green,

Figure 5.2 *Thectardis,* a possible sponge fossil from Mistaken Point, Newfoundland, Canada.

and Shields 1997). These assignments are somewhat equivocal, however, and an entirely reasonable reading of the fossil record suggests that sponges with preservable spicules did not appear until the early Cambrian (Xiao, Hu, et al. 2005). This suggestion is not surprising if sponge spicules appeared independently in different groups (Sperling, Pisani, and Peterson 2007).

FOSSIL EMBRYOS

There is no good reason to expect animal embryos to be preserved as fossils—they are, after all, simply minute bags of cells sometimes enclosed by a membrane. But when gentle acids were used to free beautifully preserved algae and organic-walled microfossils from Ediacaran cherts and shales of the Doushantuo Formation in south China (Xiao, Knoll, et al. 2004; Xiao et al. 1998; Yuan, Li, and Cao 1999), minute balls of cells were found as well. Their pattern of cell division and form suggested that some could be fossilized animal embryos, opening a new window into both the record of the earliest animals and, of equal importance, of their earliest developmental stages. The arrangement of cells within the developing embryo reflects the characteristic patterns of cell division within the particular clade to which a developing embryo belongs (chap. 4), and the patterns of cell division and form in the Doushantuo fossils appear similar to animal embryos (Donoghue and Dong 2005; Kouchinsky, Bengtson, and Gershwin 1999; Lin et al. 2006; Steiner et al. 2004). Although paleontologists were surprised by these discoveries, their acceptance was eased by earlier reports of fossil embryos, albeit with very different preservation, from Cambrian rocks (Bengtson and Zhao 1997).

The oldest putative metazoan embryos are found just above the Marinoan glaciation, and they continue through the rest of the unit (Yin et al. 2007), but most of them are found high in the formation, where a diversity of embryos exhibiting a wealth of cellular structure have been described. Many of these fossils are sufficiently simple that they could belong to a wide variety of clades, but sponges and cnidarians are particularly likely (J. Y. Chen et al. 2000, 2002, 2009; Xiao and Knoll 2000).[1] Some of these fossils, however, have been described as bilaterian embryos, a claim that has been disputed. Even if bilaterians were present, their broadcast spawning habits probably significantly reduced the probability of preservation of embryos from those clades (Gostling et al. 2008; Hagadorn et al. 2006; E. C. Raff et al. 2006). Experimental studies with modern embryos have established that fossilization is possible in a reducing environment when a fertilization envelope is preserved, although the fidelity of internal cellular structures is variable (E. C. Raff et al. 2006). Early cleavage stages are most likely to be preserved, consistent with the evidence from the Doushantuo, whereas hatched embryos and larvae, lacking an envelope that can encapsulate a geochemical environment favorable for preservation, are less likely to be fossilized (Dornbos et al. 2005, 2006; Xiao and Knoll 1999). Preservational considerations pose many problems in distinguishing the original morphology of these fossils from subsequent alterations (Dornbos et al. 2006; Schiffbauer et al. 2012; Xiao and Knoll 2000). The recent application of synchrotron X-ray tomographic microscopy provides a nondestructive means of imaging the internal structure of these fossils and of distinguishing some taphonomic artifacts (J. Y. Chen et al. 2009; Donoghue et al. 2006; Hagadorn et al. 2006).

The diversity of forms identified in the Doushantuo makes the age of the fossil horizons a critical factor in establishing the divergence times of many metazoan clades. The Doushantuo Formation overlies a Marinoan age (635 Ma) glacial deposit known as the Nantuo tillite (Condon et al. 2005). The Dengying Formation overlies the Doushantuo Formation, where a few, rare Ediacaran fossils have been recovered, and its uppermost part yields fossils. Establishing the age of these horizons precisely is difficult, however. Integration of U-Pb geochronology from China, Oman, and Namibia (chap. 2), stratigraphic correlations within south China, and $\partial^{13}C$ data all suggest that the fossil embryos are younger than the Gaskiers glaciation (580 Ma) and thus broadly correlative with soft-bodied Ediacaran fossils (Condon et al. 2005). Results from radiometric dating of whole-rock phosphorite samples from the Doushantuo in south China using two different systems (Lu-Hf and Pb-Pb), however, suggest older dates of 602 ± 48 Ma and 599 ± 4 Ma (Barfod et al. 2002). Dates from black shales between the upper and lower phosphorite beds yield Pb-Pb dates of 572 ± 36 Ma (Y. Q. Chen et al. 2009), which are thought to be the more reliable. Although the younger dates seem more likely, the discrepancies have not been resolved, and the Doushantuo fossil embryos could date from as early as 600 to 590 Ma or later than 580 Ma (fig 5.3).

Although the age of the Doushantuo fossils remains uncertain, if they are metazoan embryos, they would help us understand the sequences of developmental evolution. For example, the large sizes of Doushantuo embryos suggest direct development, contradicting hypotheses that the earliest metazoan larvae had planktonic feeding stages (Nielsen 1987, 2001). Indeed, it now seems likely that planktotrophic larval development evolved independently from four to eight times in the latest Cambrian and Early Ordovician (Peterson 2005; Signor and Vermeij 1994). Fossil embryos disappear from the Neoproterozoic record about 550 Ma, roughly coincident with the oxygenation of shelf waters discussed in chapter 3, although some reappear in the Cambrian. Cohen, Knoll, and Kodner (2009) have drawn attention to the similarities in size, ornamentation, and internal structure between the Doushantuo fossils and the resting-stage cysts formed by some animal groups, suggesting that many of the large ornamented Ediacaran microfossils may represent metazoan resting stages produced in response to the harsh environmental conditions of the early Ediacaran. The proposal of Cohen and colleagues has been met with serious objections, however (Moczydlowska et al. 2011), and it is probably premature to conclude that those fossils are metazoan resting stages.

EDIACARA MACROFOSSILS

The Ediacaran macrofauna includes a range of forms: centimeter- to meter-long fronds, disks, and other architectures with more complex shapes, some generally resembling modern animal groups (Fedonkin et al. 2007; Gehling et al. 2005; Jenkins 1992; Narbonne 2005; Vickers-Rich and Komarower 2007; Xiao and Laflamme 2009). Although many display apparent bilaterial symmetry, none show evidence of appendages or sensory features, and, with two possible exceptions, none display signs of feeding and or even of digestive systems. Such a lack of diagnostic metazoan anatomical characters has led to

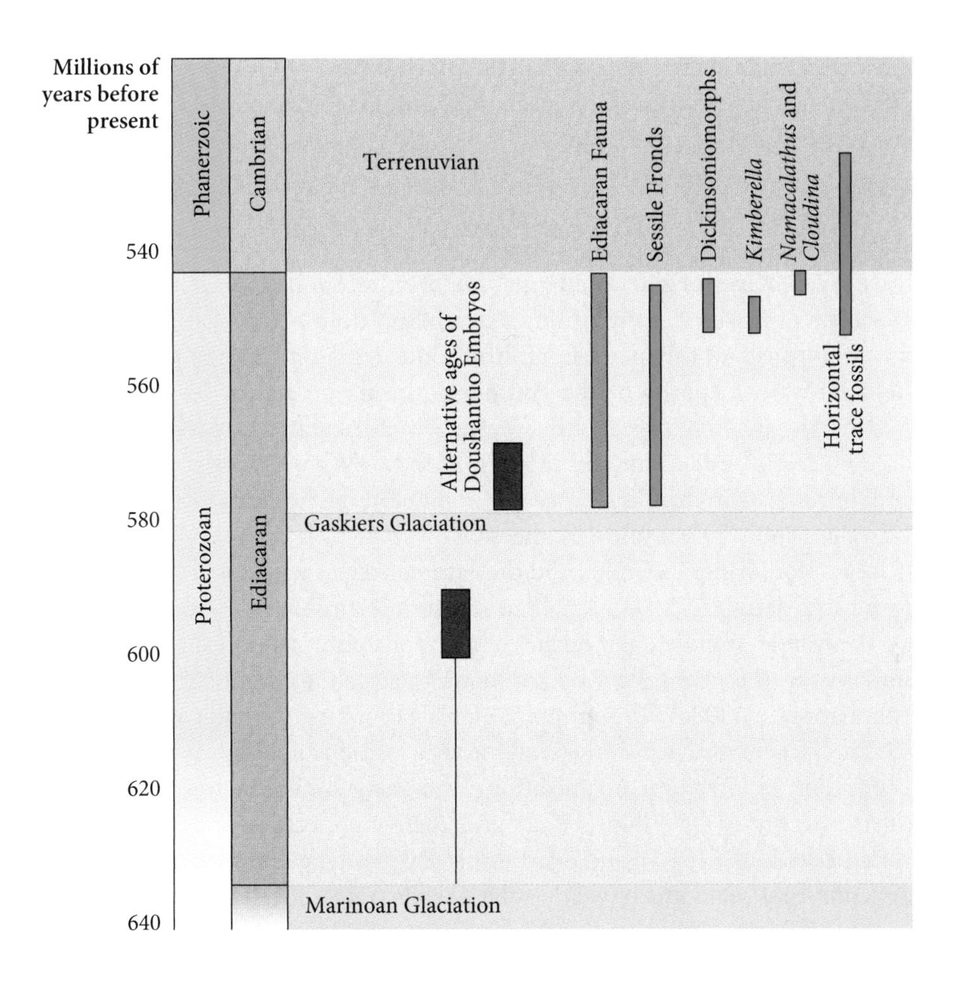

Figure 5.3 Major events of the Ediacaran Period. The oldest evidence for animals is sponge biomarkers that predate the Marinoan glaciation. The age of the Doushantuo fossils is unclear. Some radiometric dates suggest an age of 590 to 600 Ma, but more recent dates suggest that the fossils postdate the Gaskiers glaciation and are of the same age as the earliest Ediacaran fossils from Newfoundland.

a persistent controversy about what kinds of organisms these fossils represent. Do they form a single clade? Are some of them early representatives of arthropods, annelids, and other bilaterian clades? Indeed, are they animals at all, or were these impressions produced by lichens, fungi, or even prokaryotes? Even the age of these fossils has been a source of controversy. Until the mid-1990s, it was believed that the Ediacaran fossils lived tens of millions of years before the advent of Cambrian fossils, with a long gap between the two. We now know from radiometric dating that the oldest Ediacaran fossils date to about 578 Ma, just after the Gaskiers glaciation, and that some persisted to the very base of the Cambrian, 542 Ma.

Ediacaran Faunal Assemblages

Three assemblages of Ediacaran age have been recognized, each with distinctive suites of fossils: (1) the Avalon assemblage, dominated by a group called rangeomorphs and the meter-long, frond-like *Trepassia* and best known from the deep-water deposits of the Avalon Peninsula of eastern Newfoundland (578–565? Ma) (fig. 5.4); (2) the White Sea–Ediacaran assemblage, containing the "classic" Ediacaran taxa in sections in the White Sea north of Arkhangelsk in Russia and from the Ediacara Member in the Ediacaran Hills of South Australia (560?–542?); and (3) the shallow-water Nama assemblage, which includes more complex fronds and the earliest skeletonized fossils, best known from rocks dating to 549–542 Ma in southern Namibia.

The fronds, spindles, and bushes of the rangeomorphs, so-called from their similarity to the genus *Rangea,* dominate the Avalon assemblage (Laflamme and Narbonne 2008a). This group contains a diversity of fronds, some recumbent and others attached to the seafloor via a bulbous holdfast, but all sharing a common architectural organization. The Rangeomorpha were first established as a group by Pflug (1972) and their composition extended by the work of Jenkins (1985) and Narbonne (2004). The internal structures of rangeomorphs were finally well resolved when a new locality was found near Spaniard's Bay in Newfoundland. The Spaniard's Bay fossils have beautifully preserved frondlets with petals alternately branching off a central axis, and they show four levels of fractal branching (Narbonne 2004) (fig. 5.5). The diameter of the smallest frondlets (only 0.05 mm) seems to preclude their having housed polyps. Several frondlets form a module, and the modules combine to form the larger fronds. A variety of fossil types with these fractal architectures are now known, including *Avalofractus,* with beautifully preserved frondlets; large fronds such as *Charnia*; the bush-like *Bradgatia*; and the spindle-shaped *Fractofusus* (Clapham, Narbonne, and Gehling 2003; Hofmann, O'Brian, and King 2008; Laflamme and Narbonne 2008a, 2008b; Narbonne et al. 2009) (fig. 5.6). The phylogenetic affinities of these fractally constructed organisms remain a mystery (e.g., Peterson, Waggoner, and Hagadorn 2003). Although some have linked them to cnidarians, most paleontologists who have studied them agree that they represent architectures unlike any other metazoan group. The rangeomorphs likely form a natural clade, extending through all three Ediacaran assemblages. Although the rangeomorphs comprise some 74% of

Figure 5.4 Outcrop photograph of beds with elements of the Avalon assemblage, at Mistaken Point, Avalon Peninsula, Newfoundland. The surface in the foreground is covered with abundant fossil specimens preserved beneath a thin layer of volcanic ash dating to about 565 Ma.

most Avalon assemblages, a few other forms are included among the twenty genera that have been described from this assemblage (Shen et al. 2008). None of the taxa were skeletonized, and there is only one report of possible trace fossils from these localities (A. Liu, McIlroy, and Brasier 2010).

A great challenge is to "reanimate" the flattened fossils into their original three-dimensional forms. Frequently, several fossils are superimposed, obscuring details of their form and structure. *Charnia* (fig. 5.7) is one of the more common elements of the Avalon assemblage and illustrates some of the difficulties in reconstructing the original form. Studies by Antcliffe and Brasier (2008) and Laflamme and colleagues (2007) provide the latest, somewhat contrasting views of *Charnia. Charnia* lived attached to the substrate by a bulbous holdfast embedded in the sediment from which arose a central stalk

Figure 5.5 A specimen of *Avalofractus* showing the frondlets that serve as the basic building block for the rangeomorphs from the Avalon Peninsula, Newfoundland. Notice the three orders of fractal branching, with the smallest branches < 150 μm in diameter. Both sides of a frondlet are identical. From the Trepassy Formation, Spaniard's Bay, Newfoundland. Photograph courtesy of Marc Laflamme.

and a frond that extended into the water column (Laflamme et al. 2007). The frond was constructed with a central axis and alternating primary branches with secondary branches composed of the rangeomorph modules described

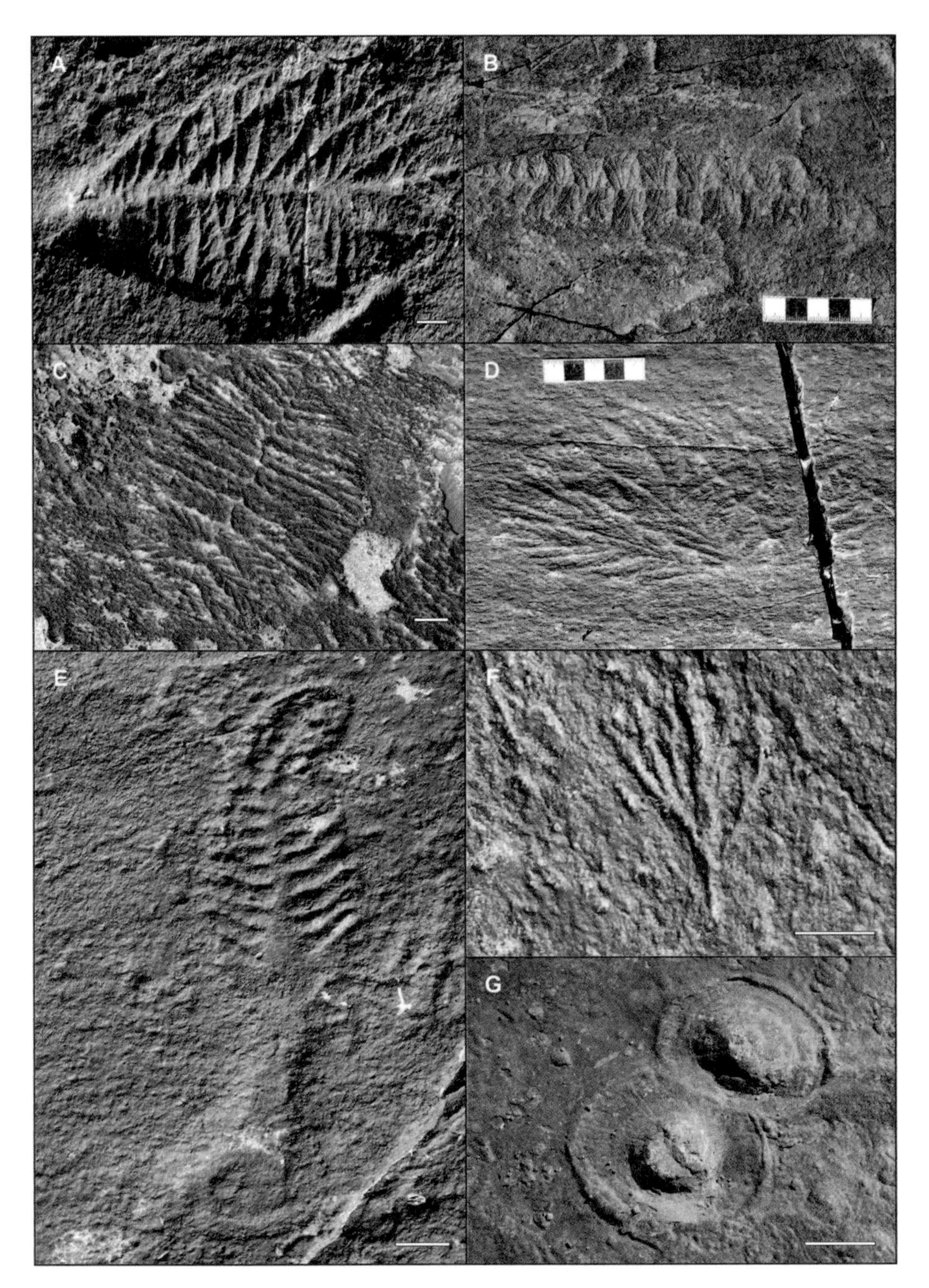

Figure 5.6 Representative specimens of the Avalon assemblage, from Newfoundland. (A) *Beothukis.* (B, C) *Fractofusus.* (D) *Bradgatia.* (E) *Charnodiscus.* (F) *Parviscopa.* (G) *Aspidella.* All are field photographs except for (A), which has been retrodeformed to remove the tectonic deformation of the shape of the fossil. Scale bars 1 cm. Photographs courtesy of Marc Laflamme.

above. Specimens of the closely related genus *Trepassia* are as much as 2 meters in length (Narbonne et al. 2009). Although it has been suggested that *Charnia* is allied to the cnidarian sea pens, the two growth patterns are in fact quite different. *Charnia* evidently grew by insertion of new branches at the apical end (away from the central stalk), the opposite of the growth patterns of pennatulaceans. Pennatulaceans are, in any case, fairly derived cnidarians; the difference in growth patterns would seem to eliminate any connection between them and any rangeomorphs (Antcliffe and Brasier 2008).

The Newfoundland fossils are exposed on more than one hundred large bedding planes, some of which can be traced for several kilometers. On each of these bedding planes, the fossils were buried by volcanic ash, preserving the original ecological relationships among thousands of specimens in what paleontologists call census assemblages. By mapping the relationships among the fossils, their ecological relationships have been reconstructed (Clapham, Narbonne, and Gehling 2003). Most of the seafloor was initially colonized by microbial mats, which in turn were invaded by flat-lying forms that acted as pioneer species and stabilized the mat. A tiered community of upright forms—including bushes, fronds, combs, and plumes—then replaced the recumbent forms during a mid- to late successional stage. More than 90% of the individuals in a locality belong to the lowest tier (with heights of fewer than 8 centimeters), including the spindle-like *Fractofusus*. The intermediate and upper tiers have progressively fewer specimens. There is also some evidence that similar species clustered together, which may say something about the patchiness of their resources. Overall, the species richness, abundance, and spatial patterning of these assemblages were similar to those documented in modern shallow-water invertebrate communities (Anderson, Conway Morris, and Crimes 1982; Clapham and Narbonne 2002; Clapham, Narbonne, and Gehling 2003; Laflamme and Narbonne 2008a, 2008b; Narbonne 2005). The diversity of rangeomorph morphologies appears to have been driven by adaptations to exploit resources at different levels above the seafloor, and even within the erect fronds, four different architectures appear to have been present (Laflamme and Narbonne 2008a, 2008b). The small size of the terminal units (0.05 mm) in rangeomorphs renders filter feeding implausible. Furthermore, the Newfoundland rangeomorphs were living too deeply to have been photosynthetic. We are left with the possibility that rangeomorphs fed by absorbing nutrients directly (Laflamme, Xiao, and Kowalewski 2009). These uncertainties emphasize just how different members of this clade were from extant animals. Rangeomorphs are not restricted to the Avalon region and related localities (fig. 5.8); *Charnia* is also found in South Australia and the White Sea faunas, and a later rangeomorph, *Rangea* itself, occurs in the White Sea–Ediacaran and Nama assemblages.

Figure 5.7 *Charnia* is one of the more enigmatic of the Ediacaran fossils. (A) Photograph of the holotype specimen of *Charnia masoni,* from Charnwood Forest, United Kingdom. The specimen is 208 mm long. (B) Reconstruction of *Charnia* based on Antcliffe and Braiser (2008). Photograph (A) courtesy of Marc Laflamme.

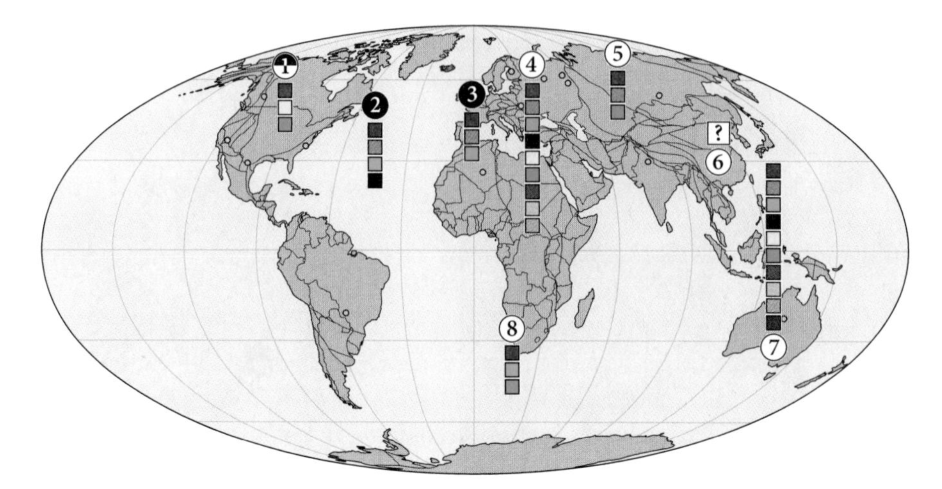

Figure 5.8 Map of the location of important Ediacaran localities: (1) northwest Canada (Avalon and White Sea assemblages); (2) Newfoundland (Avalon assemblage); (3) Charnwood Forest (Avalon assemblage); (4) Winter Coast, White Sea (White Sea assemblage); (5) northern Siberia (White Sea assemblage); (6) Yangtze Gorge, south China (White Sea assemblage); (7) Ediacara Hills, South Australia (White Sea assemblage); (8) southern Namibia (Nama assemblage). Colors correspond to those used in fig. 5.13. Dark circles denote deep-water assemblages; open circles denote shallow-water assemblages.

The White Sea–Ediacaran assemblage represents the "classic" Ediacaran fauna of discs and fronds (Fedonkin et al. 2007; Gehling et al. 2005; Glaessner 1984; Glaessner and Wade 1966; Jenkins 1992; Jenkins, Ford, and Gehling 1983). It is the most diverse of the three assemblages, encompassing about seventy-seven genera (Shen et al. 2008). There is a greater variety of forms in this assemblage than in the other two suites, from the rangeomorphs *Charnia* and *Rangea* and the frond *Charniodiscus* to seemingly bilaterally symmetrical forms such as *Dickinsonia* and more truly bilateral forms including *Spriggina, Parvancorina, Kimberella,* and *Yorgia* (fig. 5.9). In the White Sea area, the assemblage was deposited in relatively shallow-water sands (i.e., above the depths reached by storm waves). Classic assemblages have also been recognized in similar shallow-water clastic settings in northern Norway, the Urals, and Siberia and occasionally in more offshore deposits.

Dickinsonia (fig. 5.9C, D) is one of the most distinctive of the Ediacaran fossils, with specimens ranging from almost circular to long and ribbon-shaped; all are roughly blaterally symmetrical with a series of usually close-set transverse furrows that are interrupted by a central longitudinal furrow (Fedonkin 2003; Gehling et al. 2005; Valentine 1992). Seilacher interpreted *Dickinsonia* (and many other Ediacaran forms) as having a quilted structure,

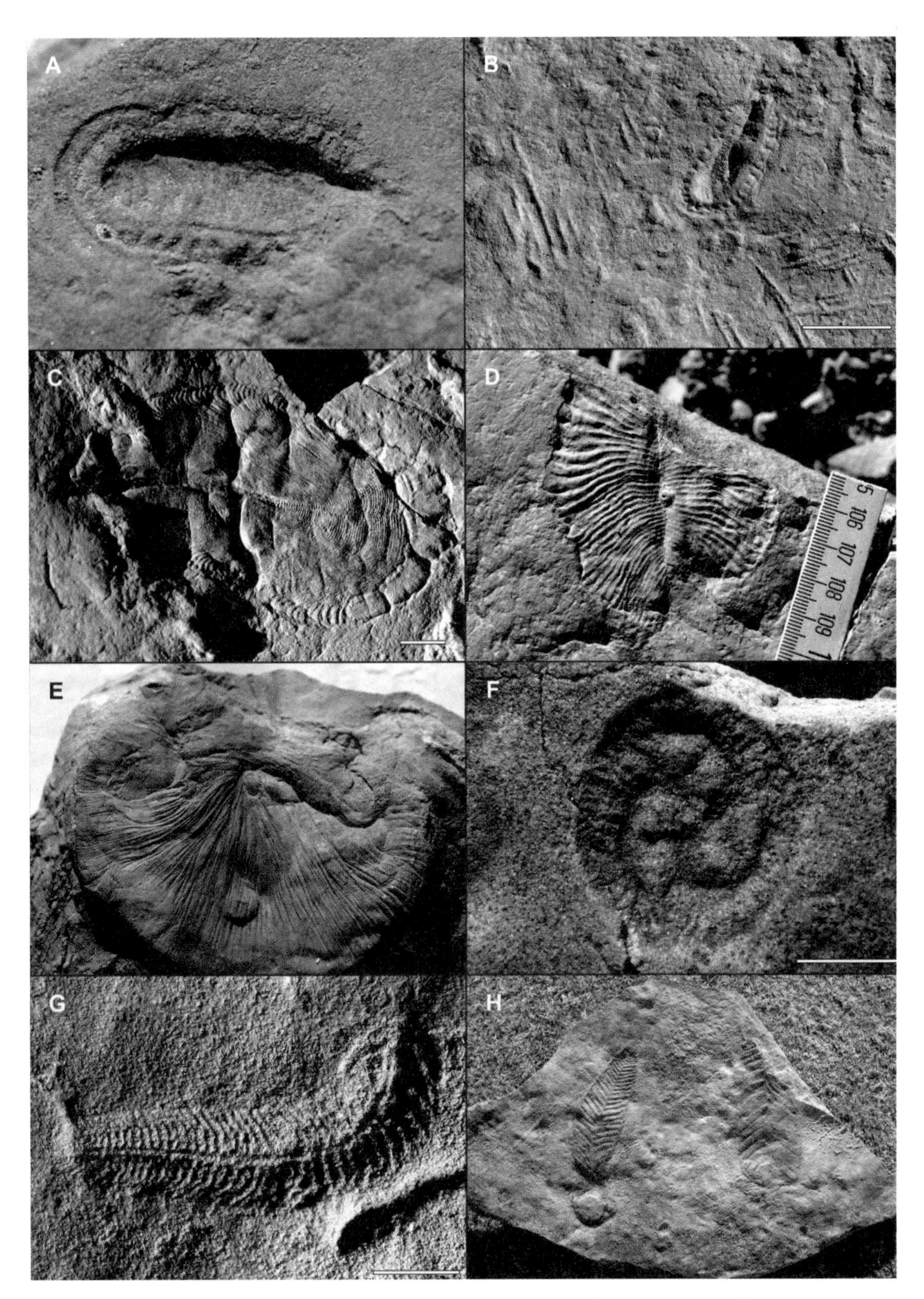

Figure 5.9 Representative specimens of the White Sea–Ediacaran assemblage. (A) *Kimberella.* (B) *Kimberella* with associated scratch marks, interpreted as radular feeding on microbial mats. (C–D) *Dickinsonia.* (E) *Inaria,* probably produced when the bulbous holdfast of a frond was plucked out of sediment during burial. (F) *Tribrachidium.* Paratype. SAM P12898. (G) *Spriginna.* Holotype. (H) *Charnodiscus.* Photographs A, B by Douglas Erwin; C–G by Marc Laflamme; H by Jim Gehling. Specimens G–H belong to the South Australian Museum.

with presumably fluid-filled spaces, called pneus, between the lateral furrows (Buss and Seilacher 1994; Seilacher 1989). Compared with many of the other Ediacaran fossils, *Dickinsonia* impressions have several unique features, including apparent shrinkage marks and concentric contractions around the periphery of some specimens that have been interpreted as indications of a muscular body; in one instance, a series of ghost-like impressions lead to a well-preserved specimen at the end of the series (Ivantsov and Malakhovskaya 2002). Such fossils appear to preserve the impressions of a single individual moving across microbial and algal mats, presumably feeding by absorbing nutrients through the underside of the organism (Sperling and Vinther 2010). Similar impressions are known from *Yorgia,* a relative of *Dickinsonia* with a lopsided anterior region (X. L. Zhang and Reitner 2006).

The curious, eight-armed *Eoandromeda* has been recovered as carbonaceous compressions from black shales of the Upper Doushantuo Formation and the Miaohe biota in the Yangtze Gorges of south China and as casts and molds from the Ediacaran biota in South Australia (Zhu et al. 2008) (fig. 5.10). It is one of the few Ediacara macrofossils preserved in two such different styles. The tubular, spiral arms with transverse ridges suggested a ctenophore affinity to Zhu et al. (2008), and a new study with additional Chinese specimens led to suggestions that they were likely stem ctenophores because they lack some features of the crown group (Tang et al. 2011).

Kimberella (fig. 5.9A, B) is an oval fossil, probably with a slug-like shape in three dimensions and with a crenulated margin, a wavy central area, and a complex region at the presumed anterior that probably includes a mouth. Specimens from the White Sea often show an extensible "proboscis" in that region. The wavy area may represent postmortem contractions of transverse muscles. The marginal crenulations have been interpreted as respiratory folds, perhaps analogous to the ctenidia of mollusks (Fedonkin, Simonetta, and Ivantsov 2007; Fedonkin and Waggoner 1997). First collected in South Australia, *Kimberella* is best represented at White Sea localities where hundreds of well-preserved specimens have been found in fine-grained rocks underlying a volcanic ash bed that is confidently dated to 555 Ma (Martin et al. 2000). The finer-grained siltstone and mudstones at the White Sea localities preserve greater morphological detail than the coarser sandstones of the Ediacaran Hills. At many localities, *Kimberella* is associated with fan-shaped bundles of parallel scratch marks (a trace fossil known as *Radulichnus*; fig. 5.9B) that appear to indicate a habit of grazing on microbial mats (Ivantsov 2009; Seilacher and Hagadorn 2010). The furrows that comprise the *Radulichnus* traces are reasonably believed to have been created by a large number of teeth set in parallel rows, similar to a molluscan radula, and strongly suggest that *Kimberella* was a bilaterian, perhaps a stem protostome, a stem lophotrochozoan, or

Figure 5.10 Carbonaceous compression fossil of *Eoandromeda,* interpreted as a stem group ctenophore. Specimen JK 10909 in the Institute of Geology, Chinese Academy of Sciences, Beijing. Scale bar in mm. Photograph courtesy of Feng Tang.

even a stem mollusk. Some *Kimberella* crawling trails have been described as well.

The vast deserts of Namibia in southwestern Africa have yielded the last of the three Ediacaran assemblages, the Nama assemblage (fig. 5.11). The sandstones, siltstones, and carbonates of the Nama Group include some of the youngest rocks of the Ediacaran Period. The diversity of fossils is much less than in the White Sea–Ediacaran assemblage, with only about seventeen genera (Shen et al. 2008). Because the rocks in Namibia have had the least study of any of the assemblages, that fauna may prove to be more diverse with more intensive fieldwork. The oldest skeletonized fossils known are also found there (Cohen et al. 2009; Grant 1990; Watters and Grotzinger 2001), along with a diversity of tubular body fossils that are not yet understood. One region in

Figure 5.11 Representative specimens of Nama assemblage fossils. (A) *Swartpuntia.* (B) *Rangaea.* (C) *Pteridinium.* (D) *Ernietta.* Photographs A–C by Douglas Erwin; D from Shuhai Xiao.

particular, the Farm Aar localities in southern Namibia, includes some outstanding exposures containing trifoliate fronds of *Pteridinium* and the sac-like *Ernietta,* commonly engorged with fine sand. These genera originally lived in shallow subtidal sands, suggesting that each possessed a flexible and elastic wall, likely proteinaceous and perhaps even collagenous.

Interpretations of *Pteridinium* are particularly contentious because it is often preserved in a jumble of specimens. It was initially reconstructed from incomplete material as an attached, epifaunal frond with three vanes radiating from a central axis, with each vane composed of a set of tubes (Jenkins

1985, 1992), but a well-preserved suite of complete specimens from Farm Aar appears to indicate that the three vanes come together to form a point much like a canoe. *Pteridinium* has been suggested to have lived infaunally in sand rather than attached to the seafloor as were most other Ediacaran species (Grazhdankin and Seilacher 2002; but see Droser, Gehling, and Jensen 2006). Such a sand-filled bag is an unusual morphology, albeit one possibly shared by *Ernietta* and a few other forms.

Early researchers at these Ediacaran localities identified what they believed to be abundant long tubular trace fossils, which were interpreted as formed by shallow-burrowing bilaterian worms. Now, however, it appears that this conclusion was mistaken. Instead, both the White Sea–Ediacaran and the Nama assemblages include a kind of fossil newly recognized as a common constituent of the Ediacaran faunas: tubular body fossils. In other words, those fossils are not burrows, but instead represent the organisms themselves. One of them, recently named *Funisia dorothea* from the Ediacaran of South Australia, forms dense (1000/m^2) aggregations of tubes, suggestive of the growth of a single cohort. The organisms grew both by branching and by adding new units to the end of the tube, giving them a segmented appearance (fig. 5.12). Their

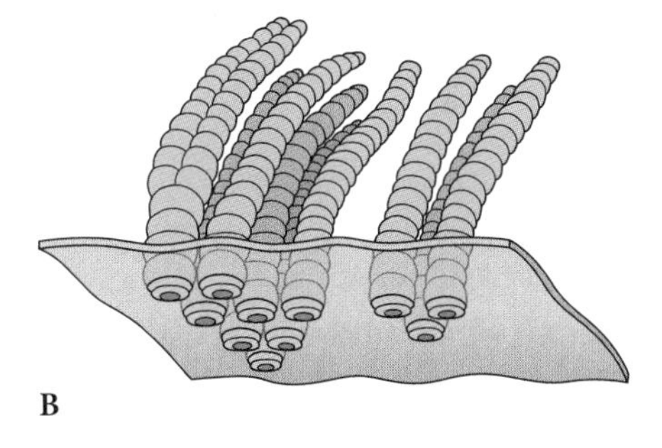

Figure 5.12 *Funisa dorthea,* an example of one of the tubular body fossils found in the late Ediacaran in South Australia, preserved as external casts. (A) Photograph of the holotype, South Australian Museum P40725, showing a variety of preservational types. (B) Reconstruction, showing the holdfast beneath the mat. Scale bars 2 cm. From Droser and Gehling (2008).

phylogenetic affinities are unclear, but they may have been stem-group cnidarians or sponges (Cohen et al. 2009; Droser and Gehling 2008).

Preservation

An outstanding problem is how these soft-bodied organisms were so well preserved during the Ediacaran. Was there something unusual about conditions at that time that enhanced the probability of preservation? Or, was there something about the organisms themselves that facilitated their preservation? The exquisite preservation of the purported Doushantuo embryos and the fidelity of many of the Ediacara forms surely represents different preservational histories but emphasizes the generally excellent quality of preservation during the Ediacaran. Indeed, one can argue that preservation may have been better during this interval than during some parts of the early Phanerozoic. Modes of preservation of Ediacaran fossils are varied; they include molds of the external surfaces, positive casts and composite molds of the upper and lower surfaces, and molds of some internal structures (Gehling 1988, 1999; Glaessner and Wade 1966; Wade 1968). The soft-bodied Ediacaran fauna has been entombed in volcanic ash in Newfoundland (Narbonne 2005), preserved through authigenic carbonate cementation in Siberia (Grazhdankin et al. 2008), and protected beneath microbial "death masks" in many localities (Gehling 1999), while some taxa are filled with siliciclastic sediments. In Namibia and the White Sea, *Pteridinium, Ernietta,* and some forms are preserved in three dimensions (Crimes and Fedonkin 1996; Gehling 1999; Ivantsov and Grazhdankin 1997). The abundance of external molds must mean that many Ediacaran organisms were composed of a sufficiently tough external integument to support the surrounding sediments until early mineralization and cementation of the sediment preserved their shapes (Darroch et al., 2012; R. Norris 1989; Seilacher 1989, 1992). Although much additional work is needed on the meaning of preservational differences between different forms, the variety of preservational styles within a single assemblage suggests that Ediacaran organisms were not all constructed from similar elements or a similar design (Gehling 1991, 1999; Jenkins 1992; but see also Retallack 1994; Seilacher 1984, 1989).

Widespread microbial mats may have enhanced the preservation of many Ediacaran specimens by reducing the diffusion of fluids and thus the rate of organic decay (Gehling 1999; Gehling et al. 2005; Hagadorn and Bottjer 1997; Seilacher 1999). Evidence for microbial mats comes from interpretations of particular sedimentary textures and structures such as "elephant skin" surfaces and wrinkle marks (or *runzelmarken*) that were widespread during this time. The marine sandstones of the Ediacaran and early to middle Cambrian were evidently embedded within flourishing microbial communities. In 1999, Gehling proposed that deposition of sands, stirred up by storms, buried the living organisms, thus preserving a snapshot of their communities. Subsequent growth of a new microbial mat community then sealed off the underlying beds, isolating anoxic or dysoxic pore waters in the sediments from the oxygenated waters above and thus producing a reducing chemical environment. Sulfate-reducing bacteria generated sulfide, which combined with iron to form pyrite (FeS_2). This mineral produced a "death mask," aiding in the preservation of the Ediacaran organisms. Thus, although many of the Ediacaran fossils may have possessed a tough integument, widespread microbial mats are likely to have aided their preservation at many localities. Once metazoan burrowing and bioturbation of the substrate increased in the Cambrian, the mats were commonly broken up, and this preservational window closed (Gehling 1999).

Biogeography

A question raised by the differences among the Avalon, White Sea–Ediacaran, and Nama assemblages (fig. 5.13) is whether they represented three different biogeographic realms, three successive evolutionary phases, or faunas from three different environments. In a test of the biogeographic hypothesis, the distributions of genera at more than thirty different localities were analyzed (Waggoner 2003), but the results revealed no coherent biogeographic pattern (despite their names) and suggested that they could represent different temporal assemblages. An alternative interpretation is that each assemblage represents a different depositional environment, implying a high degree of environmental and ecologic specialization among the Ediacaran fauna (Grazhdankin 2004). In fact, there is probably both a temporal and an environmental component to the three assemblages (Narbonne 2005). Moreover, extensive facies differences in fossil distributions within the Ediacaran Hills of South Australia suggest environmental controls as well (Droser, Gehling, and Jensen 2006). In the absence of tighter geochronologic control, a definitive answer to this problem is not yet available; a reasonable interpretation is that the three Ediacaran assemblages differ in age, as initially suggested, but also document distinctive environmental specializations among elements of the fauna during the later Ediacaran. It is also possible that preservational differences can explain some of the differences between assemblages (Laflamme et al. 2007).

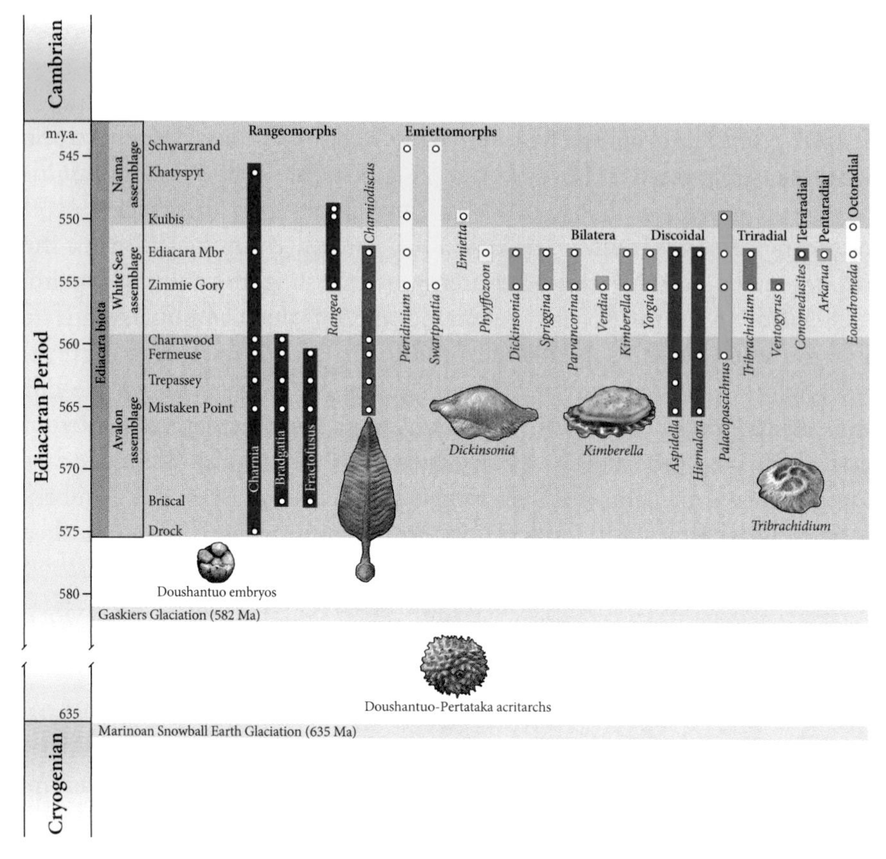

Figure 5.13 The timing of the Ediacaran fossil assemblages and the duration of representative genera of the major groups of Ediacara macrofossils (shown in colors). Adapted from Xiao and Laflamme (2009).

Construction of Ediacaran Organisms

Select Ediacara fossils appear to share sufficiently similar features with modern phyla that some paleontologists assigned them to living clades: *Dickinsonia* to the annelid worms, *Parvancorina* and *Spriggina* to the arthropods, and *Arkarua* to the echinoderms, for example. These assignments now seem fanciful. Others, such as the triradial *Tribrachidium,* are more difficult to associate with any living clade on any basis. Members of the Ediacaran fauna, however, exhibit various regularities of growth and form, producing morphological clusters within the Ediacaran fauna, although we do not yet know whether these represent phylogenetic groups (Xiao and Laflamme 2009). As part of an effort to characterize patterns of origination during the Ediacaran and Cambrian, which will be discussed further in chapters 6 and 7, Marc Laflamme developed a preliminary classification of the Ediacara macrofossils that recognizes twelve different groups, several of which are almost certainly polyphyletic: the rangeomorphs, erniettomorphs, dickinsoniomorphs, arboreomorphs (*Charniodiscus* and allies), bilateralomorphs, radialomorphs (*Mawsonites*), triradialomorphs (*Tribrachidium* and related forms), tetraradialomorphs (*Conomedusites*), pentaradioalomorphs (*Arkarua*), kimberellomorphs, tubular fossils, and poriferans (Erwin et al. 2011; Laflamme, forthcoming; see also fig. 5.14 and the appendix). To summarize the discussion, the rangeomorphs dominated the Avalonian assemblage and united forms with alternately arranged frondlets in a fractal structure. This group is likely to be monophyletic. The erniettomorphs *Ernietta, Pteridinium,* and *Swartpuntia* have been interpreted as quilted tubes alternately arranged along a midline. This group appears to have been constructed of stiffer material than *Dickinsonia* (Gehling et al. 2005). *Dickinsonia* shares some of the features of the erniettomorphs but exhibits a higher degree of bilateral symmetry and anterior-posterior differentiation, and there is the trace evidence of movement. These features are consistent with the recent proposal that the feeding mode of *Dickinsonia,* by external digestion of microbial mats with the sole of the organism, is most consistent with a placazoan affinity (Sperling and Vinther 2010). *Dickinsonia* thus bridges the morphological gap to a larger, more heterogeneous assemblage of roughly bilaterally symmetrical forms, including those with apparent segmentation such as *Spriginna* and *Yorgia.* Beyond these forms are a host of others that are more difficult to categorize, including the many discoidal forms that likely represent holdfasts for fronds, the newly described tube-like organisms, and others with a variety of symmetries.

Phylogenetic Affinities

The identification of these twelve constructional groupings does not resolve the most long-standing and contentious issue associated with these fossils: their phylogenetic affinities (reviewed by Runnegar 1995). For example, *Charnia* is essentially a phylogenetic Rorschach test: initially described as an alga, this genus has since been viewed as a pennatulacean soft coral (Glaessner 1984), a lichen (Retallack 1994), a member of the Fungi (Peterson, Waggoner, and Hagadorn 2003), a colony of the prokaryotic myxobacteria (Steiner and Reitner 2001), and a representative of an extinct higher-level clade (Seilacher 1992). Although most of the Ediacara macrofossils have been viewed at one time or another as being of cnidarian grade, some workers have argued for protostome or deuterostome affinities (see above). Furthermore, competent

A
B
C
D
E
F
G
H
I
J
K
L
M

Figure 5.14 (left) Twelve different groups of Ediacara macrofossils have been defined by Laflamme, nine of which are likely to be distinct clades. (A) Rangeomorphs: *Avalofructus.* Specimen NFM-F754. (B) Close-up of an individual frondlet. (C) Erniettomorphs: *Ernietta.* Close-up of a frond. (D) Dickinsoniomorphs: *Dickinsonia.* Specimen SAM P42158 a+b. (E) Arboreomorphs: *Charniodiscus.* Specimen SAM P40786. (F) Bilateralomorphs: *Spriggina.* Specimen SAM P29801+2. (G) Kimberellomorpha: *Kimberella.* (H) Triradialomorphs: *Tribrachidium.* (I) Tetraradialomorphs: *Conomedusites.* (J) Pentaradialomorphs: *Arkarua.* Holotype SAM P26768. (K) *Mawsonites.* SAM P41276. (L) *Palaeopascichnus.* SAM P36854. (M) A sponge, *Thectardis.* Photographs A, B, D, E, F, H, I, and L are of specimens from the collections of the South Australia Museum (SAM) and photos are courtesy of Marc Laflamme. Photograph C courtesy of Shuhai Xia. Photographs G, M by Doug Erwin. Photograph K courtesy of Jim Gehling.

specialists have not been able to agree on whether Ediacarans represent a single clade or instead are a mix of organisms representing different clades, nor do they agree on whether some or all of the forms have phylogenetic links to Cambrian and younger clades or are instead completely distinct. Rehearsing the various debates about the phylogenetic affinities of Ediacaran fossils tries the patience even of enthusiasts. Hence, we will confine our discussion to outlining two extreme views: first, Adolf Seilacher's suggestions that these forms represent extinct or at least nonmetazoan clades; second, an alternative view that many Ediacaran organisms have close affinities to crown clades.

In 1984, Seilacher made the provocative proposal that many Ediacarans shared a quilted construction, much like an air mattress. Such an architecture is so unlike any living multicellular group that he suggested that it represented a failed experiment with multicellularity (the Vendozoa hypothesis). Initially, Seilacher proposed the Vendozoa as an extinct kingdom, but he later suggested that they were an extinct metazoan clade, encompassing all Ediacaran genera (the vendobionts), although he has since acknowledged that at least some Ediacaran fossils seem to represent true metazoans (e.g., he has suggested that *Tribrachidium* was a sponge-grade organism and that *Kimberella* was a mollusk) (Buss and Seilacher 1994; Seilacher 1989, 1992, 2007a). Seilacher recognized two different groups of vendobionts: one encompassing the serially quilted forms, including the erniettamorphs, *Dickinsonia* and allies, and *Charnia* and related forms; and a second encompassing fractally quilted forms, including what we have previously described as the rangeomorphs (fig. 5.15). The serially quilted forms grew by insertion of new quilt structures, whereas the fractal forms expanded and subdivided into new frondlets (Laflamme, Xiao, and Kowalewski 2009). The recovery of an apparently serially quilted vendobiont from the Denying Formation of China (mid-Ediacaran in age) provides some

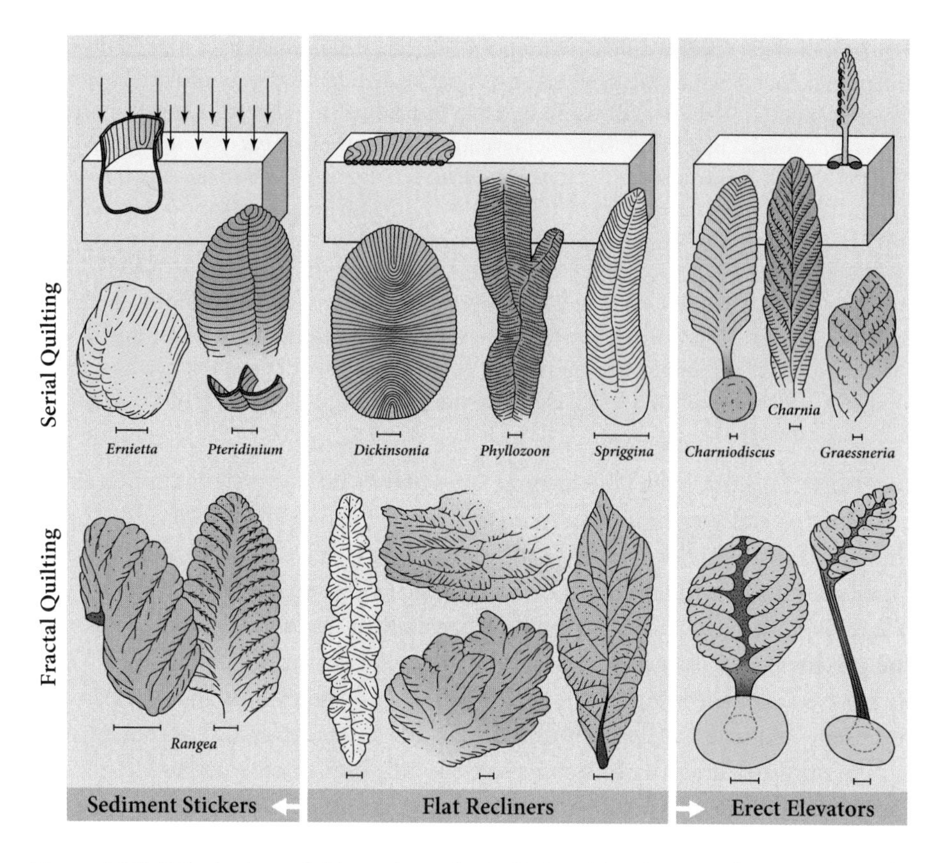

Figure 5.15 Seilacher's vendobionta hypothesis distinguishes between forms with serial quilting and those with fractal quilting, with several similar lifestyles represented among each of the different architectures. Arrows in upper left show sediment burying an *Ernietta*. Adapted from Seilacher (1992).

support for a vendobionta hypothesis (Xiao, Shen, et al. 2005). The Chinese forms have a tube-like cross section but with open distal ends, forms that are difficult to understand unless their internal composition was gel-like. In 2007, Seilacher again modified many of his assignments. The core vendobiont clade, he suggested, is an extinct group allied to rhizopodean protists such as the Xenophyophoria [as originally suggested by Zhuravlev (1993)]; thus, it is not even multicellular. Living Xenophyophoria possess a flexible wall and produce large, sausage-shaped compartments that are filled not just with protoplasm, but with up to 50% fine-grained sand. Seilacher suggests that the sand-filled Ediacaran forms such as *Ernietta* represent larger members of the same clade. Seilacher's 2007 version of the vendobiont hypothesis is perhaps strongest in dealing with erniettomorphs and related forms, but weaker for rangeomorphs and *Dickinsonia*. There are, moreover, few characters to suggest a phylogenetic link between the serially quilted forms and the rangeomorphs, and the syncytial structure required by all Seilacher's proposals is contradicted by strong evidence for cellular structures in many presumed vendobionts, a criticism that would also apply to his xenophyophoria hypothesis.

The apparently bilateral symmetry of *Dickinsonia, Spriginna, Parvancorina,* and some other forms led them to be assigned to a variety of bilaterian clades, including annelids and arthropods, while the small, circular *Arkarua* was placed with the echinoderms. Glaessner was a fervent proponent of this viewpoint, and other Australian paleontologists have tended toward a similar approach (Gehling 1991; Glaessner 1984; Jenkins 1992). Furthermore, many Ediacaran fossils were described as representatives of highly derived clades within those phyla, necessitating that much of the metazoan diversification within crown phyla had happened before the Ediacaran faunas appeared, but had been unrecorded by fossils. Phylogenetic assignments depend on the identification of specific synapomorphies, and as our knowledge of the Ediacaran fauna has expanded over the past several decades, synapomorphies with derived clades have been increasingly difficult to identify. For example, as described earlier, Antcliffe and Brasier have shown that the growth patterns of *Charnia* are inconsistent with it being a pennatulacean cnidarian. Similarly, the dickinsoniomorph morphology has not been reconciled with annelids, and other assignments, such as *Parvancorina* to the arthropods, are largely based on perceptions of overall similarities of form rather than on detailed phylogenetic character analyses. This research tradition has nonetheless identified characteristics, such as the possible muscular contractions in *Dickinsonia,* that support metazoan affinities if not a specific phylogenetic placement. Sperling and Vinther (2010) have argued that these characteristics are consistent with a placazoan affinity for *Dickinsonia* and its allies.

Figure 5.16 summarizes potential phylogenetic positions of elements of the Ediacaran assemblage under a variety of assumptions. We see little support for claims that the Ediacaran fauna represents a single clade. Rather, it appears to be a heterogeneous assemblage encompassing sponges, a possible ctenophore (*Eoandromeda*), possible cnidaria, two groups of probable bilaterians (*Kimberella* and, perhaps more uncertainly, the dickinsoniomorphs), and several major clades of uncertain placement, including the rangeomorphs and erniettomorphs. Generally speaking, preservational variation can explain many of the differences among fossil morphologies, and given the variety of facies and the challenges of preservation, this problem cannot be overestimated. Clearly, the primary difficulty in establishing the affinities of many Ediacaran fossils is the lack of informative morphological characters, and there seems little likelihood of agreement breaking out soon.

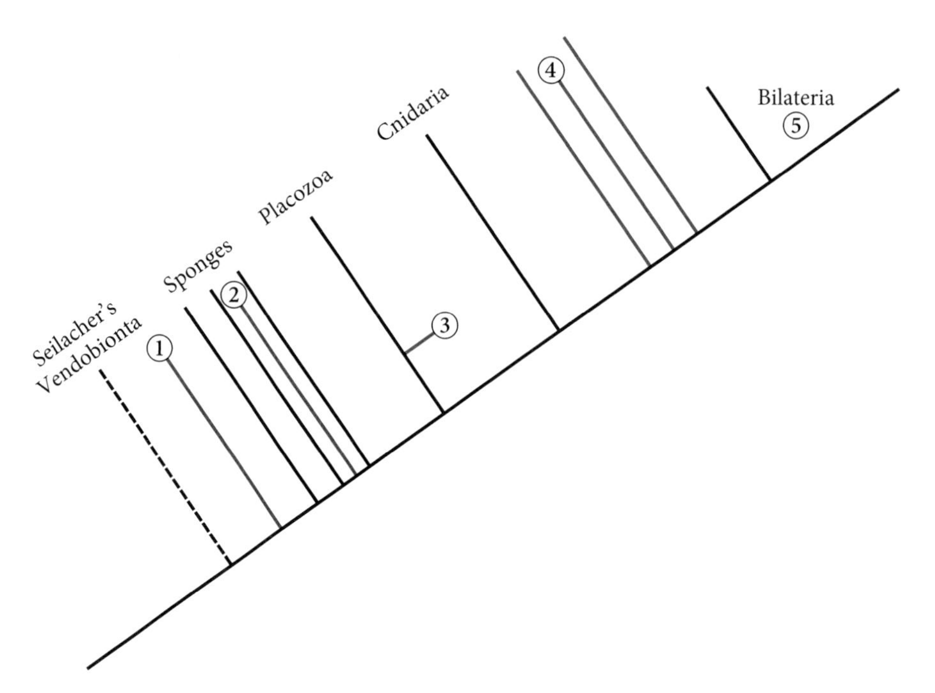

Figure 5.16 A metazoan phylogeny showing a variety of possible placements of different groups of Ediacara macrofossils. Seilacher's vendobionts would be an independent clade from the Metazoa, and are shown at the base of the tree. Position 1 is clades at the base of the Metazoa, possibly including rangeomorphs; position 2 denotes sponges, such as *Thectardis*; position 3 corresponds to the Placazoan hypothesis for *Dickinsonia,* as suggested by Sperling and Vinther (2010); position 4 would place at least some of the Ediacarans as extinct clades between Cnidaria and the Bilaterian LCA; position 5 reflects probable bilaterian Ediacarans, such as *Kimberella.* The array suggests that the Ediacaran organisms do not represent a single clade but have a variety of phylogenetic affinities instead.

Because we will forever lack molecular sequences of the long-extinct Ediacarans, we will never have direct evidence of the nature of their genomes. Perhaps the best we can do at present is to examine the genomes of those crown groups that might have originated in the general area of the metazoan tree where we believe the Ediacara macrofossils might belong, looking for hints as to what Ediacaran genomes might have been like and what that might imply about their histories and lifestyles (Erwin 2009). The two crown phyla that seem closest are Placozoa and Cnidaria, and there are complete genome sequences available for each (Putnam et al. 2007; Srivastava et al. 2008). As we discuss in chapters 8 and 9, each of those organisms contains genes that are devoted to the development of bilaterian tissues and organs not possessed by any known members of those phyla, with cnidarian genomes containing more regulatory genes than the placozoan. It would appear that if Ediacarans arose in the general neighborhood of the tree from which those phyla branched, their genomes could have contained the molecular tools employed in the development of even more complex morphologies than they exhibit. In other words, the insights from comparative genomic and developmental studies of modern organisms reveal that there is nothing in the morphology of the Ediacara macrofossils, other than *Kimberella* and the dickinsoniomorphs, that requires their placement higher than cnidarians on the phylogenetic tree.

Ecology

The ecology of the Ediacara organisms has been as perplexing as their phylogenetic affinities. There is much we can learn about the ecological aspects of the fauna simply by studying the structure of individual species, as well as their abundances, distributions, and patterns of association. A critical ecological observation is that no indications of predation, infaunal burrowing, or scavenging have been identified in any Ediacaran assemblage. The issue of motility is far more controversial. In the early 1990s, Jenkins's interpretations suggested that mobility was widespread (Jenkins 1992), yet mobile organisms were missing from the interpretations of other workers. The scratch marks *Radulichnus,* almost certainly feeding traces, demonstrate the establishment of grazing on microbial mats by 555 Ma, and they are usually attributed to *Kimberella.* In addition, trails composed of multiple impressions of dickinsoniomorphs have been identified in the White Sea and Australia, and specimens of *Yorgia* from the White Sea suggest a mobile organism (Dzik 2003). The most plausible interpretation of these fossils is that they were feeding on the underlying mat and had acquired mobility by about 555 Ma.

A variety of other feeding strategies has been proposed for Ediacaran organisms, ranging from suspension feeding to chemosymbiosis (fig 5.17). Photoautotrophy is excluded for rangeomorphs, at least, by their presence in life position in fairly deep waters. The epifaunal tiering of many fronds in the Avalon assemblage are suggestive of suspension feeding patterns, but the lack of apparent feeding structures is inconsistent with this feeding strategy; the same difficulty eliminates predation as a feeding strategy. There is a tiering of animals within the microbial mats: undermat miners, mat stickers (early skeletonized fossils that lived embedded in microbial mats), and mat encrusters can be distinguished (Clapham and Narbonne 2002; Clapham, Narbonne, and Gehling 2003; Droser, Gehling, and Jensen 2006). Rangeomorphs and erniettomorphs may have acquired nutrients by absorption of dissolved organic

Figure 5.17 A reconstruction of a late Ediacaran community assemblage, showing the diversity of architectural styles and ecological roles. Reconstruction by Quade Paul.

carbon (DOC). As discussed in chapter 3, DOC appears to have been much more abundant in Ediacaran seas than in younger oceans, making this feeding strategy, known as osmotrophy, feasible at that time. Today, osmotrophy is limited to bacteria, sponges, and some animal larvae, but the high ratios of surface area to volume of Ediacarans such as *Rangea, Ernietta*, *Swartpuntia*, and *Pteridinium* are consistent with an osmotrophic feeding behavior (Laflamme, Xiao, and Kowalewski 2009; Sperling, Pisani, and Peterson 2007).

Whatever Ediacaran organisms were doing, resources were not scarce, for the abundance of many organisms was quite high. The disk-shaped *Aspidella* from Newfoundland probably represents a holdfast for a frond-like organism, and densities are on the order of 1000–4000 individuals per square meter (Gehling, Narbonne, and Anderson 2000). Such numbers may not reflect a single living population, however, because a single bedding plane may include impressions of the central rachis of fronds from multiple levels, not all of which were necessarily alive at one time. Bedding plane surfaces in Newfoundland often contain tens to hundreds of larger specimens of rangeomorphs. Similar densities have been recorded for other Ediacaran assemblages, including those of *Pteridinium* and *Ernietta* from Namibia. Abundances are generally similar to those found in many early Paleozoic marine ecosystems (Clapham, Narbonne, and Gehling 2003; Crimes and Fedonkin 1996; Dzik 1999) or for that matter modern benthic marine communities.

Although attention is naturally drawn to the body fossils of the Ediacaran biota, study of broad exposures of Ediacaran rocks in South Australia and the western United States reveals that textured organic surfaces (TOSs) are widespread in most siliciclastic environments (Gehling and Droser 2009). These TOSs include large numbers of macroscopic organisms as well as microbially induced sedimentary structures and are far more common than body fossils. Tubular body fossils that have previously been described as traces are also abundant (Droser and Gehling 2008). The organisms that constructed these features were evidently important components of Ediacaran ecosystems, yet they have only recently been recognized, revealing an obvious lacuna in our understanding of the paleoecology of these ecosystems.

Techniques developed to quantify the increased morphological diversity (what paleontologists term *disparity*) found during the Cambrian explosion have recently been applied to Ediacaran organisms. Differences in morphologic disparity between the Avalon, White Sea, and Nama assemblages have been compared to differences in the number of taxa (taxic diversity). We have already seen that of the three faunas, the White Sea–Ediacaran assemblage had the greatest generic diversity, but surprisingly, maximal disparity occurs in the earlier Avalon assemblage (Shen et al. 2008). As in the Cambrian radiation (chap. 6), morphologic disparity increased more rapidly that taxic diversity.

Other persistent problems are why most Ediacaran fossils disappeared so abruptly near the Ediacaran-Cambrian boundary and how they are connected to the rest of metazoan evolution. Several Ediacaran-style fronds have been reported from the Cambrian, including *Maotianoascus* and *Stromatoveris* from the early Cambrian Chengjiang fauna (Dzik 2002), an assemblage of *Swartpuntia*-like fronds found interbedded with Cambrian trace fossils in the Uratanna Formation of Australia (Jensen et al. 1998), and the pennatulacian-like *Thaumaptilon* from the Burgess Shale (Conway Morris 1993). Although these fossils are similar in overall shape, it is not sufficient to demonstrate a phylogenetic connection (M. Laflamme, pers. comm., 2010). It may be that such forms were convergent due to environmental conditions. The microbially bound sediments of the Ediacaran disappear in the early Cambrian, and some paleontologists have linked this difference in preservation to the disappearance of many Ediacaran fossils. As discussed below, however, the changes in substrate type appear to have been gradual, extending at least into the mid-Cambrian. The question of whether the disappearance of many Ediacaran elements near 542 Ma indicates a real biodiversity crisis or is an artifact of preservation remains unresolved.

MICROFOSSILS

Microfossils form the bulk of the Proterozoic fossil record and they undergo significant changes during the Ediacaran (Huntley, Ziao, and Kowalewski 2006; Xiao, Knoll, et al. 2004; Zhou, Brasier, and Xue 2001; Zhou et al. 2007). After the end of the Marinoan glaciation at 635 Ma, simple, round, relatively smooth organic-walled microfossils known as leiosphere acritarchs were replaced by a particular group of acanthomorphs that were much larger (> 100 μm) ornamented forms with much shorter geologic ranges. Known as the Doushantuo-Pertatataka assemblage, a key feature of these acanthomorphs is the presence of one to many spiny processes that protrude from their surfaces. These acanthomorphs in turn disappeared between 580 Ma and 551 Ma (probably about 560 Ma) and were followed by a return to simpler, spheroidal leiospherid acritarchs in the latest Ediacaran.[2] Ornamented acritarchs reappeared in the earliest Cambrian and underwent an extensive diversification coincident with changes in the ecology of marine ecosystems. The term *acritarch* is a bit of a taxonomic garbage can and includes fossils that may represent a variety of different groups but are generally regarded as encompassing a variety of algae. Their large size and associated structures may indicate that some of these forms actually represent fungi, choanoflagellates, or mesomycetozoans. If so, the trophic complexity of post-Marinoan seas would have been substantially increased.

THE FIRST ANIMAL SKELETONS

There is no evidence for mineralized skeletons among the classic Ediacara macrofauna, but it would be unusual to find lightly skeletonized fossils in the clastic sandstones in which most of these fossils are preserved. Ediacaran limestones and dolomites tell a different story, however. In the last few million years of the Ediacaran, a variety of early skeletonized animal fossils appeared and are well known from wonderful exposures in southern Namibia (R. A. Wood 2011). These fossils include the possible tabulate coral *Namapoikia* (R. A. Wood; Grotzinger, and Dickson 2002) as well as the conical *Cloudina* (Amthor et al. 2003; Grant 1990; Hofmann and Mountjoy 2001; Hua et al. 2005) and its enigmatic associate, *Namacalathus* (Grotzinger, Watters, and Knoll 2000; Watters and Grotzinger 2001). *Cloudina* is built of a stack of lightly calcified, funnel-like tubes 2 to 7 millimeters in diameter (fig. 5.18). *Namacalathus* is particular interesting. One of the authors of this book, Erwin,

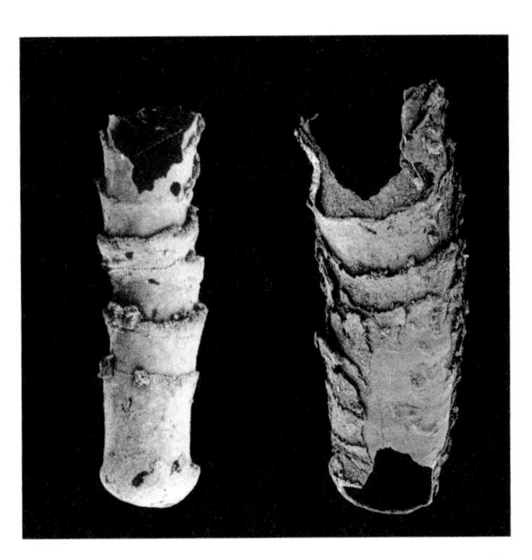

Figure 5.18 *Cloudina* from the late Ediacaran of southern Namibia, one of the earliest skeletal forms.

was doing fieldwork in southern Namibia with colleagues from MIT and Harvard in the late 1990s and collected samples from rocks near Zebra River with a large assemblage of cups, tubes, and rings. These fossils convinced his colleagues and him that they had stumbled upon a diverse assemblage of early skeletal fossils. It was only through exhaustive study that these different bits were revealed as various views of a single species. Serial sectioning and computer reconstruction revealed the original form of *Namacalathus* (fig. 5.19). This lightly calcified animal had a hollow stem attached to a goblet-shaped cup with six or seven holes around the cup. Like *Namapoikia,* it may have been a cnidarian, but it is so different from anything else known that it cannot be placed with any confidence.

Both *Namacalathus* and *Cloudina* comprise a geographically widespread, very latest Ediacaran assemblage found in carbonate packstones embedded within stromatolitic units. For example, in southern Namibia these forms are often found embedded in the heads of stromatolites, within the clotted microbial structures known as thrombolites, or in the spaces between the heads where they were swept by currents. In fact, a skeletal-rich debris is often found between these heads. Unlike the small shelly fossils of the early Cambrian (chap. 6), these first skeletal fossils do not appear to have been mobile: all were what Seilacher describes as "mud-stickers," living embedded in mud or microbial mats. Such a lifestyle is consistent with a cnidarian mode of life. Claims for borings in specimens of *Cloudina* from China suggest the possibility that early predators had become part of the ecosystem (Bengtson and Zhou 1992), but not all paleontologists accept these holes as signs of predation.

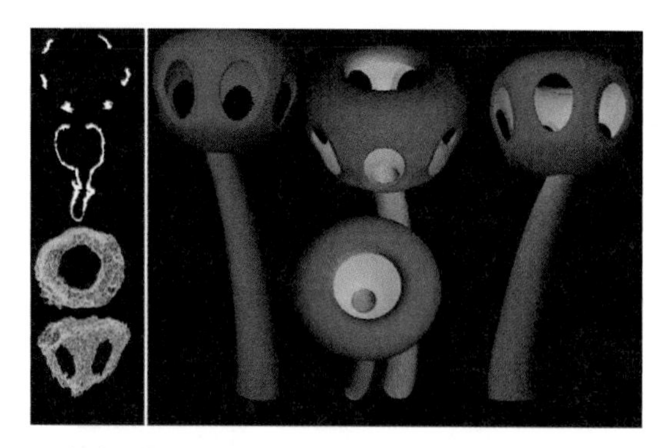

Figure 5.19 *Namacalathus* from the late Ediacaran of southern Namibia. These specimens may be 1 to 2 centimeters high. Photo and reconstruction from Watters and Grotzinger (2001).

Although these are the best known of latest Ediacaran skeletonized fossils, three other forms have been described from Siberia and Spain, and Zhuravlev and colleagues (2012) have suggested that as many as fifteen additional skeletal genera may exist in the late Ediacaran. While the Ediacaran is characterized by soft-bodied organisms, the proportion of skeletonized forms in the latest Ediacaran, as a percent of total diversity, may not be much different from either the Burgess Shale fauna or modern oceans (Knoll 2003a).

TRACES OF BEHAVIOR

One intriguing fact about the Cambrian radiation is that geologists are able to identify the evolutionary burst simply by looking at the rapid increase in the diversity and behavioral complexity of trace fossils (tracks, trails, and burrows), in the extent to which the sediment was burrowed (bioturbated), and in the disappearance of sedimentary fabrics associated with microbial binding of sediment. No fancy geochemical whizbangs or sophisticated statistical approaches needed, just a good pair of boots. Through the Ediacaran, the first traces of metazoans appear as surface trails across the top of the sediment, with some shallow mining into microbial mats. Possible bilaterian burrows have been identified as old as > 585 Ma (Pecoits et al. 2012), and penetrating vertical burrows may be as old as 555 Ma (Rogov et al. 2012) but such burrowing remains rare through the Ediacaran. In the absence of burrowing animals to disturb the sediment, rocks formed in shallow marine environments are generally finely laminated (fig. 5.20). Any disruptions to the sediment were produced by drag marks or by shrinkage as sediment dried out. The increase in bioturbation of sandstones, shales, and carbonate sediments from the Ediacaran into the Cambrian is captured by examining the changing sedimentary fabric of the rocks. A qualitative metric for describing changes in the extent of bioturbation has produced important data, particularly in logging stratigraphic section where vertical profiles expose a cross section through the rocks (Droser and Bottjer 1986, 1988; Droser, Gehling, and Jensen 1999; Droser, Jensen, and Gehling 2002; Droser, Jensen, Gehling, et al. 2002).

Such studies reveal what has been described as an "agronomic revolution," a change from dominantly microbially bound substrate in the Ediacaran, with organisms feeding on microbial resources and a sharp sediment-water interface, to a bioturbated sediment in the early Cambrian. This change is, of course, documented by the appearance of penetrating vertical burrows, by a broader range of behavioral trace patterns, and by the disappearance of microbial sedimentary features that suggest the development of a mixed boundary layer at the sediment-water interface (Bottjer, Hagadorn, and Dornbos 2000; Callow and Brasier 2009; Dornbos, Bottjer, and Chen 2005; Droser and Bottjer 1988; Gehling 1999; Seilacher 1999; Seilacher and Pfluger 1994). The change in substrate style was not abrupt and did not coincide with the onset

Figure 5.20 Diagrammatic representation of the change in sedimentary style from Ediacaran to Cambrian associated with the onset of bioturbation and the loss of microbial mats. Peter Trusler, Melbourne, from *The Rise of Animals: Evolution and Diversification of the Kingdom Animalia* by Fedonkin, Gehling, Grey, Narbonne, and Vickers-Rich. Johns Hopkins University Press, 2007.

of the Cambrian radiation in Stages 2 and 3 or with the earlier increase in bioturbation. Rather, organisms adapted to Proterozoic-style firmgrounds persisted through Stage 5, but then gradually gave way to forms with adaptations to unconsolidated sediment. The gradual change in substrate may have been closely related to the early Cambrian diversification (Dornbos 2006; Dornbos, Bottjer, and Chen 2004, 2005). For example, the mud-sticking behavior of early Cambrian helicoplacoids (see chap. 6) suggests that the great morphological disparity of such Cambrian ecosystems may in part reflect the presence of groups adapted to both kinds of substrate (Dornbos and Bottjer 2000).

The change in burrowing activity is also apparent in the morphology of discrete and identifiable trace fossils (ichnotaxa) (Crimes 1987; Seilacher 1956). The earliest accepted Ediacaran trace fossils are simple, poorly specialized, and unbranched horizontal grazing trails just a few millimeters thick, produced in sands close to the sediment-water interface. Some bedding planes are densely covered with such traces. Although they appear to be surficial traces, they are sometimes found on the underside of the overlying bed (as negative sole marks), so they must actually have been produced in the sediment or by burrowing through a microbial mat. Paleontologists have christened such fossils with names such as *Planolites, Gordia,* and *Helminthopsis.* Slightly more complex horizontal traces have shallow levees on either side, produced as the burrower pushed through the sediment and displaced material to the sides. These traces are very similar to trails or burrows produced by mollusks, and an organism with an analogous system of locomotion may have been responsible for these forms. The Ediacaran examples of such burrows have been given a variety of names, but *Archaeonassa* is sufficient to cover most if not all of them (Jensen 2003; Seilacher 1999; Seilacher, Buatois, and Mangano 2005).

One of the more complex Ediacaran trace fossils is *Helminthorhaphe* from South Australia, which has a relatively tight spiral meander connected to a more random trail. The complex spiral suggests a moderately well developed sensory system (Jensen 2003; Seilacher 1967). Another complex trace is *Radulichnus,* mentioned previously as the surficial grazing trace in microbial mats sometimes closely associated with *Kimberella* (Fedonkin, Simonetta, and Ivantsov 2007; Ivantsov 2009; Seilacher 1999). The only other Ediacaran trace fossils associated with an Ediacaran body fossil are the "footprints" associated with *Dickinsonia* and *Yorgia* in the White Sea assemblage (Ivantsov and Malakhovskaya 2002). The traces are of roughly the same dimensions as the associated body fossils and establish that these two genera were at least occasionally mobile. By the very latest Ediacaran, burrows assigned to the trace fossil genus *Treptichnus* are found. These treptichnids have a more complex burrowing system than seen earlier in the Ediacaran, with both a vertical and a horizontal component, representing movement of the organism down, up, or up and down through the sediment (Jensen et al. 2000). Indeed, as described in chapter 2, the base of the Cambrian is defined by a particularly distinctive burrow, *Treptichnus pedum* (Vannier et al. 2010). Whatever made those traces clearly possessed anterior-posterior and likely dorsoventral differentiation, anterior sensory organs, and probably a fluid or at least a tissue skeleton.

Ediacaran rocks contain a host of other structures that have been described as trace fossils, but careful studies by Jensen suggest that many tight meandering forms (often known as *Palaeopascichnus*) and others that have been interpreted as packages of fecal pellets are not actually trace fossils at all (Jensen 2003; Jensen, Droser, and Gehling 2005). Claims for penetrating vertical burrows in the Ediacaran also seem unlikely to be true trace fossils; some probably represent body fossils of anemone-like organisms [the report by Rogov et al. (2012) has yet to be thoroughly investigated]. In fact, according to Jensen, there appear to be no valid metazoan trace fossils earlier than 560–555 Ma, with the possible exception of the recently described horizontal burrows from Mistaken Point (A. Liu, McIlroy, and Brasier 2010). The absence of any reliable records of trace fossils older than 560 Ma is also a critical constraint on claims that bilaterians must significantly predate 560 Ma. The trace fossil record strongly indicates that any benthic animals older than this age were likely to have been less than about 1 centimeter in length. In any event, any older animals do not appear to have left a record of their existence.[3]

Reevaluation of the many named Ediacaran trace fossils has shrunk their apparent diversity, but they provide useful insights into the expanding behavioral repertoire of early animals. For example, a pioneering review by Crimes (1994) accepted thirty-five Ediacaran ichnogenera, but many of them are now known to represent preserved organic tubes, others were improvidently described, as discussed previously. The recognition that many of them are actually body fossils has contributed substantially to the reduction in apparent Ediacaran trace fossil diversity. Consequently, some twenty-two of Crimes's thirty-five Ediacaran ichnogenera either lack valid Ediacaran records or are body fossils (Droser, Gehling, and Jensen 2005; Jensen, Droser, and Gehling 2005, 2006; Seilacher, Buatois, and Mangano 2005). Although the diversity of trace fossils certainly increased during the Ediacaran, most were small, shallow traces as described above, and their diversity was low until the very latest Ediacaran. As a preview of events during the early Cambrian, the diversity and inferred behavioral complexity of trace fossils increased substantially. Arthropod traces, undermat miners, and traces of deposit feeding all appear during the pretrilobitic early Cambrian, but bioturbation remained low until Stage 2, when signs of vertical burrows as domiciles (e.g., *Skolithos*) and indications of multiple trophic levels appear (Mangano and Buatois 2007).

Paleontologists are justifiably concerned with the reliability of the fossil record and whether various preservational biases could significantly alter the diversity patterns. For the Ediacaran and Cambrian, there are several lines of evidence, suggesting that the record of trace fossils and bioturbation is likely to be a relatively faithful one. First, the absence of widespread bioturbation in Ediacaran and earliest Cambrian made for firmer sediments, increasing the preservation potential of shallow traces, which are the ones most likely to be destroyed (Droser, Jensen, and Gehling 2002). This change explains the numerous shallow traces preserved at that time and suggests that the absence of deeper trace fossils, which are more easily preserved, is reliable evidence of the lack of much burrowing activity. Second, microbial mats, widespread through the Ediacaran and into the earliest Cambrian (Dornbos, Bottjer, and Chen 2004), appear to have aided preservation of shallow traces. Indeed, as noted above, many shallow traces may have been formed beneath microbial mats. The preservation potential of truly surficial traces, like those that can be produced by creeping sea anemones (A. Collins, Lipps, and Valentine 2000), is unclear, but it seems to have been low.

Bioturbation certainly has important effects on biodiversity. Increased sediment mixing can modify sediment geochemistry, increase nutrient cycling, and thus increase primary productivity and microbial activity (Lohrer, Thrush, and Gibbs 2004; McIlroy and Logan 1999; Seilacher and Pfluger 1994). Callow and Brasier (2009) have constructed an integrated model of these geochemical effects. These processes in turn provide ecological opportunities for deposit feeders such as *Taphrehelminthopsis*. Such positive feedbacks can be described as ecosystem engineering when they spill over to affect other species. In chapter 8, we will return to the issue of whether the spillover effects from such ecosystem engineering provided endogenous drivers for increased diversity during the Cambrian.

Fossil evidence of animal life during the Ediacaran Period can be roughly divided into a series of stages (fig. 5.3). From the end of the Marinoan glaciation at 635 Ma until about 579 Ma, including the Gaskiers glaciation, animal life was limited to sponges, probably a variety of very small animals that we have not yet found preserved as fossils, and, depending on dating estimates, the organisms preserved in the Doushantuo Formation. Soon after the Gaskiers glacial episode, the soft-bodied fossils of the classic Ediacaran fauna appear, starting with the rangeamorphs of the Avalon assemblage and then into the far more diverse fossils of the White Sea–Ediacaran and the Nama assemblages.

Trace fossils are diversifying at this time, but remain relatively simple. The next stage of evolution commences about 555 Ma with the appearance of *Kimberella*. Near the very close of the Ediacaran, perhaps at about 545 Ma, the first weakly skeletonized marine animals appear, foreshadowing the great explosion of animal diversity in the Cambrian.

NOTES

1. An alternative interpretation is that the Doushantuo fossils are not embryos, but giant sulfur bacteria that produce symmetrical cell divisions and could have aided phosphate formation and thus their own preservation (Bailey et al. 2007). The reductive cell divisions of modern sulfur bacteria are not known to persist long enough to produce objects of 100 to 1000 cells, however (Donoghue 2007), nor do sulfur bacteria encyst, as do some of the Doushantuo embryos (L. Yin et al. 2007), and taphonomic studies of giant sulfur-bacteria do not produce structures similar to the fossil embryos (Cunningham et al. 2011). Helical spheroidal fossils reported by Xiao et al. (2007) were interpreted as postblastula embryos, although this interpretation raises the issue of why definitive blastula-stage embryos have not yet been reported.

2. In Australia, acanthomorphs first occur above the Marinoan glaciation and immediately above a large negative carbon shift and the debris from a significant extraterrestrial impact known as the Acraman impact. Despite suggestions that the diversification of the acanthomorph acritarchs occurs during recovery from the biotic effects of the impact and that the impact occurred between 580 Ma and 570 Ma, there are few reliable constraints on the age of the Acraman impact or the subsequent radiation of acritarchs. In light of the geochronologic framework presented in chapter 2, the most reasonable interpretation is that the Acraman impact occurred well before 600 Ma, with the simple, leiospherid acritarchs predominating through the remainder of the Ediacaran. For discussion of the relationship between the Acraman and acritarch diversity see Grey, Walter, and Calver (2003) and Willman, Moczydlowska, and Grey (2006).

3. Earlier trace fossils are occasionally reported (generally to considerable press attention), some of them dating far back into the Proterozoic. Jensen, Droser, and Gehling (2005, 2006) reviewed the numerous recent reports of early bilaterian traces and concluded that none withstood scrutiny; the purported fossils usually turn out to be the result of cracking and drying of the sediment, drag marks, or some other sedimentary feature. The absence of verified reports before the Ediacaran does not mean that earlier trace fossils are impossible, only that a critical eye is needed to evaluate them. Seilacher (Seilacher 2007b) contains excellent illustrations of various sorts of dubiofossils with explanations of how they were formed.

CHAPTER SIX

Metazoan Architectures of the Cambrian Explosion

The simple skeletal fossils of latest Ediacaran strata are a prelude to the main event registered in the Cambrian. Those first skeletonized forms indicate that the seawater chemistry of the time permitted the secretion and preservation of mineralized skeletons, the durability of which would be a boon both to Cambrian organisms and, some 545 million years or so later, to paleontologists. In addition, as noted previously, the trace fossils appearing in the early Cambrian were roughly an order of magnitude larger than most of their Ediacaran predecessors and included burrows that penetrated relatively deeply into the seafloor sediments. Clearly, larger and more active animals were evolving. The Phanerozoic Eon had dawned.

Unlike the Ediacaran fossils, whose internal architectures and phylogenetic relationships remain largely unknown, the Cambrian fossils are mostly stem groups of well-known Phanerozoic clades. Although the Cambrian fossils are distinct from their living relatives, our understanding of their crown groups provides a foundation for their interpretation. In all, the Cambrian explosion fauna has representatives from at least fourteen groups—nearly half of crown phyla—but there are many other Cambrian fossils that cannot be placed among living phyla. Some of these fossils are represented by isolated and disarticulated skeletal elements that do not shed much light on their body plans; for others, sufficient morphological evidence is preserved to rule out membership among any living phyla. Most of these unusual forms are likely to be stems of the superphyla to which crown phyla are assigned. That is, they represent early branches, with their own unique body plans, of groups from which living phyla arose at a later date. In this chapter, we describe the varied and often spectacular morphologies of the major groups of fossils generated during the Cambrian explosion, review their likely phylogenetic affinities, and discuss the implications that this extraordinary record of the early blossoming of morphological disparity holds for evolutionary theory.

CHARACTERISTIC EXPLOSION ASSEMBLAGE TYPES

Fossils represent the remains of organisms preserved against all odds. Clearly, the chance of any particular organism leaving any sign of its presence some half a billion years later is exceedingly small, especially for the relatively soft-bodied animals that make up a large fraction of early Cambrian fossils. Conditions on the seafloor vary greatly from place to place and time to time, and situations conducive to the preservation of dead organisms or traces vary as well. Preservation also depends on the presence or absence of other organisms; for example, algal mats can protect and thus help preserve traces, whereas bioturbators erase them. Furthermore, fossil preservation is determined by the chemical history of the sediments in which the fossils are entombed. Many Cambrian localities contain the sorts of biomineralized skeletons that form the bulk of the Phanerozoic invertebrate record, representing a range of "normal" preservational modes. As in most cases throughout the Phanerozoic, skeletal fossils dating from Neoproterozoic or Cambrian times often have had their original minerals replaced.

The record of the Cambrian explosion is enhanced by particularly striking and unusual body fossil assemblages that are characterized by distinctive styles of preservation. Many explosion fossils are represented by original organic carbon films (Butterfield 1990, 1995). This type of preservation is aptly named for the famous Burgess Shale fauna from British Columbia, Canada, and also occurs at the localities preserving the earlier Chengjiang fauna from Yunnan, China (Butterfield 1990, 1995; Gaines, Briggs, and Zhao 2008). This style of preservation involves suppression of the usual processes of microbial decay, probably by anoxic waters. These and similar faunas have contributed a large proportion of the fossils that we review in this chapter. Another particularly important type of preservation in the early Cambrian is of very small shells in phosphate deposits; the shells are either originally phosphatic or their original constituents have been replaced by phosphate minerals during fossilization. Different histories of accumulation and fossilization of organisms will preserve a different fraction of the original biota, and we are lucky to have a range of such histories represented in early Cambrian deposits. Indeed, the richness of these faunas indicates just how much we are missing from times when only "normal" modes of preservation are represented.

Small Shelly Fossils

The minute, largely phosphatic fossils called small shelly fossils (small shellies or just SSFs) in paleontological jargon are chiefly under 2 mm in their greatest dimension (fig. 6.1) (see Bengtson 2005; Bengtson and Conway Morris 1992). Although some of the SSFs are whole shells, such as those of tiny brachiopods, mollusks, and an array of cones of uncertain affinities, others are individual plates, spines, and other skeletal elements that once adorned larger animals and were preserved by the same phosphatization processes. These skeletal elements are called sclerites. The complete skeletal assemblage of the living organism is its scleritome, which is sometimes orders of magnitude larger than the individual sclerites. A new compilation of the origination patterns of the small shellies shows that they appear at the base of the Cambrian and increased steadily in diversity through Stage 1 (Maloof, Porter, et al. 2010). They probably reached peak diversities during Stage 2 and were still fairly diverse during Stage 3, when our record of the Cambrian explosion is at its height (Porter 2004).

The morphological diversity of early Cambrian small shelly sclerites suggests that they represent many different clades, although lophotrochozoans are most commonly represented. Deuterostomes are evidently largely absent, although they are present in the exceptionally preserved explosion faunas of both Stages 3 and 5. Some scleritomes were composed of more than one type of sclerite, which complicates their interpretation. Fortunately, we now have a variety of well-preserved scleritomes, with sclerites in their living positions, preserving the body plan of numbers of sclerite-bearing animals. For example, a type of sclerite named *Microdictyon* (fig. 6.2) was found and described in small shelly assemblages years before the complete animal was discovered. *Microdictyon* plates are now known to have been emplaced in the body wall above the limbs of an extinct branch of animals related to living onychophorans and arthropods, called lobopodians (J. Liu et al. 2008; Ramskold and Chen 1998). They therefore belong to the Ecdysozoa superphylum. Although ecdysozoans have the richest and most disparate fossil record of any superphylum in Cambrian rocks, they tend to be relatively rare among the classic small shelly assemblages; lobopodian sclerites are an exception.

On the other hand, other scleritomes are only hinted at by lucky discoveries of a few associated sclerites. For example, the full scleritome of a group known as cambroclaves (fig. 6.1c) is not known, but small assemblages of their sclerites have been found together at a locality in Kazakhstan, permitting a partial reconstruction of their geometry (fig. 6.3). These sclerites are mostly less than 10 mm long, and the reconstruction suggests that a mosaic of many sclerites covered at least some surfaces of the animal. Even these small insights allow

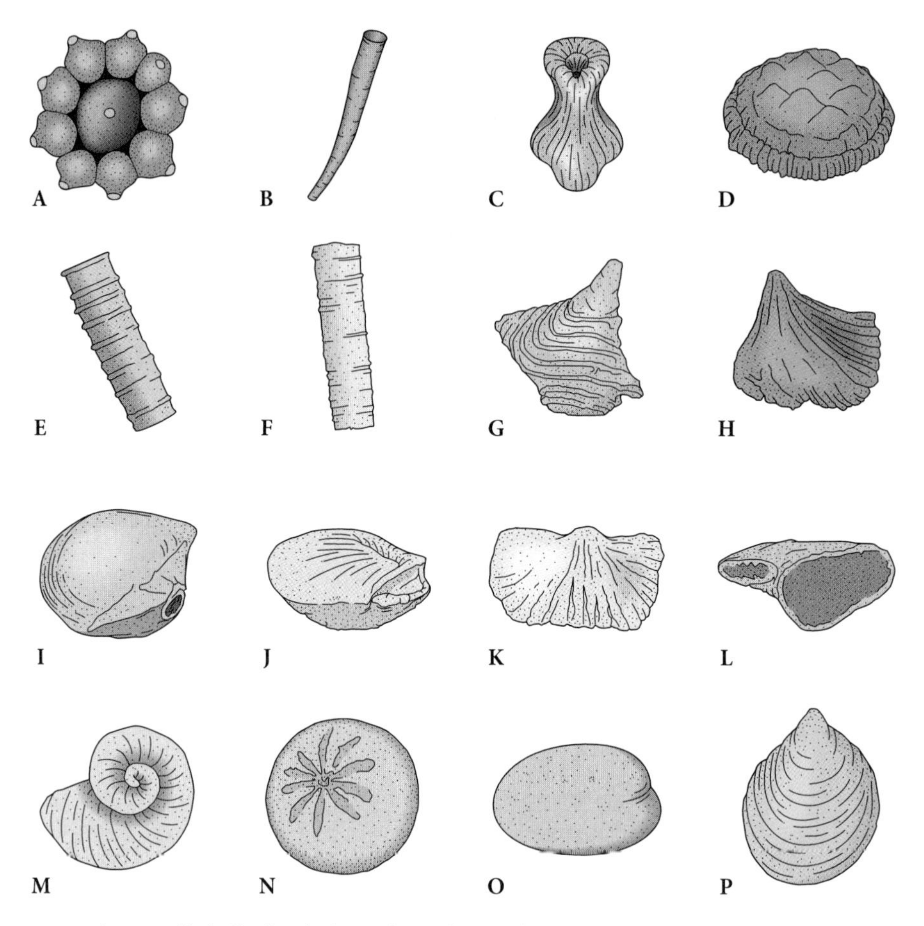

Figure 6.1 Small shelly fossils from the early Cambrian, ranging in size from < 1 mm to ~2 mm. All except (C) are from Russia; (C) is from South Australia. (A) Chancelloriid (fig. 6.11). (B) Forellellid, a closed cone with basal attachment point, perhaps a cnidarian. (C) Cambroclavid, probably a burrowing bilaterian (fig. 6.3). (D) *Hadimopanella,* a palaeoscolecid. (E) and (F) Possible annelid tubes, open at both ends; (E) hyolithellid, (F) coleolid. (G) and (H) Tommotiids, possible stem lophophorates (fig. 6.35). (I)–(K) Probable stem brachiopods; (I) of linguliform brachiopods, (J) and (K) of rhynchonelliform brachiopods. (L)–(M) Possible mollusks. (N) Mobergellid, showing internal muscle scars. (O) Rostrochonch. (P) Stenothecoid (probivalve), mollusk-like but of uncertain relations. Russian fossils after Rozanov and Zhuravlev (1992), Australian fossil after Bengtson et al. (1990).

interpretation of growth and possible life habits of this form (see Conway Morris in Bengtson et al. 1990). The spines at the inferred posterior of each sclerite bend forward (or perhaps backward) rather than outward, suggesting that the animal crawled or burrowed and might have used the spines for

Figure 6.2 *Microdictyon sinicum,* a Cambrian lobopod from the Chengjiang fauna (Stage 3). Note the pairs of small plates above each appendage. These plates had been found in the small shelly fauna decades before their body fossils were discovered and the scleritome revealed.

purchase during locomotion rather than for protection. The sclerites appear to have been originally composed of low-magnesium calcite and interlocked as shown in figure 6.3, so growth might have had to be accommodated by the secretion of new antero-posterior files of sclerites in a growth zone. It is plausible that cambroclaves were benthic bilaterians, but it will certainly require more informative fossils before they can be located with any certainty within the metazoan tree.

Although many SSFs are still poorly understood, it is clear that they are a large and phylogenetically disparate assemblage, representing a thriving fauna whose preservation was restricted to a rather narrow taphonomic window (Porter 2004). In general, the appearance of different groups of SSFs is correlated with changes in the abundance of phosphatic deposits in the sedimentary record. Such deposits are relatively rare from Stage 4 and onward, presumably reflecting a change in seawater chemistry (chap. 3). As phosphatic sediments became rarer, the diversity of small shelly fossils declined, and SSFs essentially disappear from the fossil record by Stage 5. Because there are many

Figure 6.3 Reconstruction of a small section of a *Cambroclavus* scleritome showing three interlocking files, each represented by four minute, spiny sclerites (see text). Reconstruction after Conway Morris in Bengstson et al. (1990).

phosphatic deposits of Ediacaran age that do not contain SSFs, it is likely that the small shelly fauna was rare or absent during most of the Ediacaran and evolved chiefly early in the Cambrian. The disappearance of the SSF is at least partly associated with an episode of extinction (Zhuravlev and Wood 1996), although the clades to which some of them are known to belong are well represented in later deposits. We will return to the dynamics of the SSF at the end of the chapter.

Exceptionally Preserved Faunas

Our clearest window into early Cambrian life comes from exceptionally preserved fossil assemblages in southern China that appear in Stage 3 during the peak diversity of the SSFs. These fossils preserve anatomical details of the soft parts of many different organisms in exquisite detail. The richest assemblages are from the Chengjiang fauna from Yunnan Province (J. Y. Chen and Zhou 1997; Hou et al. 2004), and there are some less diverse faunas described from nearby Guizhou Province. Although these deposits contain some mineralized skeletons (about 15% of the species would be preserved in traditional fossil assemblages of skeletons), they are deservedly famous for their collection of organisms entirely lacking in hard parts and for fossils that do have hard parts but that also show parts of their soft-body anatomy. The unusual fidelity of preservation permits imaginative portrayals of the fauna as illustrated in figure 6.4. The seafloor community depicted in that figure includes sponges, mollusk-like forms with creeping soles, onychophoran allies with stumpy legs (lobopodians), and arthropods with jointed appendages. Worms burrow into the sea floor, while the water column is inhabited by early vertebrates and large, predatory arthropod relatives. Figure 6.5 indicates the proportions of species in each major taxon known from the Chengjiang fauna: the fauna is clearly dominated by arthropods. These animals seem at once both eerily familiar and yet somehow alien, as if they were denizens of some alternate universe.

The Chinese material is complemented by the 12 My younger but similarly rich Burgess Shale fauna from Stage 5 rocks in British Columbia, Canada. As with the Chengjiang fauna, most of the Burgess Shale species are arthropods or arthropod allies, including *Opabinia* (illustrated on the cover) and

Figure 6.4 Artist's impression of a community of explosion taxa based on fossils of the Chengjiang fauna, Yunnan, China, the Burgess Shale fauna, Canada, and the Sirius Passet Fauna, Greenland. The large sponges may be 30 cm tall, and the larger swimming anomalocariid may reach 1 meter in length. Reconstruction by Quade Paul.

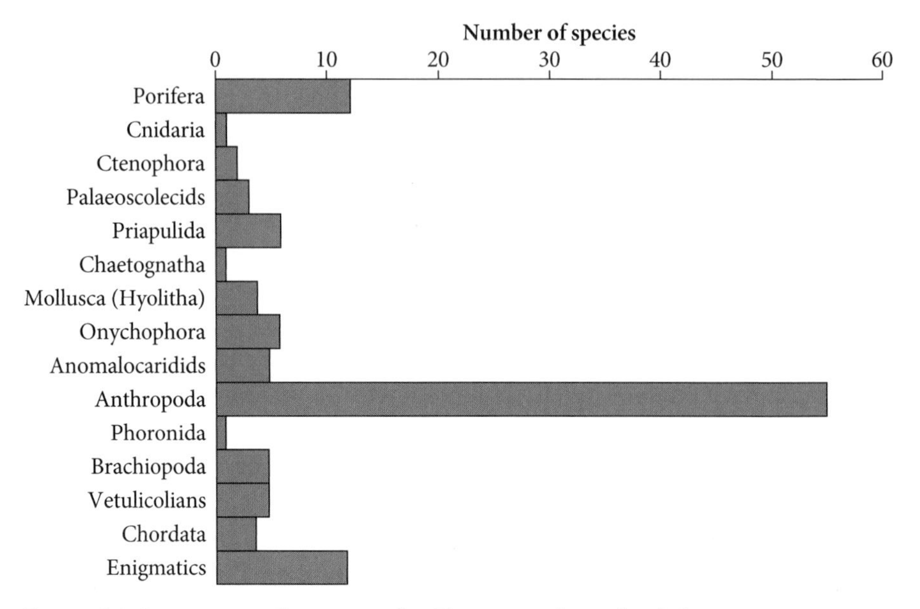

Figure 6.5 Proportions of species in the Chengjiang fauna that belong to major taxa, living or extinct. Arthropods clearly dominate the preserved fauna. The molluscs, prominent in small shelly fossils (fig. 6.1), are less important in the exceptionally preserved faunas. Palaeoscolecids, anomalocaridids, and vetulicolians are extinct. Modified after Hou et al. (2004).

the various dinocaridids, *Anomalocaris, Laggania,* and *Hurdia.* The majority of these forms can be assigned to stem groups of living phyla. Although the Chengjiang and Burgess Shale faunas have received the most acclaim from both paleontologists and the popular press, there are a variety of other assemblages with exceptional preservation in Cambrian Stages 2 to 5, particularly the Sirius Passet fauna in Greenland, the Emu Bay Shale in Australia, the Sinsk biota in Russia, and the Kaili and Guanshan biotas in China (fig. 6.6). Each of these assemblages also contains wonderfully preserved animals that have added much to our understanding of the Cambrian diversification. The remarkable discovery of a Burgess Shale–like assemblage in Lower Ordovician rocks of Morocco indicates that many of these lineages survived the Cambrian (Van Roy et al. 2010); it was the preservational window that failed.

About a dozen living phyla are represented by body fossils in the Chengjiang and Burgess Shale, including nearly all the living phyla possessing hard parts; the single exception is Bryozoa, first known from the Early Ordovician. Fossil evidence and results of molecular clock studies (chap. 4) suggest that all the stem lineages of living phyla had originated by the close of Stage 3 (indeed many are far older) (fig. 6.7). Echinoderms are evidently missing from the Chengjiang fauna, with only one problematic form among the Burgess Shale fossils, which Clausen and colleauges (2010) have interpreted as indicating a preferential diversification of echinoderm clades in carbonates rather than in the siliciclastic environments preserved in most of the lagerstätten. The "missing" crown phyla are all soft-bodied (again except for the Bryozoa), and most are very small-bodied as well. Because the living Bryozoa include major soft-bodied taxa, it is quite possible that the body plan of this phylum had originated by the time of the explosion but that lineages with mineralized skeletons had not yet evolved. Although so many phyla are present, these stem

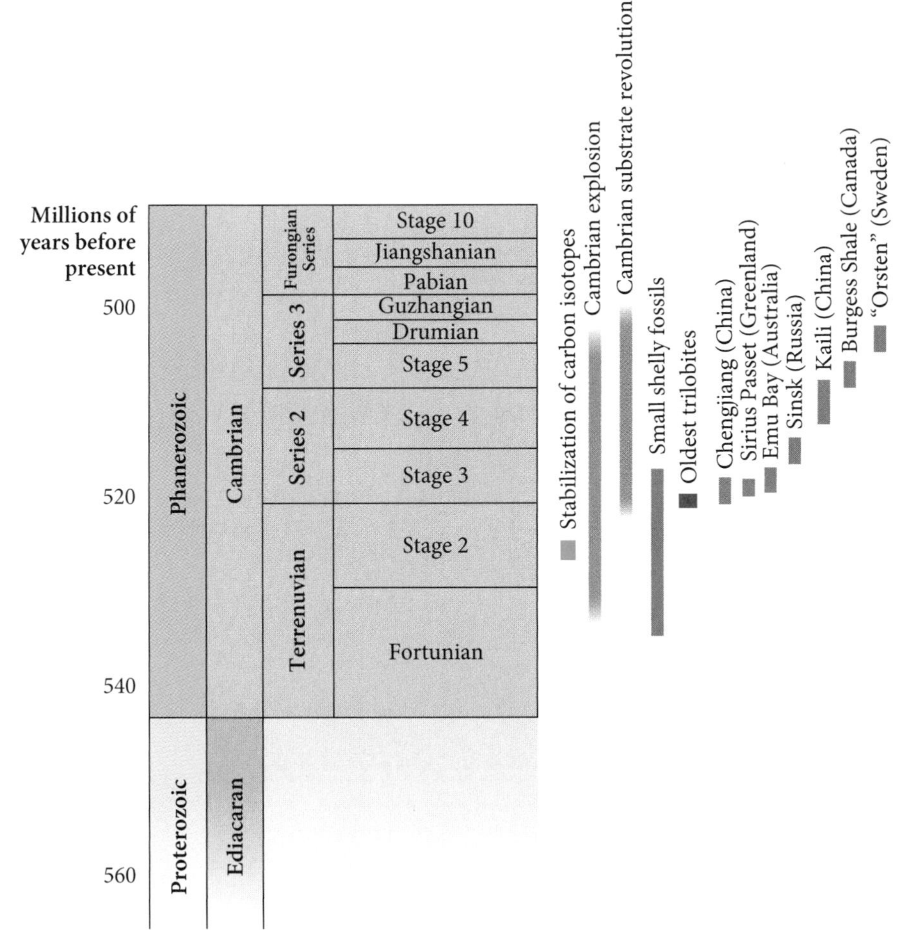

Figure 6.6 Major events of the Cambrian explosion, including the ages of the various extraordinarily preserved faunas.

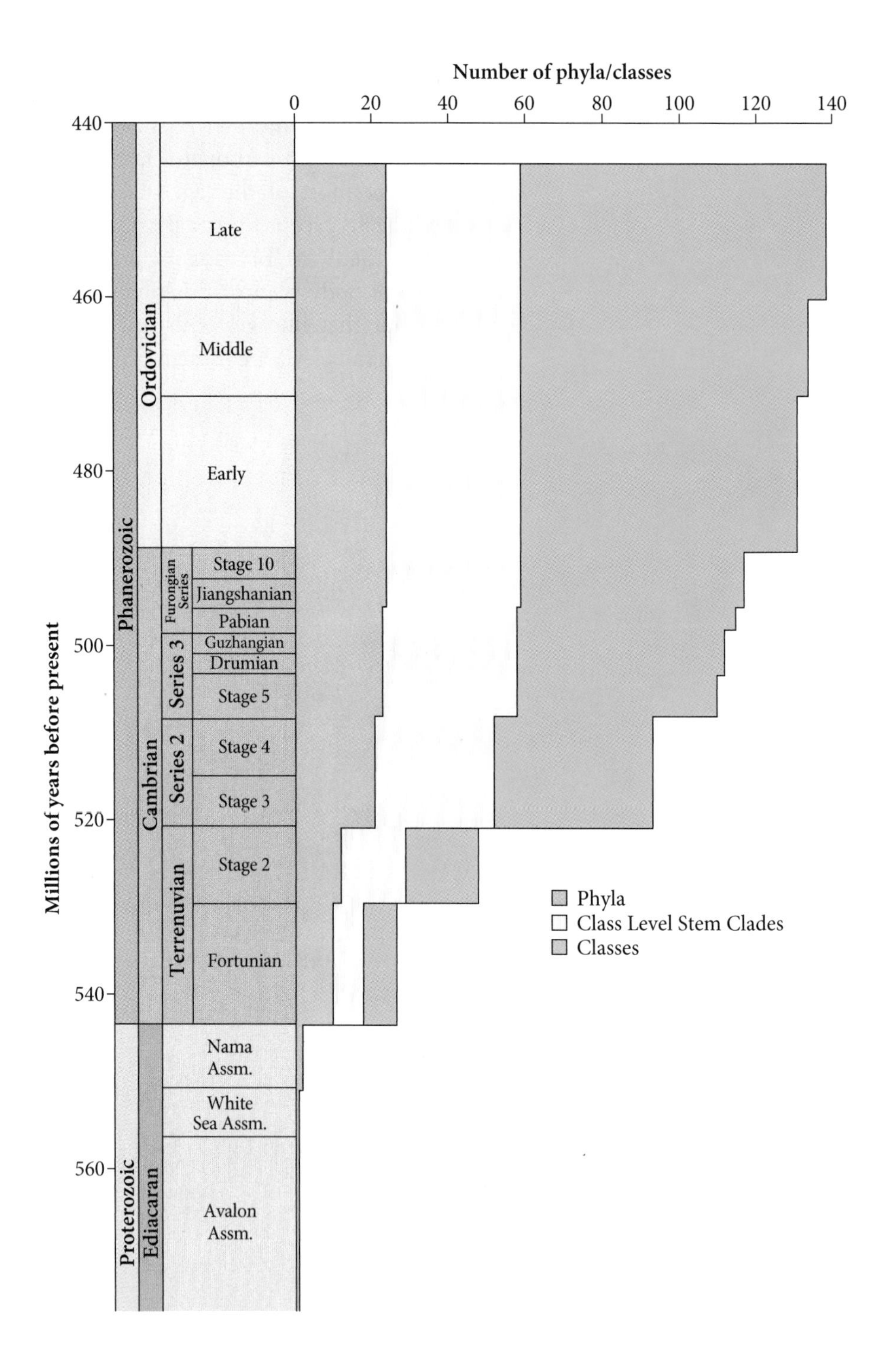
Number of phyla/classes
0
20
40
60
80
100
120
140
440
460
480
500
520
540
560
Millions of years before present
Phanerozoic
Proterozoic
Ordovician
Cambrian
Ediacaran
Late
Middle
Early
Furongian Series
Series 3
Series 2
Terrenuvian
Stage 10
Jiangshanian
Pabian
Guzhangian
Drumian
Stage 5
Stage 4
Stage 3
Stage 2
Fortunian
Nama Assm.
White Sea Assm.
Avalon Assm.
Phyla
Class Level Stem Clades
Classes

Figure 6.7 (left) Patterns of first appearances of phyla and classes as fossils from the Ediacaran through Ordovician periods, by stage through the Cambrian. The dramatic rise in the number of phyla, classes, and stem clades during the Cambrian reflects the appearance of many novel morphologies accorded high taxonomic rank under a Linnaean taxonomic system. After Erwin et al. (2011), and based on the record of appearances in the Appendix. Stem clades are class-equivalent groups.

taxa disappeared from the record long ago, which helps account for the somewhat otherworldly appearance of the Cambrian fauna when compared with living groups.

The first appearance of the Cambrian explosion fauna was geologically abrupt and quite spectacular. Because different sorts of fossil assemblages imply different preservation potentials, the first appearances of different elements of the fauna in the fossil record are tied to the presence of appropriate depositional environments. Thus, phosphorite deposits containing small shelly assemblages with the oldest fossil mollusks and brachiopods (as SSFs) occur earlier than shales and silts with soft-body preservation of the oldest known chordates and arthropods, but these appearances say little about the sequence of origination of the stem ancestors of each clade or of the times of origin of their distinctive body plans.

Because many faunal elements first found in the explosion fauna described from Stage 3 appear in Stage 5 rocks as well, especially in the Burgess Shale fauna, it is clear that once these clades were established, they persisted for millions of years. This biotic stability nonetheless contrasts with changing conditions on the Cambrian seafloor, thanks in no small part to the evolution of the Cambrian organisms themselves. The level-bottom communities of the Neoproterozoic were commonly associated with microbial mats, and those substrates were rarely disturbed by burrowing animals (chap. 5). During the early Cambrian, however, many larger burrowers and sediment feeders churned the sediments and displaced existing mat communities. The firmer seafloor sediments, bound by algal films and mats, were widely replaced by sediments capped by mixed layers, unbound and charged with water. This change affected the preservational regime in some environments, most obviously by removing the mats and mixing the sediments, and probably affected the adaptive strategies of many lineages (Dornbos 2006; Dornbos, Bottjer, and Chen 2005 and refs. therein). So, as the explosion was under way, this positive feedback was altering the evolutionary potential of the benthic environment.

FOSSIL REPRESENTATIVES OF EARLY METAZOAN GRADES AND CLADES

The general architectures of the crown phyla were summarized in chapter 4. Here we review the morphologies of the major groups of fossils that appear during the Cambrian explosion, focusing on the features that suggest affinities with living groups. There is no neat sequence of appearance of increasingly complex architectural grades with time, but rather a geologically abrupt appearance of body plans that range across levels of complexity from sponges to chordates. Although this appearance is the earliest of most of these groups in the fossil record, they clearly had a rich history of divergence and expansion of which we have few records. The explosion fauna reflects the exploration of many adaptive pathways in the evolving marine biosphere. As the burgeoning fauna interacted with physical and biotic aspects of the environment, it produced marine ecosystems whose characteristics we can recognize in the current oceans, a legacy stretching back more than half a billion years.

Basal Metazoan Clades

A suite of sponges and sponge-like organisms occupy basal positions within the metazoan tree and are important contributors to Cambrian faunas. As with many other groups, there are a variety of problematic groups whose phylogenetic affinities remain unclear.

SPONGES By the time of the explosion, sponge lineages had been around for well over 200 million years. They first appear in the fossil record as chemical biomarkers of probable demosponges dated to 635 Ma. The Ediacara macrofauna includes two possible sponges, both from the White Sea localities (Kouchinsky et al. 2012; Serezhnikova and Ivantsov 2007). Although there are reports of siliceous spicules from numerous Neoproterozoic localities, most of them can be discounted; sponge spicules first become common during the explosion itself.

Of the four living sponge clades (fig. 4.4), three are represented in explosion faunas (fig. 6.8). One clade (Calcarea) secretes calcareous spicules; the others, when spiculate, secrete siliceous spicules. Spicule formation in hexactinellids and demosponges is so similar as to be considered homologous. Molecular clock estimates indicate that the LCA of these clades originated about 700 Ma. There are numerous reports of Cryogenian and Ediacaran spicules, but most of them have been reinterpreted as pseudofossils. Reports from late

Ediacaran sections in Namibia (Reitner and Worheide 2002) are more reliable. Thus, there is a "spicule gap" of more than 200 My when spicules were likely being secreted but not preserved; presumably, the ocean water or substrate geochemistry of those times caused spicule dissolution (Sperling et al. 2010). Homoscleromorphs are not known as fossils during the explosion (but see Chap. 9).

Spicules become relatively common during the explosion, appearing in Stage 1 deposits in China and Siberia (Kouchinsky et al. 2012), and body fossils of both hexactinellids and demosponges have been recovered from Cambrian Stages 2 and 3 (Xiao, Hu, et al. 2005), although they are known from relatively few localities. These thin-walled sponge body fossils are structurally simple, seemingly adapted to quiet waters (Carrera and Botting 2008). Spicules from living hexactinellids and demosponges are siliceous, with a central organic filament but no organic sheath. Most hexactinellid spicules are six-rayed (hexactines) with three axes (triaxons), one axis of which is perpendicular to the plane of the others (fig. 6.9A). Sponges with calcareous spicules also appear in the early Cambrian and have usually been assigned to the crown class Calcarea, but recent studies have revealed a more complicated picture. Living calcareans have spicules that lack an internal filament but are sheathed in an organic layer, and most are three-rayed triaxons (triactines) or tetraxons (with four axes) (fig. 6.9B). Today, sheathed spicules are secreted extracellularly, but spicules with filaments are largely secreted intracellularly. Spicules from Cambrian Stage 4 deposits in Newfoundland show an extinct combination of these characters (T. P. H. Harvey 2010): six-rayed triaxines resembling hexactinellid spicules were organically sheathed like calcarean spicules (and indeed are preserved as carbonaceous films) and lack any obvious internal filaments (6.9C). These Stage 4 spicules recall findings of a detailed study of spicules in the best-known Cambrian sponge, *Eiffelia* (fig. 6.8E), from the Stage 5 Burgess Shale (Botting and Butterfield 2005), which has a unique variety of calcareous spicule types combined in the same scleritome. The largest spicules are hexactines but lack a perpendicular axis, whereas in successively smaller-size classes, tetraxons become progressively more abundant; these spicules were originally sheathed. Here again is an extinct combination of spicule types. Thus, some of the earliest sponges show combinations of characters— spicules in this case— that are either found restricted to different crown classes today or are unique. Either some of these characters evolved independently in different clades or the extinct sponge clades that combine characters are ancestral to living clades that have lost characters differentially. In either case, there appears to have been an early Cambrian radiation of sponge clades, some of which have become extinct and may not belong to any of the crown classes.

A

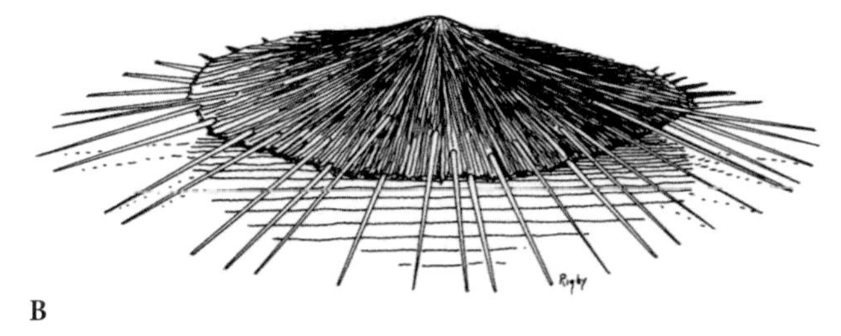

B

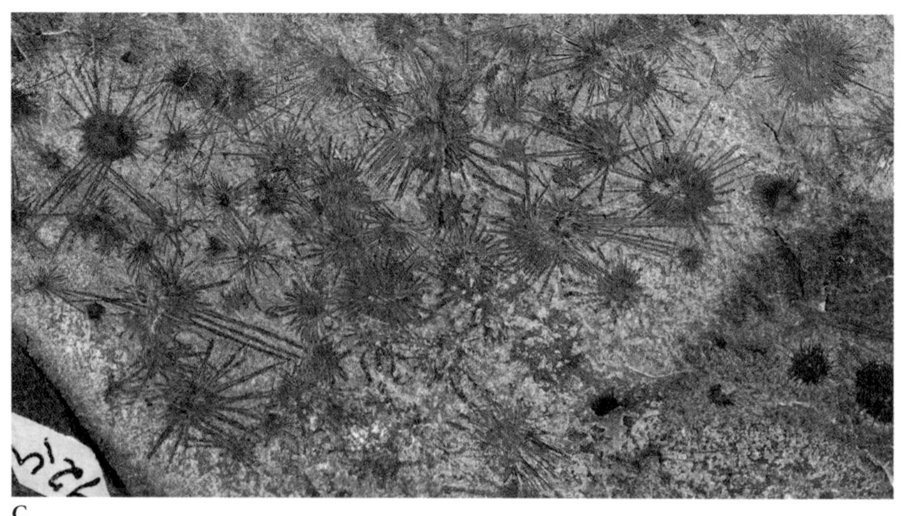

C

D

E

Figure 6.8 (left and above) Middle Cambrian sponges from the Burgess Shale. (A–C) *Choia,* a demosponge, also known from the Chengjiang fauna (see fig. 6.4). (A) and (B) *Choia carteri* USNM 66482; (C) *Choia ridleyi,* USNM 66487. (D) *Diagoniella,* a primitive hexactinellid; this genus is not known in the early Cambrian, but possible hexactinellids date from the late Neoproterozoic. (E) *Eiffelia,* USNM 66522, a calcarean, a genus also known from spicules in lower Cambrian rocks. (A, C–E) Photographs by J-B. Caron, courtesy of the Smithsonian Institution. (B) Reconstruction from Rigby and Collins (2004).

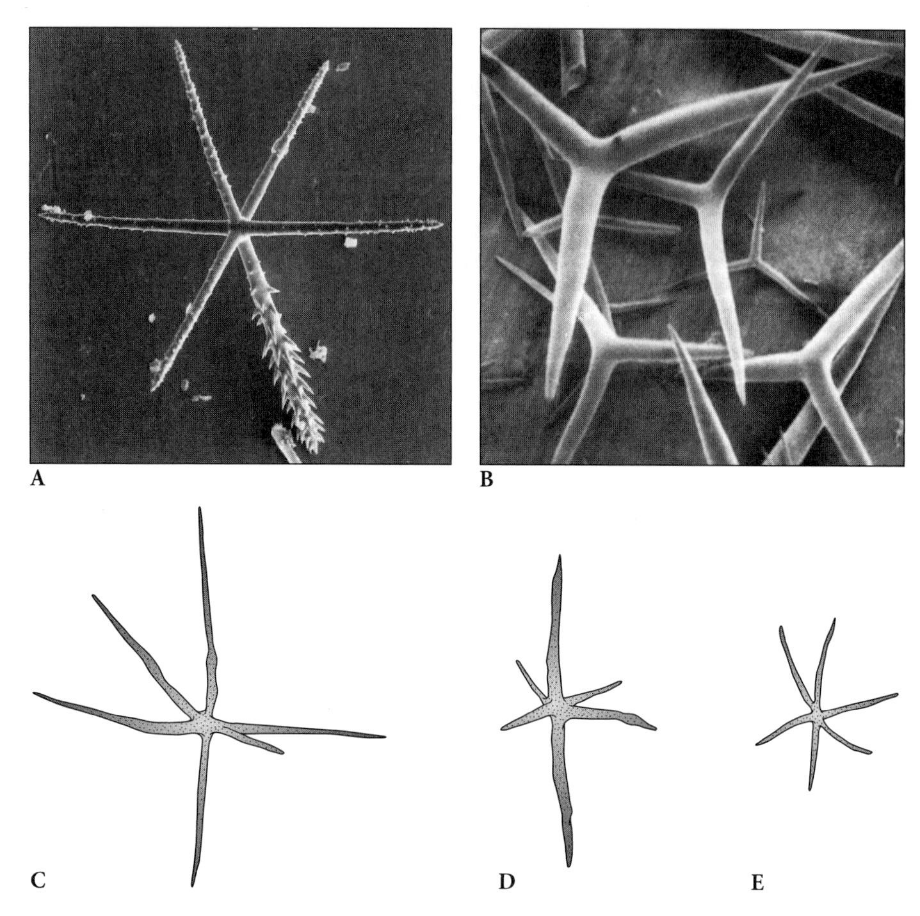

Figure 6.9 Sponge spicules. (A) A modern hexactinellid spicule, which lacks an organic sheath (of *Sympagella nux*); a hexactine. (B) Modern calcarean spicules of *Leucandria nivea,* which may be enclosed in organic sheaths; triactines. (C)–(E) Carbonaceous fossils of sponge spicules from the Forteau Formation, hexactines that were enclosed in robust organic sheaths. All to same scale; the smallest is about 150 μm across. (A) and (B) from T. L. Simpson (1984), (C–E) from T. H. P. Harvey (2010).

ARCHAEOCYATHA This notion of a Cambrian radiation of sponge-grade forms is underlined by the occurrence of the sponge-like archaeocyathans, an assemblage of heavily calcified organisms now recognized to have developed independently among several different sponge groups. Archaeocyathans built conical to cup-shaped skeletons of microgranular high-magnesium calcite and lacked spicules (Rowland 2001) (fig. 6.10). Two other fossil groups, the radiocyaths and the cribricyaths, probably evolved from different groups of sponges; the former were originally aragonitic (Zhuravlev and Wood 2008).

Archaeocyathid skeletons are penetrated by pores, and experiments with models in flowing water have shown that flow patterns through the pores are somewhat analogous to flow in living sponges, suggesting that they also fed on suspended food items by pumping water currents (Savarese 1992 and refs. therein). Archaeocyathan skeletons are commonly cemented together to form reef-like buildups on the seafloor. Their solid skeletons suggest deposition by epithelia, which are generally not well developed in sponges, although crown demosponges build calcareous skeletons with similar textures. Archaeocyathans appeared first in Stage 2 and diversified so rapidly that they appear to create a diversity peak that spoils interpretations of Cambrian benthic diversification as a logistic process. This diversification may be exaggerated by the somewhat artificial taxonomy erected for the group, however.

CHANCELLORIDA The chancelloriids are another extinct Cambrian clade that also may have had a sponge-like feeding system (figs. 6.1A and 6.11). These sessile, vase- or bag-shaped forms have hollow bodies that lack organs; the exterior of the body wall is covered with hollow, aragonitic sclerites from which spine-like rays radiate and that were attached to stalks on the body surface (Randell et al. 2005). These sclerites likely functioned as protection against predators. Chancelloriid sclerites are common in small shelly faunas

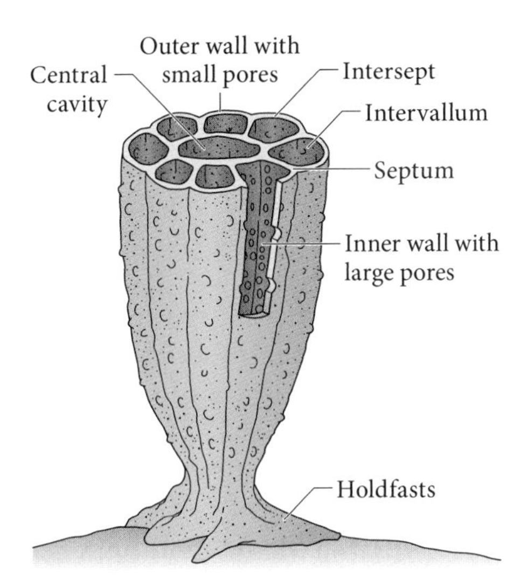

Figure 6.10 Archaeocyathans. These unusual fossils are not known from Chengjiang or Burgess Shale–type assemblages, but occur in reef-like buildups and in small shelly faunas of the early Cambrian. Once thought to represent an extinct phylum, they are now classed with sponges. A reconstruction of a regular archaeocyathid. After Hill (1972).

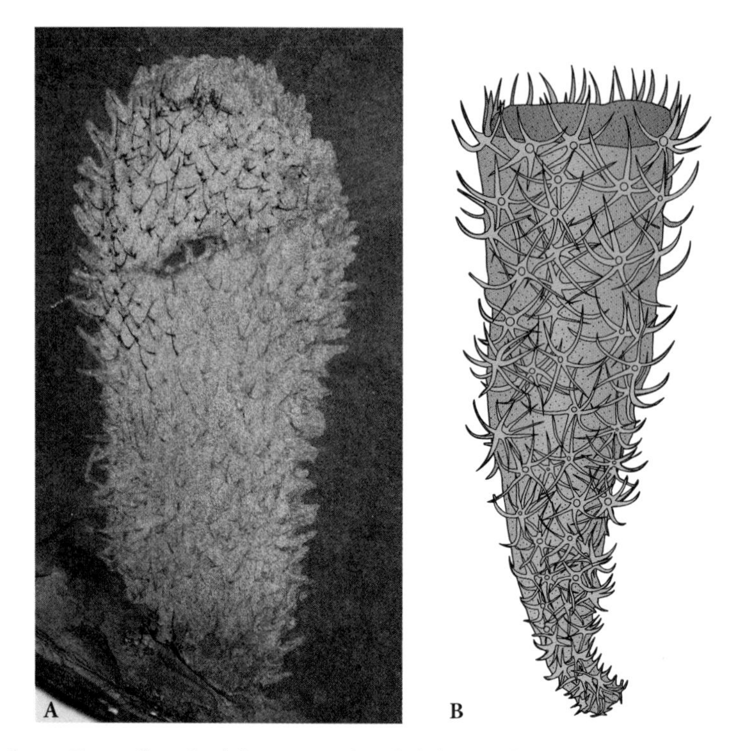

Figure 6.11 Chancelloriids. (A) An articulated skeleton of *Chancelloria* from the Burgess Shale fauna. Chancelloriid sclerites are also found among small shelly assemblages. (B) Reconstruction. Photograph courtesy of the Smithsonian Institution.

and are often found in archaeocyathan reefs. They have been interpreted as having branched from among sponge-grade organisms (Butterfield and Nicholas 1996) and were perhaps part of a radiation of sessile benthic forms, feeding on minute plankton via flagellar pumping (Sperling, Peterson, and Pisani 2009). Hollow sclerites with basal foramina are known from early bilaterians such as *Halkieria,* however (see Lophotrochozoa, below), and such sclerites have been interpreted as a synapomorphy of a distinctive clade known as the Coeloscleritophora (Porter 2008). Indeed, the similarities in skeletal microstructures among chancelloriids and halkieriids suggest a phylogenetic link between the two clades (Porter 2008). Bengtson (2005) has argued that because these characters are shared by both chancelloriids and bilaterians, they may even be eumetazoan plesiomorphies; thus, the coeloscleritophorans might be representatives of the lineage linking early nonbilaterians to bilaterians (chap. 9). Alternatively, the characters may have been convergent.

Explosion Fossils with "Radiate" Metazoan Body Plans

Although molecular clock evidence suggests the phyla comprising the Radiata (Cnidaria and Ctenophora) branched from sponges about 780 Ma, before the explosion, their fossil record before the Cambrian is equivocal (chap. 5). There are a variety of probable diploblastic-grade organisms known from the latest Ediacaran (chap. 5). The Cambrian SSFs also include a number of tube-like forms such as the anabaritids, protoconularids, and others from Cambrian Stage 1 that Kouchinsky et. al (2012) have collectively called cnidariomorphs. The "Radiata gap" between the probable origin of the clade and the first fossil representatives is less severe but just as troubling as the spicule gap. The usual taphonomic hurdles imposed by small body size and lack of hard parts may be sufficient to account for much of this gap.

CTENOPHORA Fossil ctenophores are easily identified by their comb rows and are known from Stage 3 and 5 Cambrian deposits (J. Y. Chen, Schopf, et al. 2007; Conway Morris and Collins 1996) (fig. 6.12). These fossils tend to have more comb rows than living groups, although seemingly always in multiples of eight; one possible ctenophore from the Burgess Shale is estimated to have eighty comb rows (Briggs, Erwin, and Collier 1994). This early radiation of ctenophore morphologies extends far beyond what exist today, and none of the known Cambrian ctenophores would be classed within the crown group.

CNIDARIA Anthozoans are the most basal crown Cnidarian group, and burrows attributed to sea anemones are found near, and possibly below, the base of the Cambrian (chap. 5). Some calcitic tubes and cones present in the small shelly fauna may represent Cnidarians (fig. 6.1B), although other relationships are possible. Body fossils assigned to sea anemones have been described from Fortunian-age (Stage 1) strata in China (Han et al. 2010). These beautifully preserved polyps are interpreted as stem-group hexacorals, suggesting considerable diversification of cnidarians by the base of the Cambrian. Additional cnidarian body fossils occur in the Chengjiang fauna (Hou et al. 2005) (fig. 6.13); they have twelve tentacles and are similar in symmetry to living scleractinian corals, which have a sixfold symmetry. Finally, undoubted pelagic cnidarians (Medusozoa) are reported from exceptionally preserved faunas of the Stage 4 Marjum Formation of Utah (Cartwright et al. 2008) and from Wisconsin. These forms show considerable morphological variety, with features of some specimens suggesting affinities to Cubozoa, some to Hydrozoa, and others to Scyphozoa—the three major classes of crown

Figure 6.12 Stem group Ctenophora. *Left:* Reconstruction of *Maotianoascus* from the Chengjiang fauna, with sixteen comb rows; about 6.5 mm high. *Right:* Reconstruction of *Ctenorhabdotus* from the Burgess Shale, with twenty-four comb rows; specimens range to 70 mm in height. Neither of these forms is clearly tentaculate, although preservation is far from perfect. Reconstructions by Quade Paul.

medusozoans—but none of these possible assignments are conclusive; G. A. Young and Hagadorn (2010) conclude that only Scyphozoans can be definitively recognized from the Cambrian, although some of the Utah specimens may represent hydrozoans. The nature of the morphological disparity among these fossils nevertheless indicates that medusozoans underwent a significant early radiation to produce a variety of morphologically distinctive lineages, a common evolutionary pattern among Cambrian metazoans.

Figure 6.13 (right) The sea anemone *Archisaccophyllia kunmingensis.* (A) Reconstruction of a single specimen. (B) A reconstructed life setting. (A) Reconstruction by Quade Paul. (B) Reconstruction after Hou et al. (2005).

B

Explosion Fossils with Bilateral Body Plans

It is bilaterians equipped with circulatory systems, and chiefly with fluid skeletons, that enter the fossil record en masse during the Cambrian explosion. In general, larger animals are more easily fossilized, and the clear linkage between fluid skeletons, blood-vascular systems, and larger body sizes suggests that part of the explosive appearance of these fossils represents a significant ramping up of metazoan body size independently among many different lineages. As larger bodies require more structural support, stiff integuments and mineralized skeletons evolved in response. Tough skeletons must have also provided protection against predation and certainly broadened the bilaterilan taphonomic window.

Deuterostomes: Ambulacraria

Molecular data suggest that hemichordates and echinoderms are sister groups, forming a clade named the Ambulacraria.

HEMICHORDATA The earliest hemichordate, a crown-group pterobranch, appears in the Chengjiang biota along with other deuterostome phyla (Hou et al. 2011). This fossil appears to represent a graptolite (Graptolithina), one of the more important components of pelagic macrofauna during the lower Paleozoic, lasting into the Early Devonian (a general deuterostome phylogeny is shown in fig. 6.14). The tubes of early graptolites are quite similar to those of some pterobranchs (Maletz, Steiner, and Fatka 2007), suggesting a close alliance between those groups. It is not yet clear whether graptolites are stem hemichordates or an extinct branch of pterobranchs. A possible, but undescribed, enteropneust hemichordate in the Burgess Shale fauna is currently under study.

ECHINODERMATA Well-mineralized endoskeletons of echinoderms first appear as disarticulated plates in Cambrian Stage 3. They are absent from the Chengjiang fauna, however, and are represented by only a few specimens in the Burgess Shale, suggesting that their preferred normal marine environments did not much overlap with the conditions conducive to Burgess Shale–type preservation. Clausen and colleagues (2010) have argued that the presence of echinoderms in carbonate sequences in Cambrian Stages 2 through 5 indicates a preferential diversification in carbonate environments. Echinoderm skeletons are composed of many small plates that often disarticulate after death, but with their unique stereom texture (chap. 4), even isolated

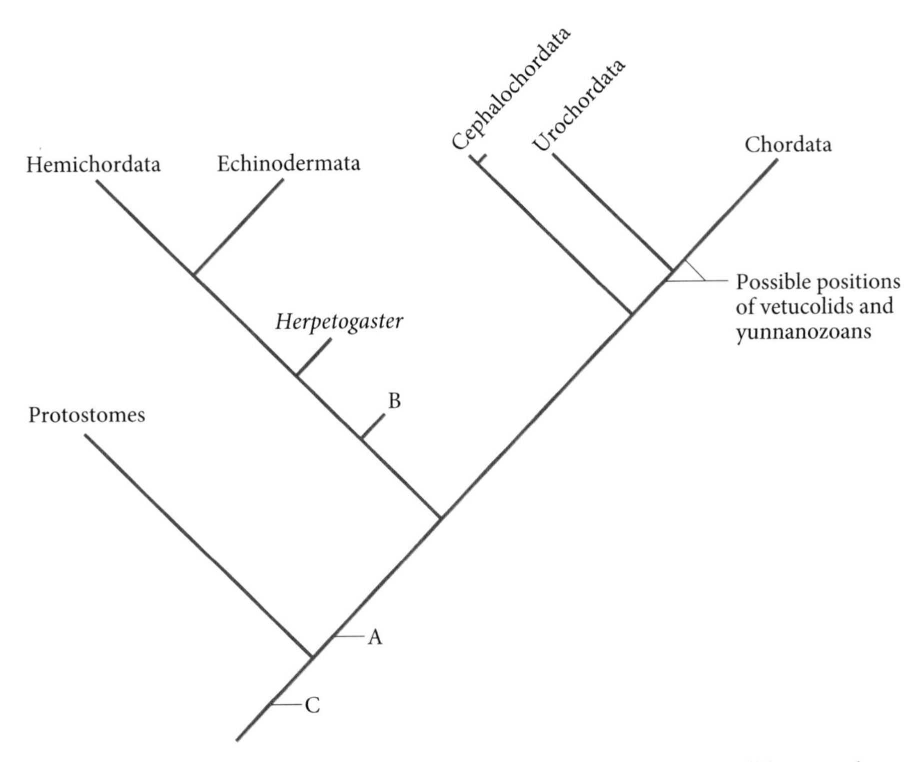

Figure 6.14 Deuterostome phylogeny based on recent molecular data: all but Acoela and Xenoturbellida are represented by fossils during the explosion. Graptolites are similar to pterobranchs, within the Hemichordata, but it is not yet clear whether they are stem hemichordates or an extinct clade of pterobranchs. *Herpetogaster* has been interpreted as a stem ambulacrarian. The vetulicolids and yunnanozoans, from the Chengjiang fauna, are shown as stem deuterostomes, although their phylogenetic position is a subject of debate. It remains unclear to what extent their phylogenetic position has been affected by stemward slippage during decay.

Alternate phylogenetic positions for the acoels are shown in: position (A) with acoels and xenoturbellids as basal deuterostomes; position (B) with acoels and xenoturbellids as basal ambulacrarians; and position (C) with acoels and xenoturbellids as stem bilaterians.

plates can be referred to Echinodermata with some confidence. These early echinoderm groups, that appeared chiefly between Cambrian Stage 3 and the Middle Ordovician, include many forms that lack the familiar pentameral arrangement of ambulacral regions found in living forms; indeed, some of those groups have only one, or two, or three, or four ambulacra. Thus Echinodermata shows the common metazoan pattern of achieving a broad range of disparate morphologies, at the level of body sub-plans, early in its history.

Many of those disparate early groups, however, are not known to have been very diverse, have relatively brief fossil records, and are relatively rare, raising the question as to how to handle them taxonomically. Under a Linnaean taxonomic approach the solution has been to recognize the disparities by assigning distinctive morphological groups to taxonomic levels that show morphological differences similar in magnitude to those shown by living echinoderm groups. Using this approach, some echinoderm workers have recognized more than twenty different echinoderm classes during the Cambrian and Ordovician periods (Sprinkle 1980; Sprinkle and Kier 1987) versus only five modern classes. The large number of relatively short-lived echinoderm classes, many with few species, reflects the systematic practice of many decades ago. As we will discuss later in this chapter, this systematic approach stands in stark contrast to the modern view of early arthropod evolution, where just as many distinctive morphologies have been discovered among Cambrian arthropods but relatively few Linnaean classes or orders have been defined. It seems likely that there were as many as thirty morphologically distinctive echinoderm clades during the Paleozoic (Sumrall and Wray 2007).

Echinoderm taxonomy has subsequently been given a cladistic treatment aimed at tracing the branching pattern of their lineages as morphological novelties appeared and accumulated rather than at identifying their disparate groups (fig. 6.15) (Paul and Smith 1984; A. B. Smith 2005; A. B. Smith and Peterson 2002). However, the early echinoderm record is clearly very incomplete and may not lend itself to detailed interpretation of branching patterns among body plans. Comparative study of the development of ambulacral geometry among echinoderm clades has led to a more general model of the pattern of evolutionary change among of the distinctive clades (Sumrall and Wray 2007). Among living echinoidern classes, all but Crinoidea have five radiating ambulacra that each terminate at a central mouth—a fully pentameral symmetry. The crinoids, and all groups of fossil echinoderms with five ambulacra, have only three ambulacra reaching the mouth, of which two split during development to produce five in adult specimens (like the edrioasteroid *Camptostroma* in fig. 6.15), so they can be described as having a 2-1-2 pattern (rather than 1-1-1-1-1). Groups of fossil echinoderms with fewer than five adult ambulacra are of course not even pentameral and are interpreted by Sumrall and Wray (2007) as having been reduced from a primitive 2-1-2 pattern through varied patterns of loss of amblacra during early developmental stages. The amblacral losses are correlated with important bodyplan differences such as characterize many of the disparate fossil echinoderm groups. The generality of this evolutionary model is appealing, but the spotty fossil record makes it difficult to test. To complicate the phylogeny further, it seems

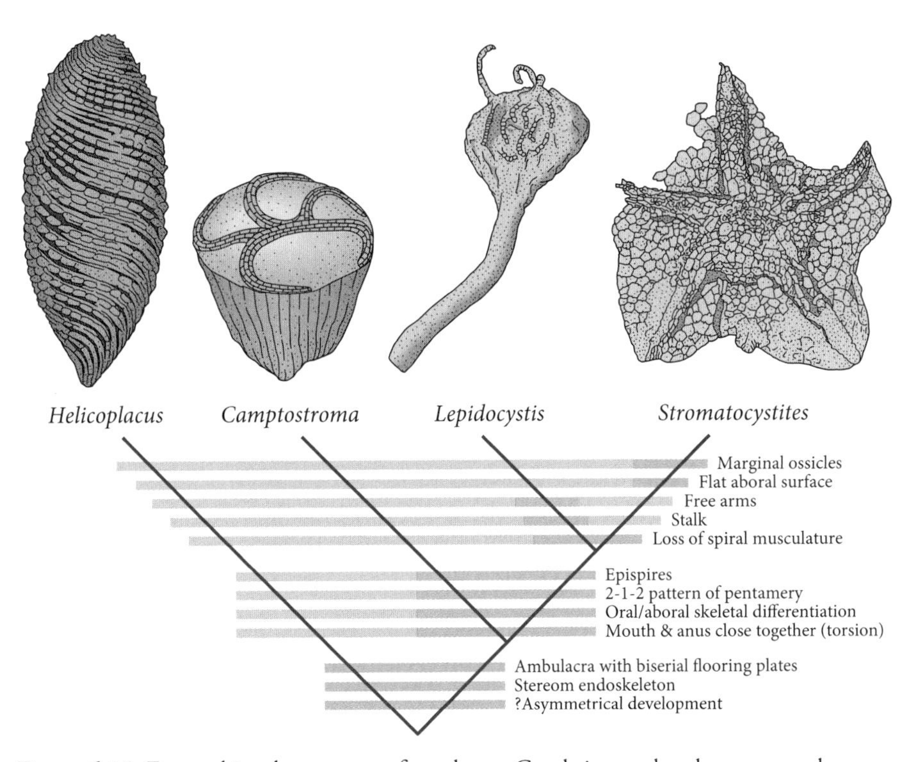

Figure 6.15 Four echinoderm genera from lower Cambrian rocks, shown according to an interpretation of their relative branching order in the echinoderm tree, with morphological criteria in support of this arrangement: *Helicoplacus* (Class Helicoplacoidea), *Camptostroma* and *Stromatocystites* (Class Edrioasteroidea), and *Lepidocystis* (Class Eocrinoidea). The topology of these genera in the tree suggests that *Camptostroma* is basal to all known pentaradial echinoderms, including all crown classes; *Lepidocystis* is hypothesized to belong to a branch, presumably arising from the eocrinoids, that gave rise to the class Crinoidea; and *Stromatocystites* is hypothesized to belong to a branch of edrioasteroids that gave rise to the crown, nonstalked classes (Eleutherozoa) as Asteroidea (starfish), Ophiuroidea (brittle stars), Echinoidea (sea urchins), and Holothuroidea. From A. B. Smith (1988).

clear that ambulacral reductions have also occurred within minor branches of some major clades. The sequence of appearance of bodyplans in skeletonized forms does not match seqential ambulacral losses, but considering the incompleteness of the early echinoderm record, this is not sufficient to falsify the Sumrall-Wray hypothesis.

The earliest Cambrian echinoderms are assigned to Helicoplacoidea (fig. 6.15) (Durham and Caster 1965). They are known only from four small-bodied (chiefly 4–7 cm) Stage 3 genera from western North America. The three best-known genera have bodies that are somewhat fusiform to top-shaped, with a skeleton of imbricating plates that have a stereom texture, arrayed in spiral columns. The columns are chiefly interambulacral, but ambulacral columns, indicating the presence of a water-vascular canal system, are shown by a few columns of much smaller plates. From the apex of the larger end, two ambulacral columns spiral down to about the widest part of the test, where they join; the resulting single ambulacral column then spirals toward the smaller end but terminates before reaching the base. Durham (1993), working with a large collection, has given the clearest description of the ambulacral structures and concluded that the mouth was at the larger end. Derstler (1981) and Paul and Smith (1984) had previously suggested that the mouth was at the junction of the three ambulacral columns. No oral framework has been detected at either site, for which deformed and collapsed states of the fossils may be to blame. If the mouth were at the junction, it would suggest that this three-canal system evolved to the 2-1-2, five-canal system found in early echinoderms by doubling of each of the paired ambulacra.

The earliest known 2-1-2 pentaradial groups are found in early Stage 4 rocks (chiefly in the Kinzers Formation of Pennsylvania). One group, often treated as a class, is the Edrioasteroidea, which were circular in outline and lay on their dorsal surfaces, evidently attached to the substrate in some cases; presumably, they were suspension feeders. The basal edrioasteroid *Camptostroma* (fig. 6.15), interpreted as a basal form, is found in the Kinzers Formation and has been hypothesized to represent the deepest known branch of all pentaradial echinoderms (A. B. Smith 1988). A later branching may have given rise to two echinoderm clades that produced the disparate pentaradial groups of the Paleozoic, one characterized by attachment stalks, the other not. One of these major clades, as represented by *Stromatocystites* (fig. 6.15), may have diversified into all the living, nonstalked echinoderm classes. If the Sumrall and Wray model is correct it would imply that the last common ancestor of those classes arose after evolution of the 1-1-1-1-1 ambulacral pattern. Another Kinzers echinoderm, *Lepidocystis,* has been hypothesized to represent a branch that gave rise to extinct stalked clades such as Eocrinoidea and to the Crinoids (fig. 6.15). Crinoids are pentaradial in the 2-1-2 fashion, with their ambulacra extending up their feeding arms (as in *Lepidocystis*; A. B. Smith 1988). For most of their existence, they were all stalked, attached to the substrate by a well-skeletonized, stem-like organ (nonstalked crinoids evolved much later). A second lineage, however, hypothesized to have branched off the edrioasteroids as well, evolved a stalk and diversified into the extinct class Eocrinoidea (fig. 6.16). Eocrinoids have arm-like structures, but they were not invested by coeloms and did not bear ambulacra; such arms are called brachioles. If the

Figure 6.16 Eocrinoids. An assemblage of fossil *Sinoecrinus lui* from the Kaili fauna, in the middle of the Kaili Formation, Jianhe County, Guizhou Province, China. This deposit is assigned to Cambrian Stage 3. Note that all the specimens have been aligned by currents and the distalmost ends of the stalks have been fractured. Photograph courtesy of Ronald Parsley, Tulane University.

phylogenetic hypothesis shown in figure 6.16 is correct, both Edioasteroidea and Eocrinoidea are paraphyletic. Testing these hypotheses will require the discovery of new fossils, however.

The carpoids, still another enigmatic echinoderm group that lacks pentaradial symmetry, is first known from Stage 5 in Bohemia. Carpoids are asymmetric, although two of the three main groups can be said to "exhibit bilateral tendencies," in Ken Caster's phrase (Ubaghs and Caster 1967, S583); indeed, no other echinoderms are remotely like them. One clade of carpoids, the Stylophora, is particularly well known to paleontologists because it was once proposed to lie in the ancestry of vertebrates (Jefferies 1986). Subsequent work has shown that carpoids are a branch of echinoderms and are not more closely allied to chordates than are other echinoderm groups. Carpoid architecture, however, clearly lies outside the borders of all other known echinoderms, and when carpoids first originated is uncertain.

Deuterostomes: Chordata

Fossils of all three living chordate subphyla (Cephalochordates, Urochordates, and Vertebrates) appear in Stage 3. The earliest known cephalochordate is *Cathaymyrus,* a poorly preserved elongate animal (about 5 cm long) from the Chengjiang (Shu, Conway Morris, and Zhang 1996), whereas a Stage 5 form from the Burgess Shale, *Pikaia,* about 4 cm long, is more complete (fig. 6.17) (Briggs, Erwin, and Collier 1994; Conway Morris 1982; Conway Morris and Caron 2012). These fossils show the seriated myotomes (muscle blocks) characteristic of living cephalochordates (and somewhat similar to those of vertebrates). Urochordates, clearly much reduced morphologically from their ancestors, are represented among Chengjiang fossils by a benthic sea squirt (fig. 6.18) (J. Y. Chen et al. 2003).

Craniate fossils are found in Stage 3 rocks near the Chengjiang localities (Shu et al. 1999). This earliest known putative vertebrate is named *Myllokunmingia* (fig. 6.19), and it appears to have a cartilaginous skull and other elements that in living vertebrates indicate the presence of tissues derived from neural crest (chap. 4) (although see the discussion below on taphonomy). Paraconodonts, which are tooth-like fossil elements, are first found in late Stage 4 and are believed to be allied to Conodonta (Bengtson 1983), a primitive craniate group (Aldridge et al. 1986); they would thus be fish-like vertebrates as well. Both molecular clock estimates and the fossil evidence suggest that Vertebrata arose during the same general round of evolutionary innovation that produced the body plans of Arthropoda, Mollusca, Brachiopoda, and so forth, the major crown metazoan body plans.

A

B

Figure 6.17 *Pikaia,* a cephalochordate from the Burgess Shale. (A) Photo of USNM 57628. Notice the chevron-shaped muscle blocks and faint notochord just above the chevron's V. (B) Reconstruction. Photograph by J. B. Caron, courtesy of the Smithsonian Institution. Reconstruction by Quade Paul.

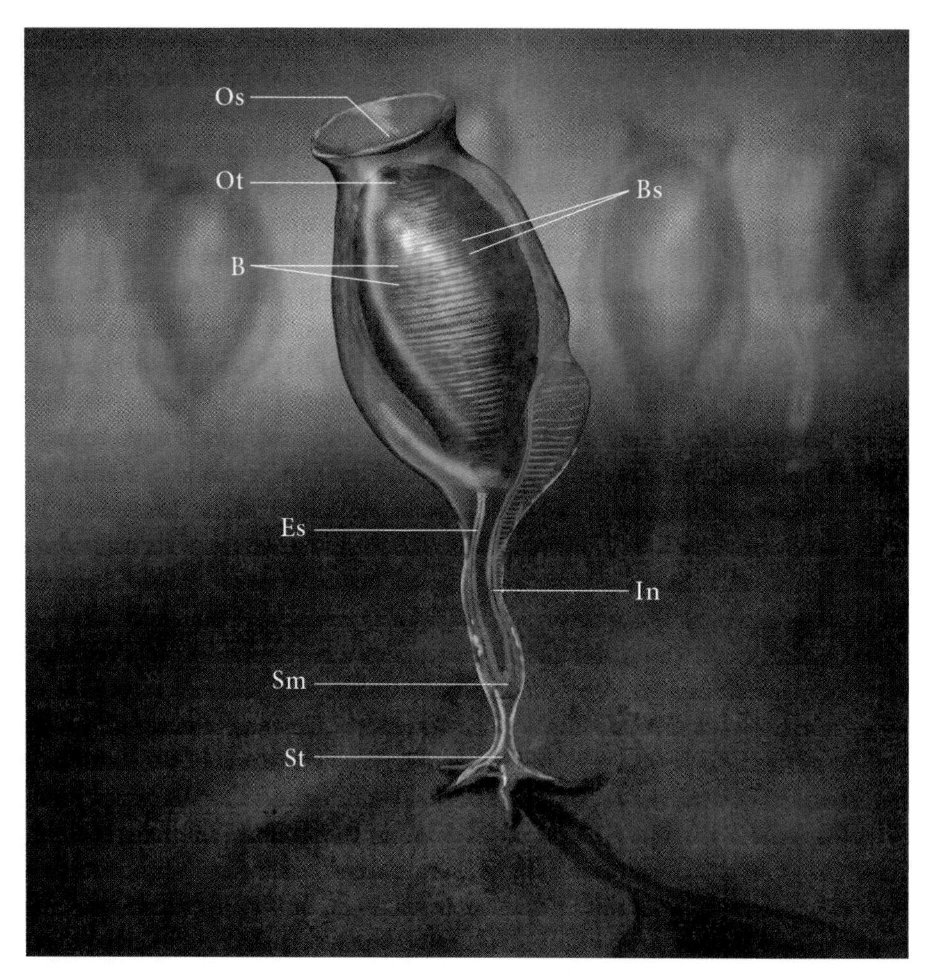

Figure 6.18 *Shankouclava,* a urochordate from the Chengjiang fauna. Reconstruction based on eight specimens. B, branchial bars; Bs, branchial slits; Es, esophagus; In, intestine; Os, oral siphon; Ot, oral tentacle; Sm, stomach; St, stalk. J. Y. Chen et al. 2003. Reconstruction by Quade Paul.

Other Possible Deuterostomes

The crown phyla maintain recognizable body plans as they are traced back to the early Cambrian. The stem groups, however, commonly have unusual synapomorphies that are not known among crown groups or that are mixtures of characters not shared by any crown group.

One such form is a tentaculate soft-bodied fossil from the Burgess Shale, *Herpetogaster collinsi* (fig. 6.20), which has been tentatively grouped with other

Figure 6.19 *Myllokunmingia,* a probable agnathan vertebrate from Haikou, Yunnan, China. Reconstruction by Quade Paul.

Cambrian fossils of uncertain affinity with which they share general morphological features. This group is interpreted as possible stem Ambulacraria (Caron, Conway Morris, and Shu 2010), that is, branching after the split between chordates and ambulacrarians but before the split between hemichordates and echinoderms (see fig. 6.14). The trunk of *Herpetogaster* appears to include a segmented region that, if coelomate, is unlike those in known deuterostomes. The tentacles contain a trace that is interpreted as either a hydrostatic canal or a vascular system. The ambulacrarian/chordate LCA had evolved pharyngotremy, and although gill pores are not unequivocally present on *Herpetogaster,* a possible location for them is behind the head. Although its preservation is reasonably good as fossils go, interpretation of this animal is very difficult and epitomizes the difficulties encountered when dealing with stem groups.

Two other possible deuterostome groups, vetulicolids and yunnanozoans, are often tentatively referred to the chordate branch or, alternatively, more basal positions as stem deuterostomes (and still other phylogenetic positions have been suggested). Vetulicolids (fig. 6.21) have bipartite bodies, with an anterior carapace-like section that may be subdivided into two or more units, sometimes appearing to be segmented, and a posterior section or trunk that appears to be narrowly segmented in some genera in a fashion recalling arthropods. They are known from both the Chengjiang and the Burgess Shale faunas

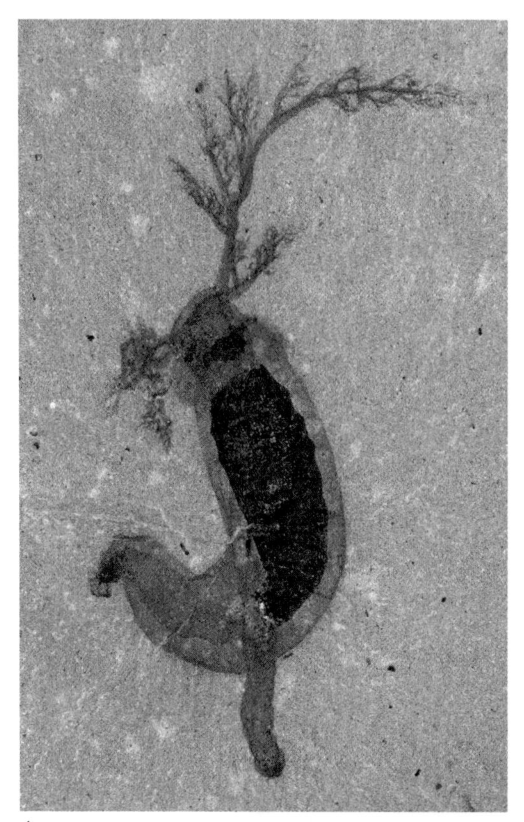

A

B

Figure 6.20 (left) A possible stem ambulacrarian, *Herpetogaster collinsi,* from the Burgess Shale; specimens are typically 3–4 cm long, including tentacles. (A) Holotype (ROM 58501) showing overall morphology; © Royal Ontario Museum; photograph by Jean-Bernard Caron. (B) Reconstruction by Quade Paul.

(Aldridge et al. 2007; Caron 2006; Shu et al. 2001). Some of the Chengjiang forms have structures interpreted as gill slits, suggesting a chordate affinity, whereas a Burgess Shale form (*Banffia,* known from hundreds of specimens) shows structures that can be interpreted as midgut diverticulae (Caron 2006), a feature lacking in crown chordates but common in protostomes. It is possible that forms assigned to Vetulicolida represent more than one group; at the very least, vetulicolid morphology is sending a mixed phylogenetic message.

Yunnanozoans (fig. 6.22), from the Chengjiang fauna, are also known from many hundreds of specimens, but they are quite enigmatic, and they have been reconstructed as stem hemichordates (Shu, Zhang, and Chen 1996) or stem chordates (J. Y. Chen et al. 1995). They have a fish- or lancelet-shaped body with a series of anterolateral structures that make sense as gills, and they have a remarkable dorsal fin. This fin is seriated or segmented, suggesting at first glance a relationship with the segmented cephalochordate or vertebrate body plans, but seems unlikely to have functioned in an analogous manner. Further, no notochord-like structure has been confirmed, which weakens a chordate assignment (see also Shu, Zhang, and Chen 1996).

Taphonomic loss and blurring of morphological characters are important reasons for the uncertainties about the phylogenetic position of these soft-bodied fossils. Studies of patterns of decay among modern chordates reveal a progressive loss of phylogenetically informative characters so that as decay advances, a specimen appears like increasingly earlier phylogenetic stages (Sansom, Gabbott, and Purnell 2010). In other words, decay is not random with respect to the phylogenetic sequence of chordate characters. These modern taphonomic experiments suggest that such sequential taphonomic biases may affect the interpretations of Cambrian chordate fossils.

Ecdysozoans

Ecdysozoans are among the most spectacular of the explosion fossils, and although the fossil taxa appear to be about as morphologically complicated as their living counterparts, they are nearly all stems of crown phyla or in some cases appear to represent even earlier branches from which the stems of crown phyla branched in turn. Because even "soft-bodied" ecdysozoans have cuticles that are usually toughened, they are almost certainly more easily preserved

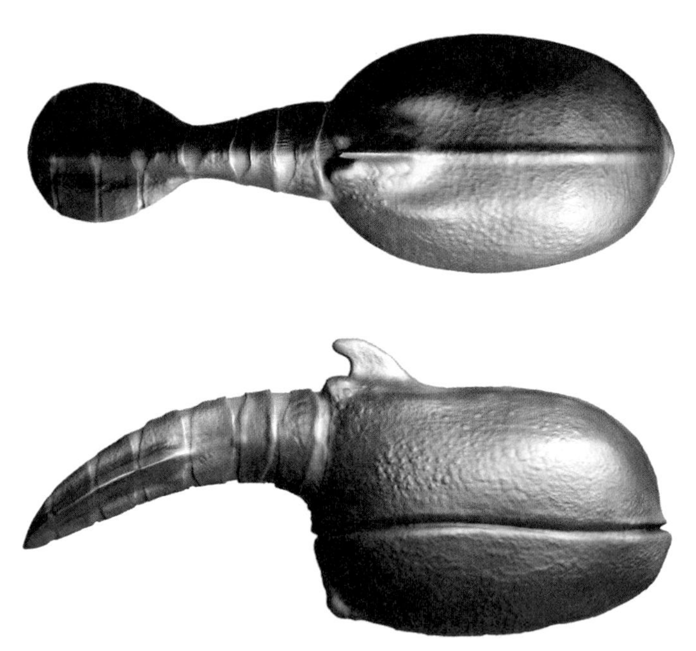

Figure 6.21 Vetulicolids, possible deuterostomes. (A)–(B) *Vetulicolia* from the Chengjiang fauna. (A) Specimen, about 9.5 cm long. (B) Reconstruction. (C)–(D) *Banffia* from the Burgess Shale. (D) One side of a complete specimen, ROM 49914, about

C

D

8.5 cm long. (D) Reconstruction of external appearance. *Banffia* is also known from the Chengjiang fauna. (A) from Hou et al. (2004), (C) © Royal Ontario Museum; photograph by Jean-Bernard Caron, (B, D) reconstruction by Quade Paul.

Figure 6.22 Yunnanozoans, deuterostomes of uncertain affinity. Specimens from Chengjiang fauna; upper specimen about 2.5 cm long.

than, say, those lophophorates that lack mineralized skeletons. Recall from chapter 4 that the ecdysozoans are divided into two large clades, the Panarthropoda, which includes the lobopods and the arthropods, and the Cycloneuralia, which includes the Nematoda and Scalidophora.

SCALIDOPHORA Two of the crown pseudocoelomate ecdysozoan phyla—Priapulida, from the Chengjiang fauna and from the Burgess Shale (Wills 1998), and Loricifera (Peel 2010), from Sirius Passet fauna in northern Greenland—are identified from presumed stem lineages in the Cambrian. Together with a third living phylum—Kinorhyncha, not known from the fossil record—these living phyla are grouped into the clade Scalidophora. There are additional stem groups in the Cambrian that are often also identified as Scalidophora. An array of forms in both the Chengjiang and Burgess Shale assemblages (fig. 6.23) have been identified as priapulids. These fossils range to more than 15 cm in length. Today, some priapulids are shallow burrowers that live in the upper few centimeters of very soft substrates; the larger forms known from the explosion may have shared this mode of life. A review of pseudocoelomates of the Chengjiang and associated faunas (Maas et al. 2007) has made the point that many features used to identify Cambrian forms may represent plesiomorphies, and, if so, assignment to crown taxa is premature. Living priapulids nevertheless have highly conserved genomes that suggest a low, even basal position within Ecdysozoa, and the Cambrian forms may well represent their stem group(s) (Webster et al. 2006).

Embryos assigned to Scalidophora are preserved in phosphatic sediments from early Cambrian to Early Ordovician faunas. The best known species belong to the genus *Markuella* (fig. 6.24A, 1–3), which is considered to represent either a stem scalidophoran or a stem priapulid (Dong et al. 2011). Late-stage embryos are most commonly preserved, but X. G. Zhang, Pratt, and Shen (2011) report some cleavage-stage forms that indicate radial cleavage, as in priapulids. Radial cleavage has been suggested as the basal cleavage state of ecdysozoans on the basis of molecular phylogenetic relationships (Valentine 1997).

PALAEOSCOLECIDAE The Palaeoscolecidae is an extinct group of uncertain affinity with thin, worm-like bodies covered with cuticular sclerites of calcium phosphate (fig. 6.24B). This group is interpreted as pseudocoelomic and is likely to have had ecdysozoan affinities: T. H. P. Harvey, Dong, and Donoghue (2010) consider them likely stem priapulids, whereas Conway Morris and Peel (2010) propose a stem cycloneuralian position. A diversity of mineralized sclerites of palaeoscolecids is found in Cambrian Stage 3, and body fossils occur in the Sirius Passet and Chengjiang biotas. Most of the worm-like forms among the living pseudocoelomates are about the right size to be responsible for the small trails and burrows recorded in the Neoproterozoic. When the phylogenies of these groups have been definitively sorted out, we should have a better idea of which, if any, of the small-bodied ecdysozoan crown groups were likely to have been members of the Neoproterozoic fauna.

Panarthropoda

Three crown phyla—Onychophora, Tardigrada, and Arthopoda—seem closely related on morphological criteria and were grouped as the Panarthropoda in a pioneering cladistic classification by Nielsen, Scharff, and Eibye-Jacobsen (1996). Molecular data indicate that Onychophora and Arthropoda are sister groups, whereas Tardigrada may be the sister of those two (Campbell et al. 2011). The addition of some of the more flamboyant animals in the Cambrian explosion fauna to phylogenetic analyses, however, produces a more complex picture. Many of these unusual forms have morphologies suggesting affinities to onychophorans, and others have characteristics of arthropods. There are

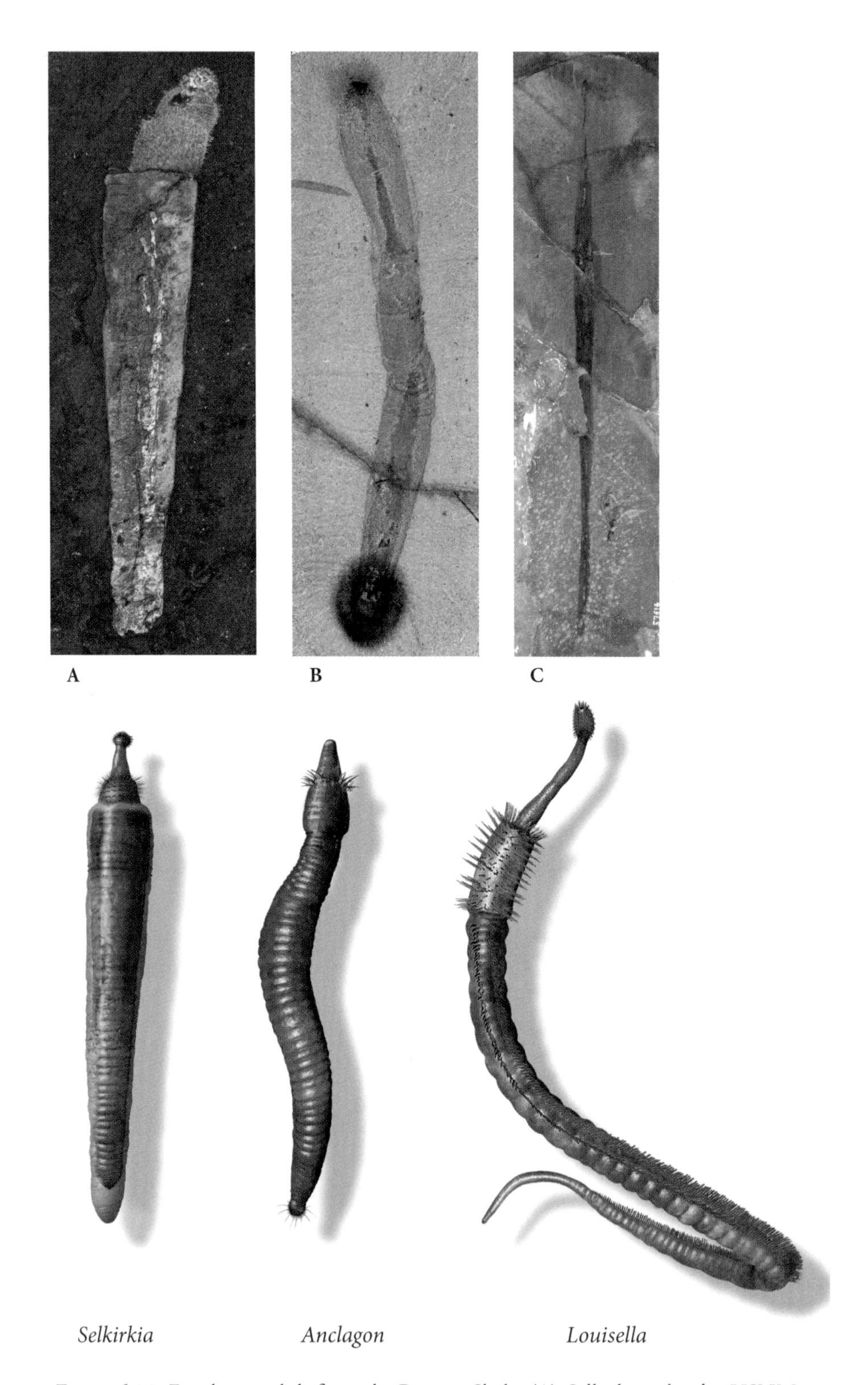

Figure 6.23 Fossil priapulids from the Burgess Shale. (A) *Selkirkia columbia* USNM 83941a. (B) *Ancalagon* USNM 57646. (C) *Louisella,* USNM 57616, anterior portion,

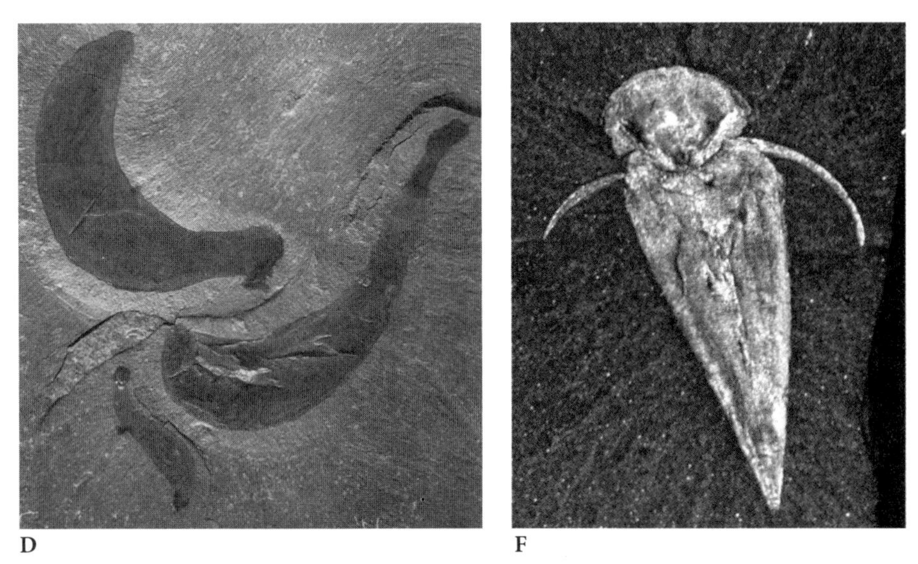

with proboscis everted. (D) *Ottoia* USNM 188609. Three specimens, with the specimen on the right with the proboscis everted. (E) Reconstruction of *Ottoia.* (F) *Haplophrentis,* a hyolithid from the Burgess Shale. Specimens of *Haplophrentis* are preserved in the gut of several specimens of *Ottoia.* These priapulids range from about 2 cm to 30 cm in length. (A)–(D), (E) Photographs courtesy of the Smithsonian Institution, (E) reconstruction by Quade Paul.

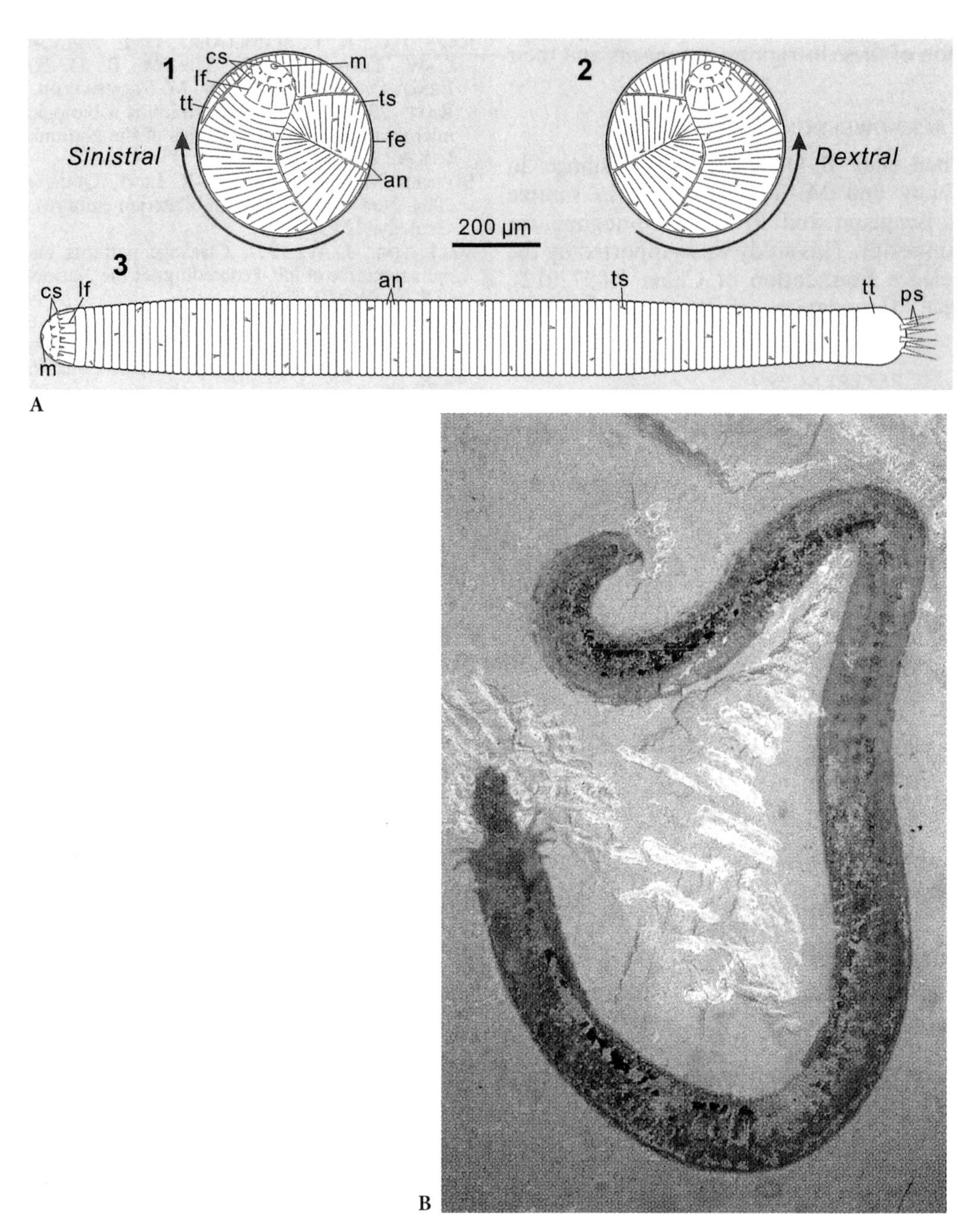

Figure 6.24 Fossil scalidophorans from the Cambrian. (A) 1–3: Reconstructions of *Markuelia,* a scalidophoran embryo from Guizhou, China. There are two types, 1 and 2, that coil in opposite directions within their fertilization envelopes; in 3, an embryo has been reconstructed as uncoiled. (B) Fossil specimen *Circocosmia,* about 32 mm in length. Both of these groups are likely to have been pseudocoelomate. (A) from X. G. Zhang, Pratt, and Shen (2011); (B) from Hou et al. (2004).

significant numbers of relatively well-preserved animals that do not fit within any crown phylum—including such interesting forms as *Opabinia,* a reconstruction of which graces the cover of the book —and the phylogenetic relationships among these early Cambrian forms have not yet been fully resolved. The earliest panarthropod trace fossils are in Cambrian Stage 2. Although some lobopod body fossils appear in Stage 2 and there are reports of arthropod body fossils, virtually all the stem arthropod clades first appear during Cambrian Stage 3.

LOBOPODIANS AND ONYCHOPHORA Living Onychophora, with their paired antennae and lobopodial (stubby, rounded, and annulated) walking legs, are morphologically similar (recall from chapter 4 that all living onycophorans are terrestrial, in contrast to the marine fossils). By contrast, the morphological diversity of Cambrian fossils with lobopodial appendages is much greater (figs. 6.2 and 6.25), generating considerable controversy over their relationship to crown panarthropods (Hou and Bergström 1995; J. Liu et al. 2008; Ma, Hou, and Bergstrom 2009; Ramskold and Chen 1998). *Aysheaia,* from the Burgess Shale (fig 6.26C) was the first Cambrian onycophoran to be recognized. Many of the explosion lobopodians are "armored," bearing a diversity of plates or spines. Although the earliest confirmed body fossils of lobopods are not found until Cambrian Stage 2, J. Liu and colleagues (2008) argue that the small shelly fossil *Zhijinites* in Cambrian Stage 1 is likely a lobopodian sclerite, and plates of *Microdictyon* have been reported from similar-aged rocks in Siberia (Kouchinsky et al. 2012). Cambrian forms all have elongate bodies with lobopod-type limbs encircled by putative hemal channels, but, with a single exception, they differ from modern onycophorans in the location of the mouth, the variable nature of limb morphology, the presence of terminal claws on the limbs, the variety of spines and plaques on the trunk, and the general lack of prominent, muscular paired antennae. (Small antennae have been reconstructed on some Cambrian species, but they are quite unlike the muscular antennae of living onychophorans.)

An exceptional Cambrian form is *Antennacanthopodia gracilis* from the Chengjiang fauna (fig. 6.26A), which has two pair of antennae (Ou et al. 2011), the larger and more anterior of which resembles the antennae of crown onychophorans. *A. gracilis* also has possible paired eye spots, a lightly sclerotized cuticle, and terminal walking pads on the limbs rather than claws. Other features are more similar to living onychophorans than to those of the lobopods. Internally, *A. gracilis* shows relatively large body spaces, presumably hemocoels, arranged much like those in oychophorans. Ou and colleagues (2011) reasonably speculate that *A. gracilis* may be related to a lobopodian branch that gave rise to crown Onychophora.

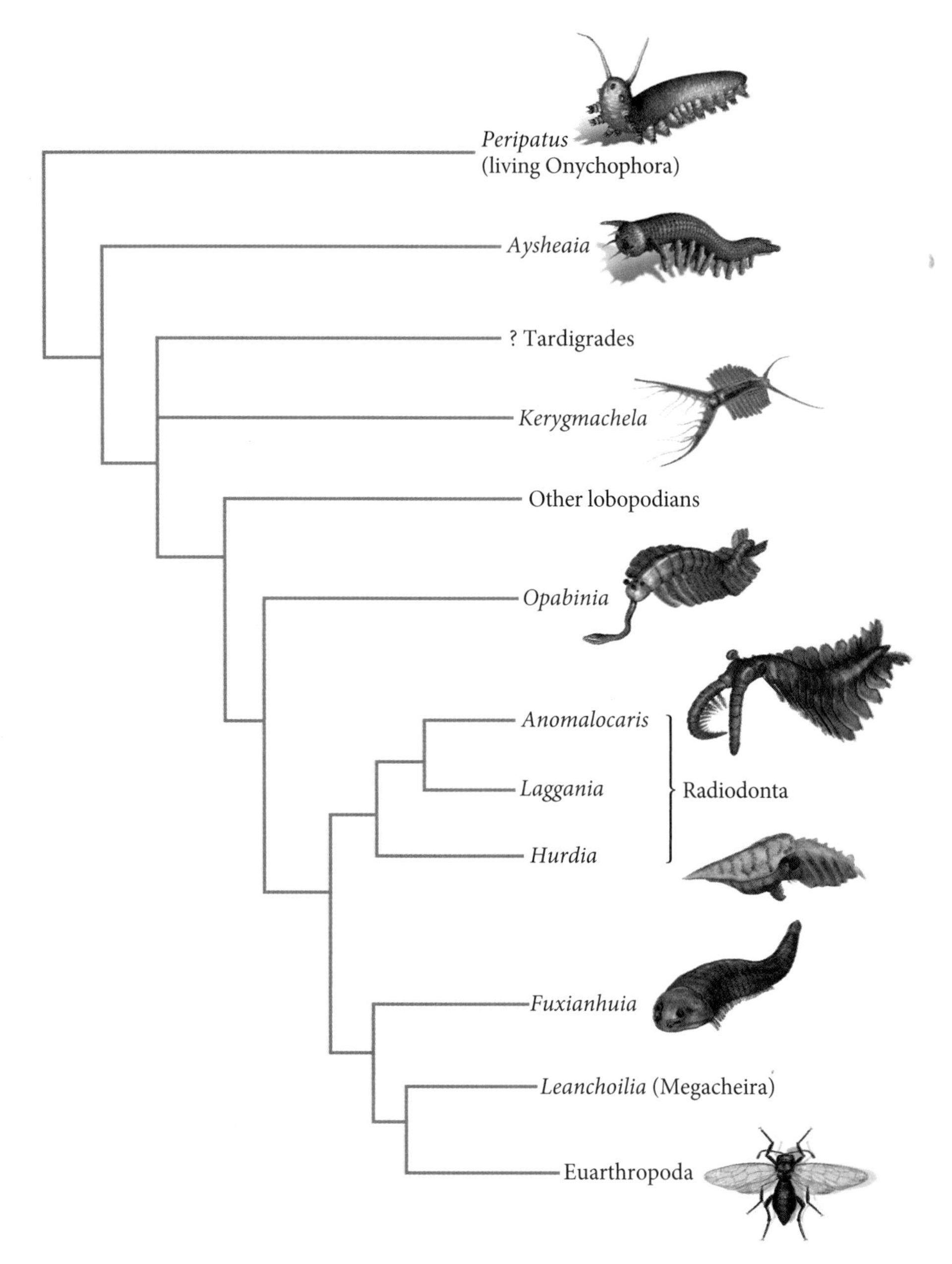

Figure 6.25 Lobopodians and Onychophora. This figure shows a consensus of phylogenetic results for Panarthropoda, including lopobodians, Onycophora, tardigrades, and Euarthropoda. The living onycophoran *Peripatus* is at the base of the tree, with a paraphyletic suite of Cambrian lobopods along the main branch. The position of tardigrades is unclear; molecular analyses suggest that they may be the sister group to Onycophora + Arthropoda. Reconstructions by Quade Paul.

Graham Budd of Uppsala University has been prominent among those who have argued for a lobopodian origin for arthropods, citing both anatomical comparisons between crown groups and the fossil morphologies that he has interpreted as having features linking onychophorans with stem plus crown arthropods (see below) (Budd 1996, 1999). The description of a spectacular new lobopodian from the Chengjiang fauna, *Diania cactiformis* (fig. 6.26B) (J. Liu et al. 2011), has added a new dimension to the question of arthropod origins. *Diania* is heavily armored, with the spiked elements encasing the limbs, suggesting jointing and at least some incipient segmentation. The trunk armor, the elements of which are quite narrow, suggests a different sort of flexibility. It is easy to imagine that the jointing that characterizes the arthropod carapace evolved from some such integumental armor while integrating internal segmental features within the jointed skeletal architecture to produce the arthropod body plan (J. Liu et al. 2011). The phylogenetic placement of *Diania* has been the subject of considerable debate, however. In their original paper, J. Liu and colleagues proposed that *Diania* was truly intermediate between Cambrian lobopods and euarthropods. This hypothesis now seems unlikely because several new phylogenetic analyses place *Diania* well within other Cambrian lobopods (Mounce et al. 2011), emphasizing both the difficulty of resolving phylogenetic relationships among the Panarthropoda and the importance of rigorously developed phylogenies.

Many Cambrian lobopod fossils are poorly preserved and incomplete, and key morphological characters may be difficult to discern. Advances in our knowledge of their morphological details and their interrelationships thus depend on continuing discovery of new fossils that reveal more features, but the general impression of the variety of body shapes and ornamentation of Cambrian lobopods is correct. The early phylogenetic hypothesis represented in figure 6.25 has already been challenged by subsequent studies. Figure 6.27 presents a panarthropod phylogeny, with the Cambrian lobopods represented as a series of paraphyletic lineages with both the tardigrades and the onycophorans as the living remnants of a once more disparate assemblage of forms. Although the topology shown in this figure has been adopted by many workers, we suspect that further phylogenetic studies may alter the picture, perhaps significantly. If anything, the Cambrian forms seem morphologically more complicated than living onychophorans.

BRANCHES BETWEEN LOBOPODA AND CROWN ARTHROPODA Figure 6.27 also illustrates that between the Cambrian lobopods and the crown-group arthropods lies a broad array of fossils that appear to have arthropod affinities: they seem to have segmented exoskeletons, and most have jointed append-

A

B

C

Figure 6.26 (left and above) Disparate lobopodians. (A) The Stage 3 lobopod *Antennacanthopodia gracilis,* which has two pairs of muscular antennae; length about 1.5 cm. (B) The Stage 3 lobopod *Diania cactiformis,* nicknamed the walking cactus, reconstructed in lateral view; length about 7 cm. (C) The Stage 5 *Ayshaeia pedunculata* USNM 83942A; length about 2.5 cm. (A–B) Reconstructions by Quade Paul, (C) courtesy of the Smithsonian Institution.

ages of some sort. Accordingly, the crown group is called the Euarthropoda to distinguish it from such stems. The more basal groups lack some of the synapomorphies of crown groups but have synapomorphies of their own. Perhaps near the base of this assemblage, close to the lobopodians, are two forms, *Kerygmachela* (fig. 6.28) and *Pambdelurion,* that are interpreted as having flexible (as opposed to jointed) sclerotized appendages that can be viewed as derived from lobopods (see Budd 1999). These forms are gilled, however, and *Pambdelurian* has a radial mouth, surrounded by sclerotized plates that resembles anomalocaridid mouths (see below). In these (and other) respects these forms differ from canonical lobopodians. These "gilled lobopods" are commonly cited as evidence suggesting that arthropods evolved from among the lobopodians.

Opabinia may represent another step toward arthopodization. With its curious single-stalked anterior appendage, multiple eyes, and odd posterior blades or flaps (fig. 6.29), *Opabinia* is so unusual that it provoked laughter when first described by the late Harry Whittington at a scientific meeting in

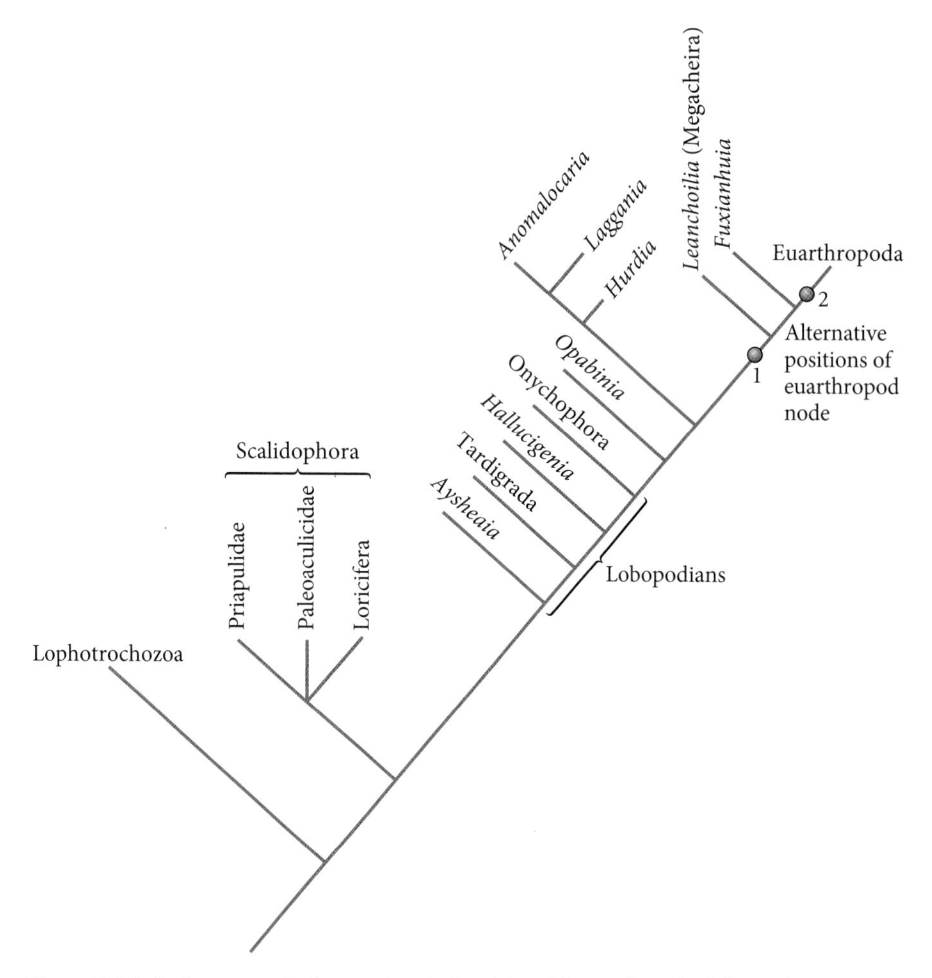

Figure 6.27 Ecdysozoan phylogenetic relationships. Notice that the lobopodians are a paraphyletic grade, of which both the extant tardigrades and onycophorans appear to be remnants. Based on Edgecombe (2010).

the 1960s. The animal is about 5 cm long. *Opabinia* must have chiefly been a swimmer and is probably allied with the anomalocaridids (fig. 6.30), which are known from several genera grouped into the clade Radiodonta. The anomalocaridids can be fairly large, and some may have been 1 meter in length. Like *Pambdelurion*, anomalocaridids have a radial mouth, preserved as a circlet of sclerotized plates (thirty-two in some forms) around the mouth opening. Paired claws flank the mouth anteriorly. The large body size and rather spectacular and fearsome-looking claws of anomalocaridids, which vary significantly among genera, have caused them to be featured prominently in artworks and displays of Cambrian faunas. Despite their stature as paradigmatic examples of the Cambrian explosion, anomalocaridids persisted at least into the Ordovician, where beautiful (and gigantic) specimens have recently been recovered (Van Roy and Briggs 2011). The jointed claws could presumably be flexed to capture prey, which could then be conveyed to the mouth circlet, which was probably specialized for crushing exoskeletons. In both *Opabinia* and the radiodontids, swimming was likely accomplished by movement of the lateral flaps, to which gills are affixed, along the side of the body. Radiodontids and *Opabinia* belong to paraphyletic assemblage of stem groups between the lobopodians and the Euarthropoda.

Resolving the phylogenetic relationships among these groups is critical to understanding the evolution of arthropod limbs and thus to understanding the origin and early diversification of the arthropods (Budd and Telford

Figure 6.28 Reconstruction of a "gilled lobopod," *Kerygmachela,* from the early Cambrian Sirius Passet fauna of Greenland; scale bars 20 mm. Reconstruction by Quade Paul.

A

B

Figure 6.29 (left) *Opabinia,* a stem arthropod and possibly allied to the anomalocaridids, from the Burgess Shale. (A) Fossil in lateral view, USNM 57683. (B) Reconstruction. (A) Courtesy of the Smithsonian Institution, (B) reconstruction by Quade Paul.

2009; Edgecombe 2010). A glance back through the figures in this section will identify some of the problems currently facing researchers in this area. For example, the lobopodians all share fairly simple, unspecialized legs, yet *Opabinia* and the anomalocaridids lack legs but have paired, lateral flaps that, particularly in *Opabinia,* have gills along the upper aspect of the flap (X. L. Zhang and Briggs 2007). Beyond the Radiodonta, however, well-sclerotized jointed appendages reappear. Are arthropod appendages homologous to those of lobopods, as Budd has argued? Are they homologous to the lateral flaps of the Radiodonta? Or are they entirely novel structures? This debate is far from settled (Budd 1996; Budd and Daley 2012; Waloszek et al. 2007; X. L. Zhang and Briggs 2007), illustrating something of the complexities of understanding the evolutionary pathways among these groups. Still other stem arthropods, almost certainly closer to crown arthropods than are radiodontids, are found in Burgess Shale–type assemblages. Two such groups (which may be closely allied) are the Canadaspids (fig. 6.31A), which are bivalved and have sometimes been assigned to crown Crustacea, and *Fuxianhuia* and related taxa (fig. 6.31B). They differ from crown groups chiefly in details of the limbs (see Edgecombe 2010).

Another stem group well represented among lower Cambrian lagerstätten are the Megacheira, the "short great appendage arthropods" such as *Fortiforceps* (fig 6.31C). This group is characterized by relatively large, paired appendages that arise on the anterior portion of the head, although it is commonly difficult to be sure precisely where they insert because the heads are partly obscured beneath a shield and the boundaries of anterior head segments are not evident. In living arthropods, anterior head appendages are innervated from the brain, which has three lobes: the procerebrum (innervating eyes), the deuterocerebrum, and the tritocerebrum in antero-posterior order. Arthropod eyes are on the most anterior head segment—the ocular segment—and are innervated from the most anterior lobe. The most anterior appendages, which are antennae in crustaceans and chelicerae in chelicerates (fig. 6.32), are found on the second segment only and are innervated from the middle lobe. No living arthropod of any type has appendages that arise on the ocular segment. Living onychophorans, however, have two brain lobes, and their antennae are on the ocular segment. The eyes are situated at the base of the antennae, and both are innervated from the anterior of the brain lobes. Thus, if crown

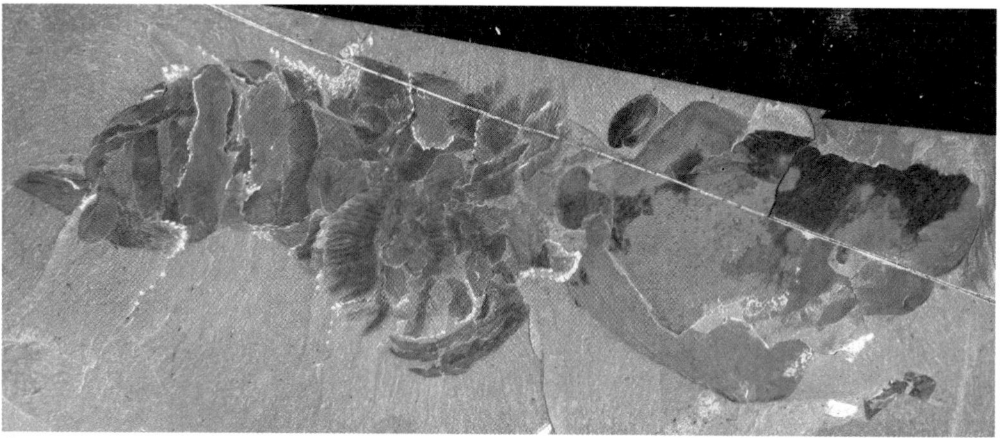

A

B

Figure 6.30 The anomalocaridids *Hurdia* and *Anomalocaris* (Radiodonta) from the Burgess Shale. (A) *Hurdia victoria* USNM 274159. (B) Reconstruction. (C) *Anomalocaris* ROM 51211. (D) Reconstruction. (A) Courtesy of the Smithsonian Institution, (B, D) reconstructions by Quade Paul, (C) © Royal Ontario Museum; photograph by Jean-Bernard Caron.

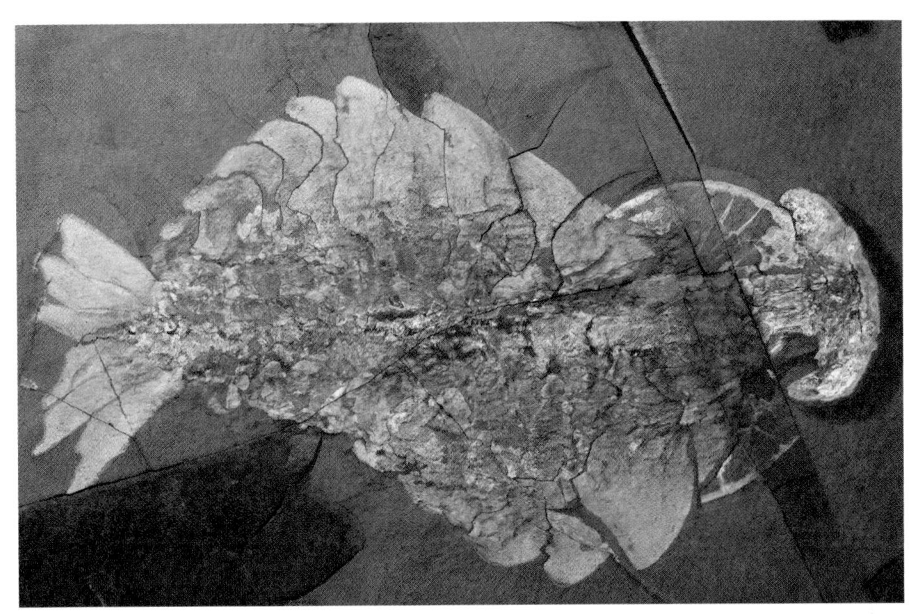

C

D

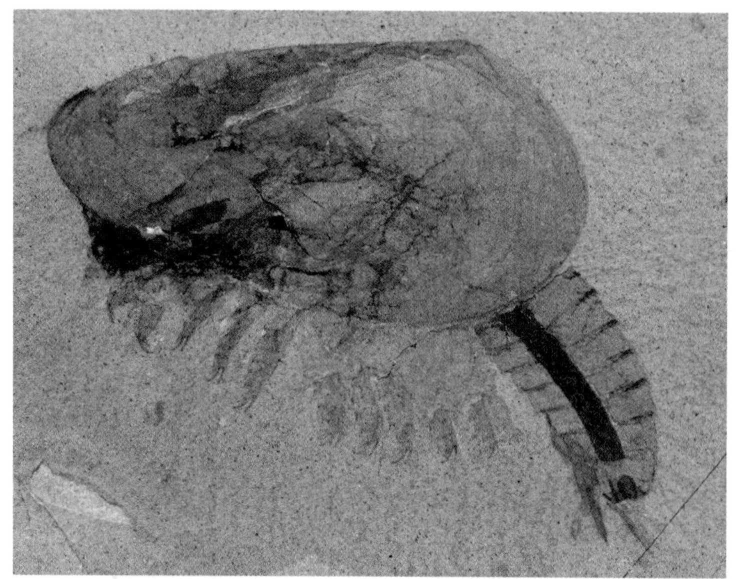

A

B

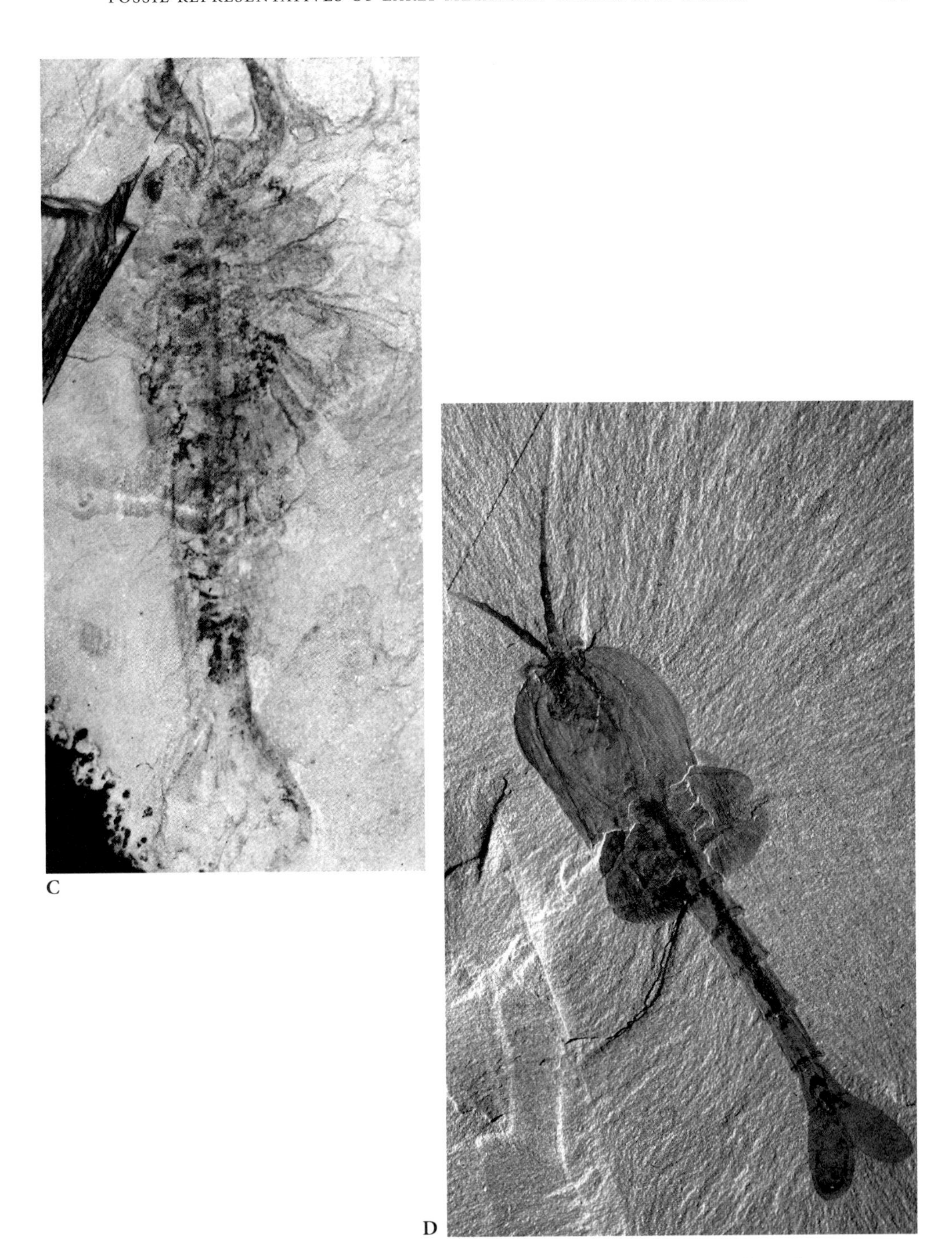

Figure 6.31 (left and above) Stem arthropods. (A) *Canadaspis* USNM 57703, from the Burgess Shale. (B) *Fuxianhuia.* (C) *Fortiforceps,* a "great appendage" form. (D) *Waptia fieldensis,* from the Burgess Shale. (A), (D) Courtesy of the Smithsonian Institution, (B), (C) from Hou et al. (2004). (B) Reconstruction by Quade Paul.

Figure 6.32 *Sanctacaris,* a stem-chelicerate from the Burgess Shale. ROM Specimen. Photograph courtesy of Derek Briggs.

arthropods and onychophorans are both derived from among the lobopodians, significant evolution of the brain and of ocular appendages occurred at some point to produce their different derived features.

Budd (2002) proposed an ingenious hypothesis to address this evolutionary problem. The "great appendages" could be interpreted as arising on the ocular segment, just as do the antennal appendages of living Onychophora. If this interpretation were correct, Budd conjectured that the appendages on the ocular segments of those Cambrian forms were innervated from the procerebrum, unlike crown arthropods. In other words, the switch from anterior innervation of anterior appendages occurred after the Megacheira (or other arthropods with similar appendage origins) had branched from the arthropod stem. Study of new fossil material has since shown that at least some of the great appendages arise just behind the eyes and so may not be on the ocular segment. Therefore, the simplest explanation for the evolution of these features is that the condition found in crown arthropods arose earlier than the branch leading to the Megacheira. We are nevertheless left with the question as to when the crown arthropod–style brain and anterior innervations arose: could it have simply been after the LCA of Onychophora and Arthropoda, within the lineage, unknown as fossils, that gave rise to the basal arthropods? Also, what were lobopod brains like? It seems doubtful that we will ever know for certain.

Before the redescription of the Burgess Shale fauna began in the mid-1960s, trilobites were the most famous and best-studied Cambrian arthropods (fig. 6.33). They are among the best-studied taxa of the Paleozoic and still far outnumber all other groups in terms of numbers of known Cambrian species and genera. The earliest trilobites define the base of Cambrian Stage 3. They

Figure 6.33 Trilobites and other arthropods from the Burgess Shale fauna. Several large specimens of *Ollenoides* are scattered across the surface, with two large specimens of *Sidneyia* in the upper and lower middle of the slab. Photograph courtesy of the Smithsonian Institution.

have relatively durable carbonate exoskeletons that are more readily preserved than the tough but unmineralized organic integuments of most Cambrian arthropods, and they are common in the many fossil faunas in which only mineralized skeletons are preserved. This obvious correlation of preservability and diversity leads to the speculation that the unmineralized groups appearing in the Burgess Shale, Chengjiang, and similar faunas may have been far more diverse than we can document. Trilobites are closely related to an unmineralized sister group, the Naraoiidae, that is known only from Chengjiang and Burgess Shale–type assemblages (Hou and Bergström 1997). Whether trilobites branched before or after the LCA of crown arthropods is uncertain.

Although crustaceans do not seem to be represented among the larger, better known fossils described from the Chengjiang and Burgess Shale assemblages, the group may be represented by fragments of mouth parts found in Stage 3 or 4 drill cores from the Northwest Territories, Canada (T. H. Harvey and Butterfield 2008). These fragments indicate a mandibular feeding apparatus that is more similar to apparatuses found in crown-group crustaceans than to known stem-group taxa. The mouth parts suggest an animal as large as 5 cm that likely fed by scraping particles from benthic surfaces, filtering them and eventually comminuting them on mandibular molars in a sophisticated feeding system. In addition, a eucrustacean larva has been identified in Stage 4 rocks from south China (X. G. Zhang et al. 2010). Thus, crown crustaceans (and therefore crown arthropods) are likely to be represented among early Cambrian fossils.

To sum up these spectacular but problematic early Cambrian ecdysozoan groups, it appears that there was a major latest Neoproterozoic and early Cambrian radiation of scalidophorans, lobopods, and arthropods. The Cambrian scalidophorans appear to be stem lineages and display more morphological diversity than their crown groups. The Cambrian lobopodians are a very disparate, paraphyletic suite of stem clades that includes ancestors to the arthropods, tardigrades, and onycophorans. The burgeoning lobopod and arthropod faunas of the Cambrian, of which nearly all clades are extinct, clearly formed major elements in the economy of Cambrian communities, and judging from their great morphological variety, their ecological effects must have been felt in most corners of the marine biosphere.

Lophotrochozoans: Spiralia

The lophotrochozoan phyla with classic spiral cleavage in quartets seem to be well established as a monophyletic group, termed the spiralia, and vary in complexity from relatively simple flatworms to the derived molluscan and annelid groups but do not achieve the morphological complexity displayed by some ecdysozoans and deuterosomes.

MOLLUSCA Although early Cambrian shelled mollusks are nearly all small and commonly fragile, they are well represented in small shelly faunas beginning in Stage 1. These SSFs are coiled shells similar to Gastropoda but that are unlikely to be crown forms (fig. 6.1L and M) and small shells assigned to Bivalvia that may also represent stem groups. In addition, there are entirely extinct groups that seem likely to be mollusks or mollusk relatives: Rostroconchia, which resemble bivalves but have single shells (6.1O); various univalved clades among what was once called the Monoplacophora; and possibly Probivalvia, with two articulated shells but not much resembling Bivalvia in gross shape (6.1P). The Hyolithida, generally considered to be a separate, extinct phylum close to the Mollusca, have conical shells with their aperture covered by a second flattish shell, the operculum. The hyolithids are particularly common among small shelly faunas where they tend to be measured in millimeters; they are also found in other types of assemblages and indeed lasted 280 million years or so, and some later ones reached lengths of 40 cm. Figure 6.34 is a sketch of lophotrochozoan phylogeny, including mollusks.

ANNELIDA Although poorly represented by body fossils in the Chengjiang fauna, several distinctive annelid body types are known from the Burgess Shale (Conway Morris 1979b). None of them are assigned to living groups, although they seem quite as complex in gross morphology as crown clades (fig. 6.35). Among the small shelly faunas are a variety of phosphatic tubes, more or less tapering and open at both ends. Such tubes have been found in living position in north Greenland assemblages, vertically embedded in the substrate with the larger end upward (Skovsted and Peel 2011). The occupant is likely to have been a suspension feeder, and Skovsted and Peel suggest that the tubes were formed by stem annelids. Also, two genera of the unsegmented group Sipuncula, which is nested within marine annelids, have been found in the Chengjiang fauna (fig. 6.36) (Huang et al. 2004). Both genera have morphological features typical of living forms, especially of the family Golfingidae, and lack any striking synapomorphies. Thus, this group has maintained an exceedingly conservative morphological mode for at least 520 million years, although our sample is still quite small and we cannot be sure that disparate sipunculan morphologies have never existed. The early establishment of this unusual annelid architecture emphasizes the morphological conservatism that many clades display over hundreds of millions of years.

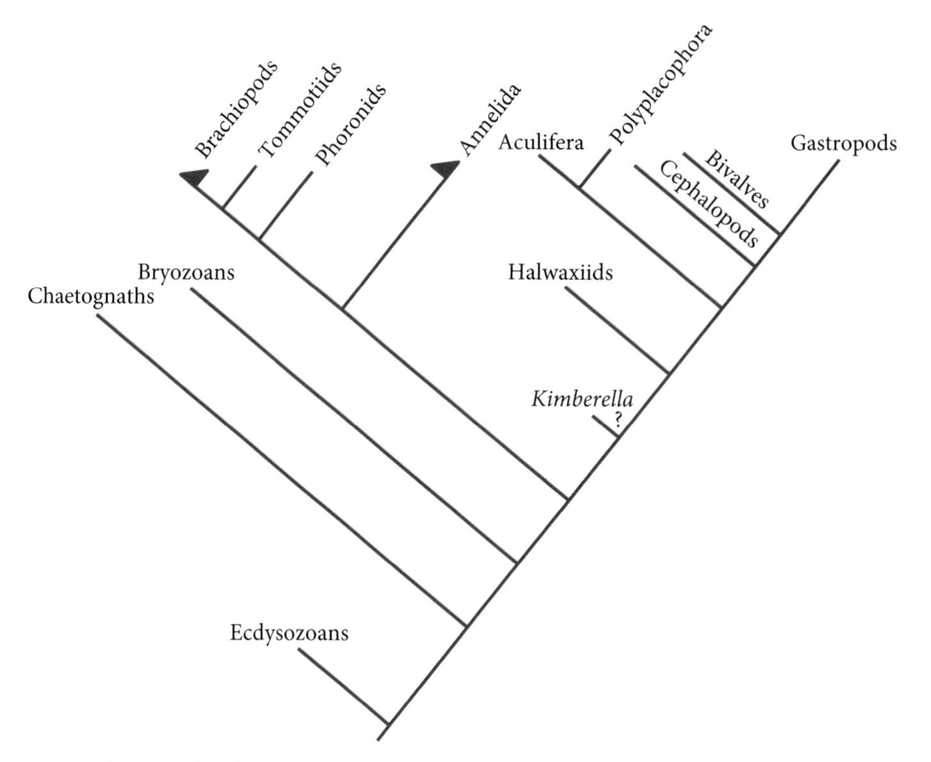

Figure 6.34 Lophophorate phylogeny, showing likely positions of *Kimberella* and the Halwaxiid clade as stem groups.

OTHER LIKELY SPIRALIANS Some enigmatic explosion taxa have unusual morphologies (by the standards of crown groups) but may also have been spiralians, for they do not appear to have molted, are quite unlike deuterostomes, and do not resemble lophophorates. One such form is the Burgess Shale taxon *Odontogriphus* (fig. 6.37), which ranges to more than 12 cm in length, and has a ventral surface that is at least partly annulated, with seriated marginal features interpreted as gills evidently placed one within each annulus (Caron et al. 2006). It is tempting to interpret the annuli as delineating the margins of coelomic compartments as in annelid-style architectures, but annulated integuments are also found in forms that lack eucoeloms, where they are usually interpreted as locomotory adaptations. Other enigmatic fossils from the Burgess Shale include *Wiwaxia,* smaller at 3 cm or so and covered by dorsal chaetae that are closely comparable to certain annelid chaeta (fig. 6.38A, B) (Butterfield 2006); and *Orthozanclus* (fig. 6.38C, D), even smaller at 10 mm

Figure 6.35 (right) Annelids from the Burgess Shale. (A)–(B) *Burgessochaeta.* (A) Photograph of USNM 57650. (B) Reconstruction. (C) *Canadia.* (A), (C) Photographs courtesy of Smithsonian Institution, (B) reconstruction by Quade Paul.

A

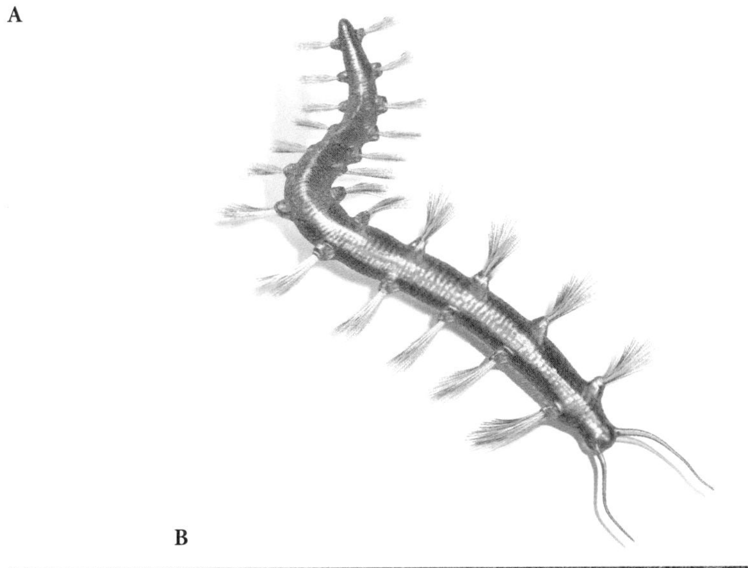

B

C

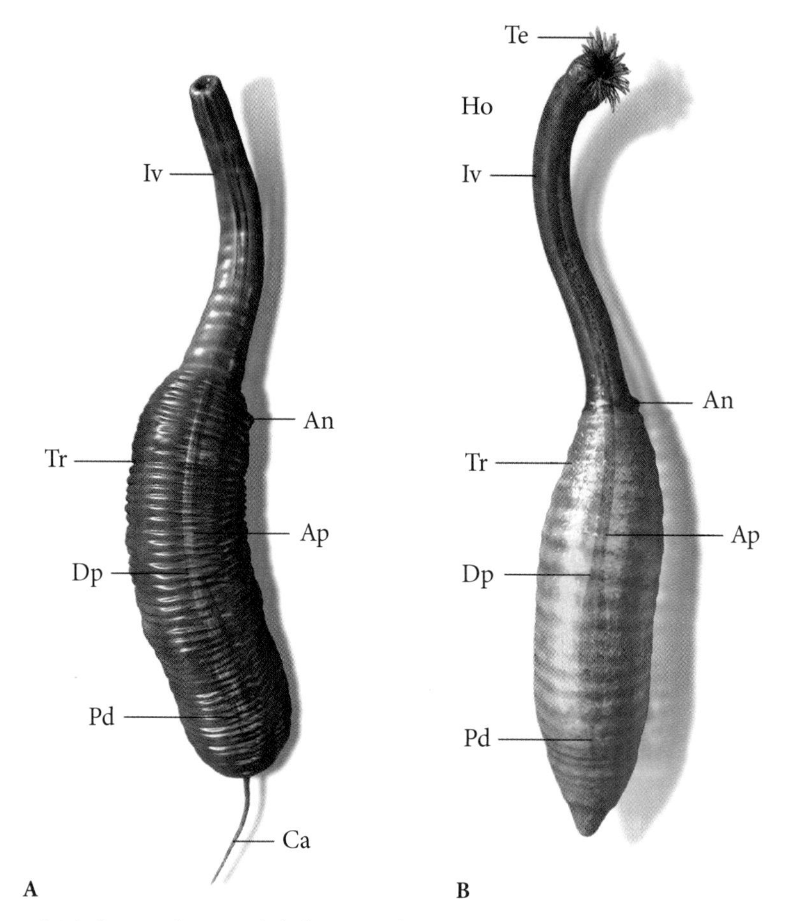

Figure 6.36 Sipunculan annelids from southwest China. (A) *Archaeogolfingia.* (B) *Cambrosipunculus.* An, anus; Ap, ascending digestive tract; Ca, caudal appendage; Dp, descending digestive tract; Ho, hook; Iv, introvert; Pd, protruding digestive tract; Te, tentacle; Tr, trunk. After Huang et al. (2004). Reconstructions by Quade Paul.

(Conway Morris and Caron 2007). Although *Odontogriphus* and *Wiwaxia* differ considerably in general appearance, they both have feeding structures composed of two or more transverse, chevron-shaped rows of teeth, possibly replacement rows, in a ventral apparatus, suggesting feeding by rasping of algal mats or sheets on the seafloor. The apparatuses are similar enough to

Figure 6.37 (right) *Odontogriphus,* a stem lophotrochozoan, from the Burgess Shale; about 6 cm. (A) Photograph. (B) Reconstruction. (A) Photograph courtesy of Smithsonian Institution, (B) reconstruction by Quade Paul.

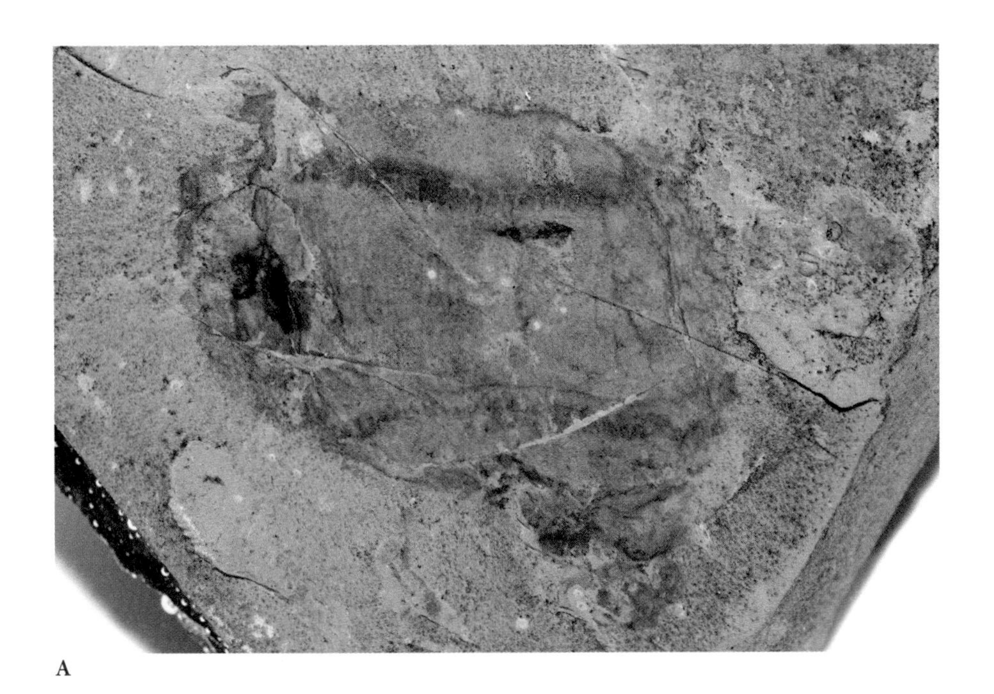

A

B

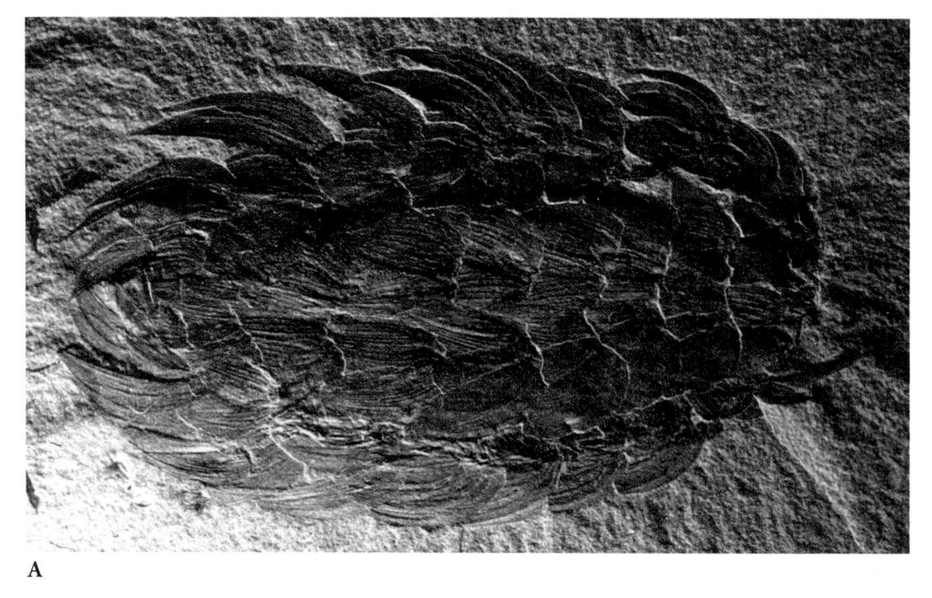

A

B

Figure 6.38 Members of the Halwaxiid clade. (A) *Wiwaxia corrugata* USNM 57635, dorsal view; about 2.5 cm. (B) Reconstruction. (C) and (D) *Orthozanclus reburrus,* a small form (to ~10 mm) with a convex dorsal cross section and with a shell-like structure anteriorly. (A) Photograph courtesy of the Smithsonian Institution, (B), (D) Recon-

C

D

struction by Quade Paul, (C) ROM 57197 © Royal Ontario Museum; photograph by Jean-Bernard Caron.

suggest a common ancestor (fig. 6.39) (see Butterfield 2006). These structures have been compared to feeding apparatuses found in most molluscan classes, which have belts of transverse tooth rows. Some crown-group annelids also have rasping feeding structures, although again they are not identical to those in these fossils. Butterfield (2006) has suggested that *Wiwaxia* may be a stem annelid and that *Odontogriphus* may be a stem mollusk or a stem that branched deeper in the lophotrochozoan tree.

Halkieria (fig. 6.40), from Stage 3 in north Greenland (Conway Morris and Peel 1990), ranges to 8 cm and has been allied to *Wiwaxia* because of its general form and its covering of flattened sclerites that are mineralized and interpreted as originally aragonitic (Porter 2008). The dorsal, shell-like structures found at both anterior and posterior ends of *Halkieria* resemble some brachiopod shells in general shape. Indeed, it has been postulated that halkieriids are a lophotrochozoan stem group from which brachiopods branched (Conway Morris and Peel 1995). Under this hypothesis, brachiopods evolved through folding and antero-posterior compression of the halkieriid body plan, although simple folding will in fact not account for basic brachiopod anatomy. Halkieriid sclerite microstructure also resembles that of the somewhat sponge-like chancelloriids (Porter 2008). The current evidence for brachiopod affinities seems very weak: the microstructural features used to link these fossils to crown groups may well be plesiomorphic. *Halkieria* has been restudied by Vinther and Nielsen (2005), who concluded that it was an early mollusk and erected a new class, the Diplacophora, to contain it. Finally, Conway Morris and Caron (2007) united *Halkieria, Wiwaxia,* and *Orthozanclus* into a group called the halwaxiids, placed within the molluscan stem. Vinther (2009) found the sclerite canals in halkieriids to be uniquely similar to canals in crown polyplacophoran (Mollusca) valves, suggesting a position as a stem-group mollusk (branching below Polyplacophora and its likely sister group Aplacophora). We are surely far from hearing the last word on the relations among these groups.

Figure 6.39 (right) The geometry of feeding structures of *Odontogriphus* (A) and *Wiwaxia* (B). The similarity of these structures indicates a close relationship, although they differ from feeding structures in mollusks (radulae) and in annelids. Notice that the second rows show more rounded denticles than the first; in mollusks, the radular rows are similar. A third row is found in some specimens of *Odontogriphus.* Both specimens with anterior to the top. (A) Backscattered image of an isolated two-row radula from ROM 57713. (B) A backscattered image from ROM 57707. (A) (B), © Royal Ontario Museum; photograph by Jean-Bernard Caron.

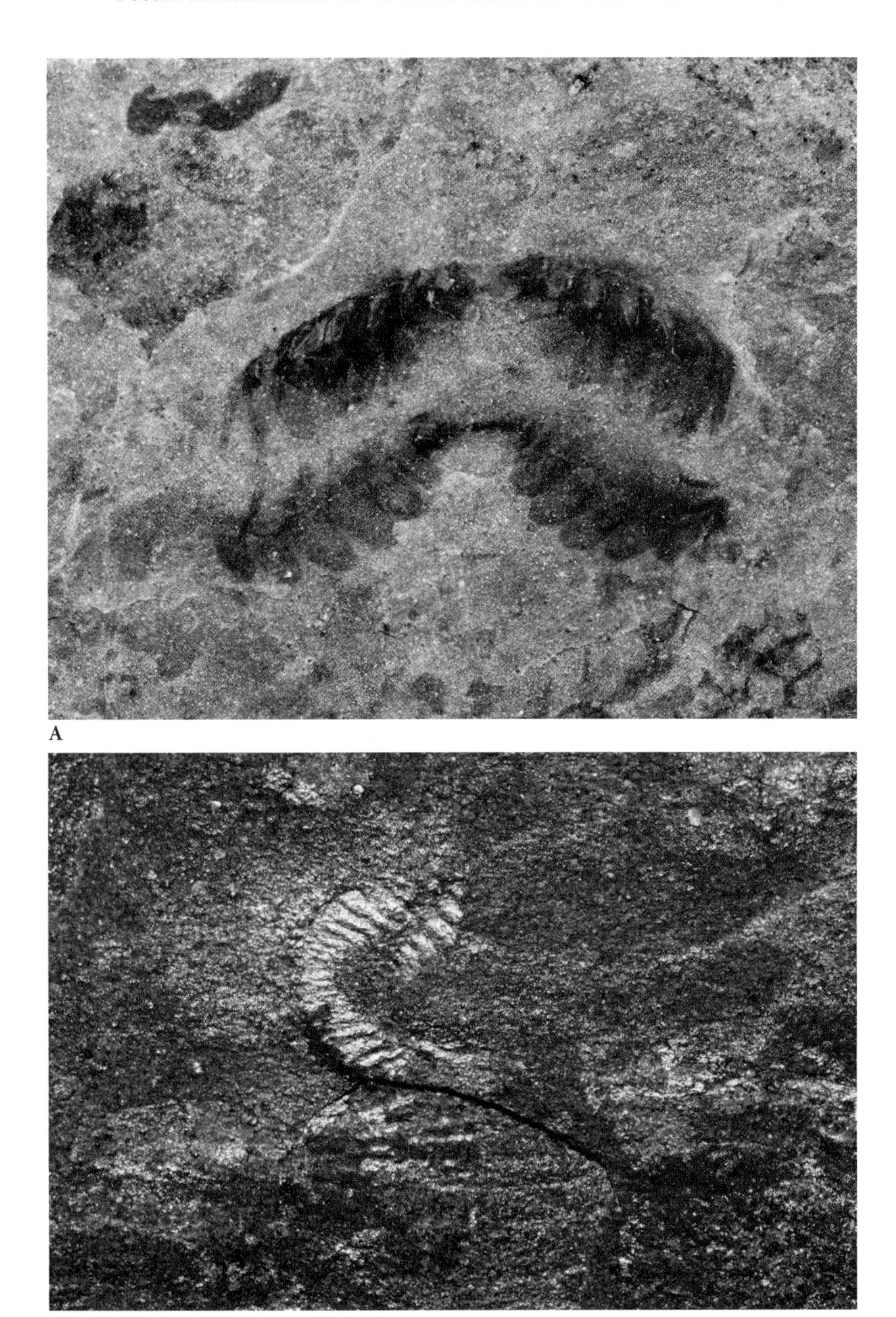
A
B

Figure 6.40 The composite scleritome of *Halkieria,* from the Sirius Passet locality in Greenland. The organism is interpreted as having a flattish sole and a converse dorsal cross section, and it reaches 8 cm in length. Notice the shell-like structures at either end. Dorsal view. Specimen MGUH 19728. Photo courtesy of Jakob Vinther.

LOPHOPHORATA All classes and fourteen (of twenty-seven) orders of brachiopods have been found in the Cambrian; ten of the orders appear in explosion faunas, and all but one of the Cambrian orders (Lingulida) are extinct. This pattern is the common one of early achievement of morphological disparity and taxonomic breadth, followed over geologic time by the loss of major clades. Much of what is now known as Stage 2 of the Cambrian was originally recognized as the Tommotian stage in Siberia and is characterized by millimeter-sized phosphatic sclerites produced by basal or marginal accretion, called tommotiids. Fully articulated scleritomes of several tommotiids have now been recovered. The tommotiid *Eccentrotheca* had a tubular scleritome with the individual sclerites fused together and was probably cemented to the substrate (Skovsted et al. 2008); it has been interpreted as being a lophophorate tube. Partial scleritomes of another tommotiid, *Paterimitra,* reveal that it is conical and was also attached to a hard substrate (Skovsted, Holmer, et al. 2009). The mineralogy of the organophosphatic shells of tommotiids is very similar to certain brachiopods (Balthasar et al. 2009). This mineralogy suggests a model in which the tommotiids are stem lophotrochozoans of various affinities, with the bivalved condition of the brachiopod skeleton originating from the progressive simplification of a benthic, tubular filter feeder with a composite scleritome; the tommotiid animal has been visualized as a stem phoronid in a tube (Skovsted, Balthasar, et al. 2009; Skovsted, Brock, et al. 2009). Tommotiid sclerites are quite diverse, so they may have achieved a considerable range of scleritome types in the early Cambrian, and the phylogenetic conjectures are quite likely to change as more specimens are evaluated.

CHAETOGNATHA The phylum of small-bodied, largely pelagic predators called Chaetognatha is likely to be represented in the early Cambrian by hook-shaped spines in functional sets that closely resemble chaetognath grasping apparatuses (Szaniawski 1982) as well as by body fossils in the Chengjiang fauna (J. Y. Chen and Huang 2002) and the Burgess Shale (Szaniawski 2005). Chaetognatha are neither spiralian nor lophophorate, but molecular evidence nevertheless suggests that they are protostomes allied to Lophotrochozoa and are possibly basal to those two groups (Matus, Copley, et al. 2006). If that topology is correct, the clade including Chaetognatha may include the LCA of living lophotrochozoans. About 20% of living species are benthic, and Casenove, Goto, and Vannier (2011) have studied the functional and morphological differences between benthic and pelagic forms. Their analysis suggests that the Cambrian species had a pelagic lifestyle. As chaetognaths have radial cleavage and are literally deuterostomous, it may well be that these features were inherited by the lophophorates, while being replaced by spiral cleavage and protostomy in the branch leading to Spiralia.

GENERAL PATTERNS

Several general patterns emerge from this discussion of the evolution of metazoan architectures during the Cambrian explosion (fig. 6.7). Members of most metazoan phylum-level clades must have been present, at least as stem groups, by the end of Cambrian Stage 3. In a few cases, crown groups had appeared at the class level (e.g., Lingulida among Brachiopoda, Bivalvia among the Mollusca, and probably Crustacea among Euarthropoda). Furthermore, numerous stem lineages are present among each of the major clades for which we have sufficiently well-preserved fossils for phylogenetic analysis. In some cases, such as with major groups within Echinodermata, these stem clades have been accorded relatively high Linnaean rank, generally as classes. In other cases, such as with Cambrian lobopods and arthropods, many of the phylogenetic relationships are still in considerable flux as new fossils are described and older discoveries are reexamined, although stem groups of roughly equivalent morphologic distinctiveness to those of the echinoderms have generally not been recognized with high Linnaean rank. Regardless of taxonomic approach, however, a remarkable degree of disparity appeared during that time.

Diversity

As partial or whole scleritomes of small shelly fossils are being recovered and reconstructed, it is evident that these fossils represent a diverse array of clades, from stem mollusks to stem lobopods, and probably stems to other groups. Although there is not yet a database cataloging the entire Cambrian diversification, a recent analysis from south China does provide a census from one region (G. X. Li et al. 2007) (table 6.1). Relatively few genera of body fossils have been identified from Cambrian Stage 1, all of which are small shellies. The number of genera of small shellies increases dramatically during Stage 2, where they still make up an overwhelming proportion of the taxa. Stage 3 includes the Chengjiang fauna and has the highest diversity (421 genera). The number of small shelly fossils has dropped from Stage 2, but they still compose 15% of the fauna (65 genera). By Stage 4, trilobites and other arthropods dominate.

Two recent global syntheses of the occurrences of biomineralized clades during the early Cambrian suggest it may be possible to discern two separate pulses of biomineralization, possibly tied to changes in seawater chemistry

Table 6.1 Generic diversity through the first four stages of the Cambrian in south China.

	Stage 1	Stage 2	Stage 3	Stage 4
Total diversity	11	169	421	66
Genera of small shelly fauna	11	164	65	11
Genera of trilobites	0	0	113	54

From G. X. Li et al. (2007). The data in G. X. Li et al. have been recompiled and updated to reflect the current Cambrian stratigraphy. The Chengjiang fauna is in Stage 3. Invalid taxa have not been included.

(Kouchinsky et al. 2012; Maloof, Porter, et al. 2010). These papers propose that first pulse occurred during Stages 1 and 2, primarily during the first 10 million years of the Cambrian. It is dominated by the appearance of total-group Lophotrochozoa (aside from the phylogenetically ambiguous chancelloriids) and includes mollusks as well as halwaxiids, hyoliths, tommotiids, and a variety of other components of the SSF. Protoconodonts, which are likely the spines of chaetognaths used in predation, are among the earliest hard skeletal parts recovered. All these forms are found near the negative carbon isotope excursion associated with the Ediacaran-Cambrian boundary. Brachiopods and related groups first appeared during Stage 2, as did the better skeletonized sponges and corallomorphs.

Although ecdysozoans and deuterostomes occurred earlier in the Cambrian (and indeed date to the Neoproterozoic), the earliest biomineralized representatives of these two clades appeared during Stage 3. These forms include trilobites and other arthropods, echinoderms, and nonmineralized chordates. This pattern is consistent with changing skeletal mineralogy driven by fluctuations in the magnesium/calcite ratio (Porter 2004; Zhuravlev and Wood 2008). The earliest biomineralized forms were aragonitic. This second phase of biomineralization occurs after a $-10‰\ \partial^{13}C$ isotopic excursion in Stage 2 and involves the first appearance of high-magnesium skeletons and a decrease in the Mg/Ca ratio. As emphasized by Maloof, Porter, et al. (2010), caution is necessary in interpreting these results. These phases may simply reflect the considerable differences in the dominant style of preservation between the two intervals: phosphatization of shells during Stages 1 and 2 followed by the exquisite soft-part preservation of many different groups in the Chengjiang fauna and similar lagerstätten of Stage 3. Neither of these studies included the sort of rigorous analysis of origination patterns that would place statistical limits on the reliability of these phases of origination.

Disparity

The most remarkable pattern to emerge from any analysis of early Cambrian metazoan diversification is the extraordinary breadth of morphologic innovation. It is evident at many different scales, from the obvious generation of morphologically distinctive groups to diversity in anatomical details. For example, one might expect that complexity and sophistication of eyes improved through the Phanerozoic, but the recent discovery of exquisitely preserved eyes from arthropods in the early Cambrian Emu Bay Shale in Australia illustrates that highly advanced, compound eyes with more than 3000 ommatidial lenses had evolved very early in the history of the clade (M. S. Lee et al. 2011). Surprisingly, many of the recovered eyes preserve a "bright zone" within the ommatidia that has higher light sensitivity and, perhaps, acuity. Such sophisticated eyes demonstrate that vision was an important adaptation in some groups.

Many morphologically distinctive groups (high disparity) in the Cambrian had relatively few species (low diversity) compared with similar groups today. This pattern is robust at the level of the fundamental body plans that originated during the early Cambrian as well as in the levels of morphologic disparities within most of those body plans. As seen in group after group, the earliest recorded phase of morphologic evolution defined the architectural themes of a body plan, whereas subsequent, postexplosion evolution was largely a process of exploiting those themes. In other words, the early species of most new major clades were widely distributed across the eventual morphologic range of the group—its morphospace boundaries—and taxa evolving later tended largely to fill in this space rather than extending it by a significant amount. Further, when the early morphospace is relatively narrow, subsequent morphology tends to be narrowly bounded as well.

Although this pattern had been known for many years—it was originally revealed by using the higher Linnaean categories of phylum, class, and order as metrics of morphological novelty (e.g., Erwin, Valentine, and Sepkoski 1987; Valentine 1980)—it received greater attention with the publication of Gould's 1989 book *Wonderful Life* (Gould 1989). Gould wrote at a high point of interest in extinct higher Linnaean ranks. Since then, the spread of phylogenetic methods has resulted in greater emphasis on the topological relationships of stem lineages within well-constructed phylogenies. The use of Linnaean ranks as proxies for disparity paved the way for quantitative morphometric studies within new cladistic frameworks. Some Linnaean taxa proved to be polyphyletic, however, and rankings certainly have a subjective component, especially when involving groups with distinct morphologies. For example, how does an order within Crustacea compare to an order within Bivalvia (Forey et al. 2004; A. B. Smith and Patterson 1988)? On the other hand, some groups initially identified as clades by cladistic methods have turned out to be polyphyletic as well, and patterns revealed in the rank-based studies have proven to be generally accurate when more quantitative approaches were introduced. As Foote (1997) pointed out, the issue is whether Linnaean categories are scientifically useful as proxies. A carefully defined Linnaean taxon indicates the rise(s), decline(s), or both of a particular collection of synapomorphies within a body plan as well as its successes and failures, which in itself has its uses.

In any event, Gould's book helped spur the development of quantitative estimates of morphologic disparity, which have largely replaced simple counts of Linnaean taxa as disparity metrics (Erwin 2007; Foote 1997). Today, a variety of quantitative approaches for the measurement of disparity are available, relying on both measurements of discrete characters (as used in phylogenetics) and on continuously variable features, such as shape, using geometric morphometrics. Evaluation of the resulting patterns of change in disparity within a phylogenetic framework also allows investigators to examine whether similar morphologies are phylogenetically related or are only similar due to convergence. These approaches are essential to testing hypotheses of evolutionary transformation. One of the challenges of disparity studies, however, is that almost all the quantitative techniques are best applied within a common body plan, where homologous characters are present. It is more challenging to examine the patterns of disparity among the Ecdysozoa as a whole, for example.

Foote (1993, 1996) identified three potential relationships between taxonomic diversity and morphologic disparity (fig. 6.41): (1) constrained morphologic increases with taxonomic diversification unrelated to disparity; (2) rapid increase in disparity, with large early steps in disparity and smaller ones later during a diversification; and (3) concordant increases in disparity and diversity (which is the pattern that Gould argued was the null hypothesis of most evolutionary biologists; a bit of a straw man). The first possibility suggests that further increases in the disparity of a group were constrained either by development or ecology. In the second case, the morphologic distance between sister taxa was larger early in the history of a clade than it was later. There are two alternative explanations for such a pattern: either the size of the morphologic transitions between taxa declined through time (perhaps as a result of increasing developmental limitations on morphologic innovation) or the success of morphologic transitions declined (likely for ecological reasons). In the third case, taxonomic diversity is a good proxy for morphologic disparity, and there is no change in the distance of morphologic transitions

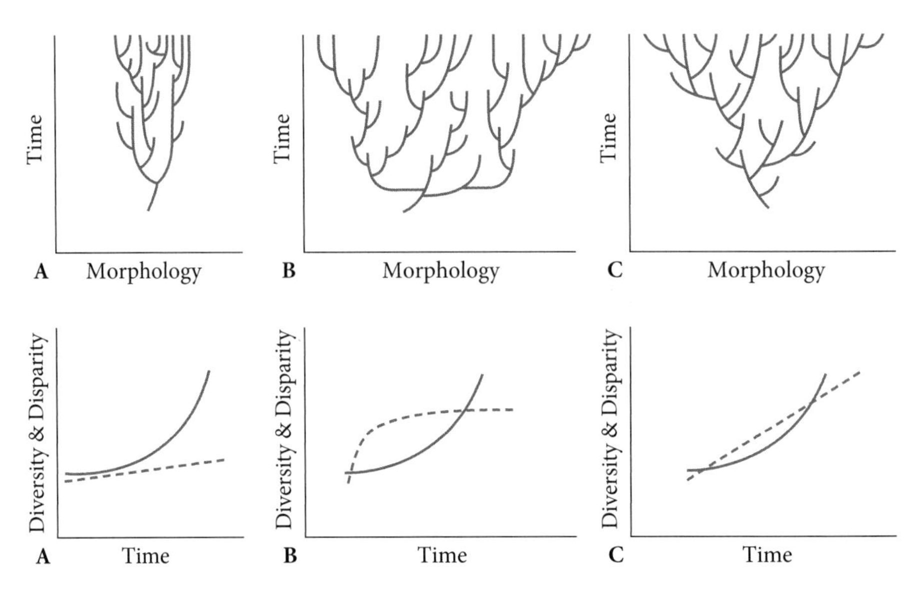

Figure 6.41 *Top:* Idealized diversity patterns (modified from Foote 1993). *Bottom:* Predicted diversity and disparity curves over time (solid line, taxonomic diversity; dashed line, morphological disparity). (A) Constrained morphological evolution with taxonomic exceeding morphological diversification. (B) Morphological diversification outstrips taxonomic diversification with large morphological steps early in the clade's history; morphological diversity may continue to increase, but steps are smaller. (C) Morphological diversification is concordant with taxonomic diversification; there is no constraint on morphological evolution nor a trend in morphological step size (redrafted from Wesley-Hunt 2005).

over time. As we discussed for the arthropods, a single sample of Cambrian arthropod disparity from the Burgess Shale occupied a larger volume of morphospace than a representative sample of living arthropods (Briggs, Fortey, and Wills 1992). This pioneering study led to a long series of discussions (reviewed in Erwin 2007). There have been numerous quantitative studies of morphologic disparity since the early 1990s, and many (but not all) have identified a pattern of great, even maximal disparity early in clade histories. Our qualitative reading of the diversification patterns for the Cambrian clades discussed in this chapter suggests that these explosion groups generally followed this same disparity pattern.

It must have become obvious as we discussed the parade of body plans and their major sub-plans that appeared during the Cambrian explosion, that establishing the phylogenetic position of these forms, while a fundamental task, is a difficult and contentious one. We have tended to use a Linnaean framework precisely because great early disparity was the common mode shown during the explosion. These Cambrian disparate clades had Ediacaran ancestors, as demonstrated by the molecular clock studies. But the molecular clock analyses also indicate that many of the crown groups first appeared during the Cambrian explosion. In chapter 8 we discuss how developmental evolution underpins morphological disparities, and in chapter 9 we speculate on the possible nature of the ancestral lineages. Here, though we want to emphasize the challenges that the evolutionary dynamics of the Cambrian and the resulting fossil record pose for phylogenetic methods. The Ediacaran silence on morphological ancestors means that characters are often subject to multiple interpretations, and phylogenetic analyses can be plagued by a variety of methodological problems, making it difficult to rigorously establish a nested series of morphological changes that define phylogenetic divergences. While this situation presents challenges, specialists in the various Cambrian groups have done heroic work in evaluating their relationships on the spotty evidence, and comparative studies in evolutionary developmental biology are showing where regulatory synapomorphies may be found. These problems are not intractable, but are hard.

PART THREE

Biological Processes

CHAPTER SEVEN

The Origin and Evolution of Metazoan Ecosystems

Central to our view of Ediacaran and Cambrian events is the involvement of organisms in constructing their environment through a positive feedback process in which new evolutionary innovations created opportunities for additional species. As early as the Cryogenian, sponges may have facilitated the ventilation of the oceans by filtering the water column, contributing to the generation of a more hospitable environment for metazoan life. New innovations in the latest Ediacaran and earliest Cambrian also propelled the generation of diversity as bioturbation of the sediment and the expansion and stabilization of metazoan components within zooplankton communities helped reengineer benthic and pelagic marine environments. Such ecological innovations fundamentally altered the complexity of marine ecosystems by generating new ecological opportunities. Novel opportunities were exploited by representatives of many different clades. There is a striking contrast in the paleoecology between the Ediacara macrofossil assemblages—with organisms of uncertain phylogenetic affinities, peculiar feeding strategies, and lacking important bioturbators or predators—and the essentially modern aspect of early Cambrian faunas that feature a full panoply of Phanerozoic clades and ecological relationships. The paleoecological challenge for the Ediacaran-Cambrian transition is to understand the pattern of ecological events, in the sense of how the animals were functioning and especially in their patterns of interaction, and the mechanisms that generated this remarkable explosion in biodiversity.

Animal-dominated ecosystems were built on a microbial foundation. Microbially dominated ecosystems appeared billions of years before the Ediacaran organisms. Their remains are well known to paleontologists as stromatolites: layered sedimentary structures constructed as microbes that trapped and bound sediment. Stromatolitic reefs occurred at least as early as 2 billion years ago. By 1 billion years ago, many clades of eukaryotic algae had diversified, and although their fossil record is not abundant, they must have been important members of Neoproterozoic shallow-water communities. Furthermore, phylogenies based on molecular data establish that many lineages of nonphotosynthesizing eukaryotes also diversified during the Proterozoic, although their fossil record is sparse. Primary producers, detritivores, predators, commensals, parasites—the whole array of activities involving energy flow within communities—are nonetheless found among the living microbial descendants of Neoproterozoic clades. The molecular clock evidence discussed in chapter 4 suggests that sponges and cnidarians were present as well, but even though the evolution of sponges added to the roster of suspension-feeding organisms, it probably did little to alter the trophic (feeding) network structures in those early communities.

At the dawn of the Ediacaran (635 Ma), the deep oceans were anoxic, sulfur rich, and perhaps iron rich in different regions. Microbially bound substrates dominated the shallow continental shelves. Because sponges inhabited the seas and their tissues need large amounts of collagen, the synthesis of which requires oxygen, the wave activity and photosynthesis along the shelves must have provided sufficient oxygen in marine waters, at least locally, from atmospheric oxygen. Microbially dominated communities remained important through the Ediacaran, when the addition of the rangeomorphs, dickinsoniomorphs, and various other elements of the Ediacaran fauna formed a predominantly benthic ecosystem. The earliest Ediacara macroorganisms—the rangeomorph soft-bodied biota—appeared by 579 Ma. The substantial vertical tiering of these Ediacarans above the sea bottom in fairly deep water is similar to that of Phanerozoic groups that feed on material suspended in the water column. It seems likely that they relied on dissolved organic material as food, however, because other rangeomorphs, like the spindle-shaped forms (*Fractofusus*) in Newfoundland, shared the same basic morphology but lay on the seafloor.

Later Ediacaran macroorganisms exhibit a greater variety of forms, and inferred ecological habits. A major ecological innovation is recorded by the mobile grazing habits of *Kimberella* on microbial mats, and the occasional movements of *Dickinsonia* and its allies also suggest heightened mobilities among the benthos. The impressions left by *Dickinsonia* suggest that it fed on microbial mats by absorbtion, moving when they had depleted the resources locally (Dzik 2003; Gehling et al. 2005; Sperling and Vinther 2010). We have no evidence for predation, and probably the late Ediacaran metazoan food web was composed chiefly of suspension-feeding sponges and osmotrophs. The expansion of trophic complexity from phytoplankton to mesozooplankton and eventually to larger animals must have occurred during the diversification of bilaterians, and mesozooplanktonic niches may have been occupied by radiates, chaetognaths, and chordates by the latest Proterozoic or earliest Cambrian time.

The complexity of marine ecosystems grew spectacularly during the early Cambrian with the appearance or expansion of predators; the evolution of shells, carapaces, and other types of skeletons that provided protection from predation; and the appearance of behaviors such as active burrowing. The advent of burrowing in the very latest Ediacaran or earliest Cambrian led to a seafloor "agronomic revolution," heralding the disappearance of the firm, microbially stabilized sediments of the Neoproterozoic and the increasing aeration and disturbance of sediments by the burrowers (Bottjer, Hagadorn, and Dornbos 2000; Seilacher and Pfluger 1994). As vigorous bioturbation raised oxygen levels in progressively deeper sediments, benthic microbial abundance would have increased substantially there. The record of acritarchs and other microfossils reveals a concurrent diversification of pelagic organisms as well. Progressive stabilization of the carbon cycle through the early Cambrian probably signals the spread of these innovations (see chap. 3) (Maloof et al. 2005).

Animals began to invade microbial reefs near the close of the Ediacaran period. Thrombolitic and stromatolitic reef complexes in southern Namibia contain the "mat stickers" *Cloudina* and *Namacalathus* as well as the oldest putative tabulate corals. These organisms were not actively involved in reef construction, but took advantage of microbially constructed structures. With the advent and rapid expansion of archaeocyathids in the first two stages of the Cambrian, most reefs came to be dominated by these mineralized sponge-grade organisms, where they were joined by a variety of calcified microbes as well as some tabulate-like corals. These reefs exhibited considerable ecological differentiation, as in modern reefs, with many different ecological guilds present, and they were also zoned, from high-energy fore-reef and reef-crest environments to back-reef lagoons. Vertical tiering of sponges and cnidarians evolved within Cambrian epifaunal marine communities. At the beginning of Cambrian Stage 3, there were many genera of archaeocyathids, but the group suffered major crises during Stages 3 and 4 before disappearing completely in the mid-Cambrian (Rowland and Shapiro 2002; Savarese 1993; R. Wood 1999).

The increased oxygen levels, bioturbation, and linkage of pelagic and benthic ecosystems were accompanied by a rapid growth of complexity in ecological networks. Indeed, the early Cambrian marine ecosystems appear to have been as complex as many modern marine ecosystems, at least in terms of their trophic relationships. The Chengjiang biota provides excellent documentation of how far the ecological diversification had progressed by Cambrian Stage 3: at least two dozen different modes of life are present, including a variety of infaunal burrowers as well as suspension, microparticle, and larger particle feeders (J. Y. Chen, Waloszek, et al. 2007), pelagic forms, and predators. Arthropod predators are heavily represented among the Chengjiang and Burgess Shale faunas. Trilobites were a dominant component of Cambrian faunas and included detritivores as well as predators (Babcock 2003; Zhu et al. 2004). Thanks to Burgess Shale–type preservation, even soft-bodied predators such as chaetognaths (among the pelagic forms) and priapulids (among the burrowers) are known as fossils. By the mid-Cambrian, extensive ecological diversification had occurred among many clades. For example, eight species of extraordinarily well-preserved echinoderms in deposits in northern Spain include two tiers of attached suspension feeders, several free-living forms, and others that were attached to skeletal debris (Zamora 2010). Overall, although the number of species within specific modes of life was far below the numbers reached later in the Paleozoic, the basic structure of Phanerozoic ecosystems had been achieved within at most 10 million years of the onset of the major bilaterian diversification.

The ecological dimensions of the Cambrian explosion are best viewed in light of general models of the processes involved in the construction of modern ecosystems. Although ecological dynamics were likely to have been different in the Ediacaran and the early Cambrian phases of diversification, both phases involved the growth of ecological networks; of particular significance were adaptations that had widespread network effects (positive spillover effects; see below). The more obvious environmental differences, discussed previously, include increased oxygen levels, the development of a mesozooplankton that linked pelagic and benthic marine ecosystems, and increased bioturbation, all arising from biological activities. There is still much to learn about such evolving systems, not least because we lack well-accepted, general models for the construction of modern ecosystems, much less those of the deep past.

MODELS OF ECOLOGICAL DIVERSIFICATION

The simplest population growth model is an exponential one in which the rate of growth is proportional to the size of the population, whether of individual organisms or of taxa (fig. 7.1A). Such a pattern is often called Malthusian growth, after Thomas Robert Malthus, who described the growth of the human population in *An Essay on the Principle of Population* in 1798 and who

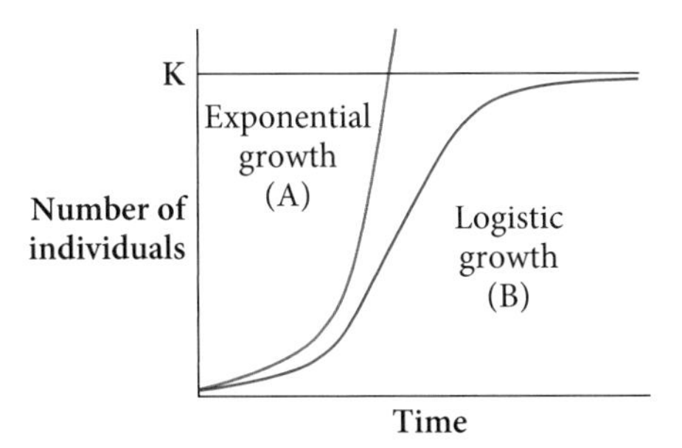

Figure 7.1 (A) Exponential and (B) logistic growth, showing the maximum carrying capacity (K) for the logistic growth model.

helped inspire both Charles Darwin and Alfred Russel Wallace to propose hypotheses of natural selection. As an expanding population approaches size limits set by the available resources, negative feedbacks increasingly limit its growth rate, and a logistic or S-shaped growth pattern is produced. When the population reaches the maximum size permitted by limiting resources, it has reached the carrying capacity of the environment and growth must stop (fig. 7.1B). Logistic growth has been invoked to explain species richness (diversity patterns) within ecological communities, assuming that diversity is locally controlled by resource limitation. In such models, local species richness controls regional diversity, permitting the extension of this approach to biogeographic regions and then to the entire global biosphere under the assumption that diversity at each of these levels is resource limited and additive. Simple richness models, however, do not address the division of resources among species: can they be monopolized by a few species with relatively large resource requirements, or are they parceled out in smaller increments among many species? In today's oceans, both situations occur, but under different environmental circumstances. Furthermore, when simple models address higher-level ecological associations such as communities or provinces, they usually do not account for many species occurring in more than one association. The number of species in a given region is not the sum of the numbers of species within each community and so on up the hierarchy of ecological associations; attempts at simple addition will invoke errors of autocorrelation. The varied ecological ranges and geographic dispersal abilities ensure that some species will spread widely and be present in many associations. Indeed, one of the important factors in the diversity of a local association is the size of the regional pool of species available to colonize that locality. If the pool is large, local species richness may rise toward a carrying capacity, or even exceed it, through consistent immigration. On the other hand, if there are barriers to consistent immigration and there is an excess of species capacity, richness may be raised by local speciation events.

Whether any given community is saturated with species is difficult to determine in the short time available for experiments and observation of modern systems, for environmental conditions fluctuate, as do species richnesses. The introduction of new species can sometimes add to community diversity, but in other cases, it may reduce it. Studies of the success of invasive species suggest that most ecological communities today do not appear to be saturated with species in the short term (Cornell 1999; Loreau and Mouguet 2000; Ricklefs and Jenkins 2011; Sax et al. 2007; but see Shurin and Srivastava 2005). Indeed, Loreau (2000) notes that there are no theoretical reasons for arguing that species interactions will limit diversity at local, regional, or larger scales. Moreover, the carrying capacities of ecosystems vary across space today (reviewed in Valentine et al. 2008) and have clearly varied over time. The evolution of the diverse bilaterian macrofauna of the Cambrian explosion must have involved an unequaled period of evolutionary creativity while filling an environment with an untapped or, much more likely, an expanding carrying capacity. The expansion appears to have been in great part thanks to feedbacks involving activities of the evolving biota itself.

Episodes of evolutionary innovation such as the Cambrian radiation can be described as involving the creation, invasion, and occupation of new adaptive zones. George Gaylord Simpson first introduced the concept of adaptive zones in his classic *Tempo and Mode in Evolution* (G. G. Simpson 1944). A Simpsonian adaptive zone is a subdivision of the environment, including both its physical and biological attributes, viewed as a set of ecological opportunities that may be exploited by appropriately adapted species. Adaptive zones are hierarchical—that is, the entire ocean is such a zone—but it can be subdivided into the seafloor and the water column, and the seafloor may be further subdivided into rocky, muddy, or sandy bottoms, and so forth. In turn, each environment may be further subdivided by organisms representing different functional categories, such as suspension feeders, detritivores, or carnivores.

The adaptive spectrum available to a population or a species is usually termed a niche. A theoretical model of a multidimensional niche, each dimension of which represents a physical or biological parameter, was first described by Hutchinson (1957, 1967). In a multidimensional space that includes all possible physical and biological parameters, the niche of any species may be represented by the multidimensional volume it occupies. Because adaptive zones can also be represented within such a conceptual adaptive space, the Hutchinsonian niche and the Simpsonian adaptive zone were regarded as coexisting in a common, multidimensional region called ecospace (Valentine

1973). This concept has found most usage among paleobiologists; because realized ecospace varies as environments and biotas change and evolve, the trajectories of physical parameters or of taxa could be traced in ecospace through geologic time. At any particular time, ecospace is divided into niches that reflect the mosaic of ecological parameters, and boundaries between mosaic pieces may be weak or strong; Simpson's adaptive zones were surrounded by barriers that were difficult for evolution to broach. When traced though time, the adaptive mosaic becomes an adaptive kaleidoscope as conditions change (Valentine 1980). For example, the environmental changes associated with oxygenation of late Neoproterozoic oceans may have altered the physical conditions of a large array of marine adaptive zones to permit the rise or invasion of a variety of evolving metazoan lineages.

When marine environmental zones were largely unoccupied by metazoans, they offered unexploited resources to clades able to acquire the necessary adaptations to access them. The possible roles of adaptive zones in producing novel morphologies have been explored by simple modeling (Valentine 1980; Valentine and Walker 1986, 1987). This model involves a large zonal ecospace subdivided into species niches. The simulation begins with only one niche occupied. Evolution within the simulation proceeds in discrete steps: at each step, a new species arises from each species present, jumping in a randomly chosen direction to land on another niche space. If the landing space is occupied by another species, the new species is not viable and disappears. The length of the jump is chosen from a probability distribution in which small jumps are common but increasingly longer jumps are increasingly rare, both because the farther the jump, the more different the new niche, and because it is deemed more difficult to adapt to larger changes than to smaller ones. This expectation leads to one more rule: the longer the jump, the larger the unoccupied area required for a successful landing. In establishing a very different adaptive type, longer-jump lineages are required to colonize larger vacant adaptive regions because their adaptations would be far from perfect in their new niches and they need to avoid too much competition or other sources of interference. Because the number of large vacant spaces declines as diversification proceeds, there is a decline in the likelihood of success of longer jumps. Correspondingly, an increasing number of spaces are filled by small jumps from adjacent spaces. The results of this model are consistent with the pattern of the appearance of taxa in higher categories observed for the Cambrian explosion and much of the subsequent Phanerozoic fossil record: an early appearance of major adaptive types, exemplified by the abrupt appearance of metazoan phyla, with most appearances of classes coming in the Ordovician and of orders later still. In other words, the record of the declining appearance of the most disparate metazoan novelties gives the impression that adaptive space was filling up. In the models, the larger vacant adaptive zones, required for the success of longer jumps, have already become occupied, so the success of longer jumps progressively declines.

This model has, of course, been tuned to produce the sort of pattern that we find to be common in the fossil record, including the pattern of early Cambrian diversifications. Used heuristically, the model suggests that ecospace availability is an important feature in the history of clades, and it is commonly the case that early in a clade's history it establishes a presence in what becomes a Simpsonian adaptive zone. The clade proceeds to radiate first into subzones and then into subdivisions of those in a series of lineage divergences, but diversification is damped because the ecospace most accessible to the clade is filled. Further changes may result from environmental shifts within the adaptive kaleidoscope, producing new conditions to which evolution must respond. Environmental change may also cause extinctions, which can free resources to permit rediversification in the newly opened ecospace.

Perhaps the most difficult part of this model as constructed is the assumption that lineages are free to make large jumps within ecospace. Simpson was faced with this problem in formulating the notion of an adaptive zone, and he hypothesized that lineages invading a new adaptive zone were preadapted. Ecological preadaptations are those that, although evolved for life in one set of circumstances, happen to provide access to a different region of ecospace and may subsequently become the starting point for diversification within a new adaptive zone. Once in the new ecospace, evolutionary enhancement of preadaptation to the new conditions would proceed apace. For a familiar example, consider the evolution of lightweight, small-bodied dinosaurs with hollow bones, equipped with feathers. Those lineages were not evolved to fly, but they had adaptations—Simpsonian preadaptations—that permitted them evolutionary access to flight; consequently, those lineages could invade and radiate into a vast and novel adaptive zone, now teeming with birds.

If the small steps that lead to the successful invasion of novel ecospace occur rapidly, it can give the appearance of an evolutionary jump in a spotty fossil record, a situation that would be most common early in metazoan history when most animals were soft-bodied and before skeletons were well developed. New conditions faced by invading lineages would favor selection for mostly small changes, which could rapidly lead along an adaptive pathway to a new body type. As we will see in chapter 8, there is strong evidence that elements within the metazoan genome have commonly served as genetic preadaptations.

From a theoretical perspective, ecospace shifts with changing environments and new ecological adaptations. This view provides an opportunity for a new round of models in which the simulation of adaptive spaces, and indeed their number, changes during the simulation. There is a long-standing debate within ecology over how niches (or their extension into adaptive zones) should be defined: do they exist independently of the species that occupy them? Convergence between organisms of very different clades is frequently a good clue that such niches or zones exist, as exemplified by the morphologic similarities between the reptilian ichthyosaurs, the piscean tuna, and the mammalian porpoises, all of which fill the role of a fast, open-ocean predator. Other niches depend in large part on interactions among species.

Ecological networks clearly expanded considerably in both size and complexity between the Ediacaran and the early Cambrian. One way to characterize this increase is by defining the modes of life occupied by marine organisms and tracing their numbers through time. A particularly useful classification of life modes has been developed by Bush, Bambach, and Daley (2007). In this scheme, marine ecospace is divided into six different tiering levels relative to the seafloor, into six different levels of motility, and into six different feeding mechanisms. These categories may be arrayed in a three-dimensional grid, a theoretical ecospace with 216 possible combinations that could be defined as distinct modes of life. Of these potential modes, only ninety-two appear to have been occupied by Phanerozoic animals; of the unoccupied modes, ninety-eight are probably not viable ways of making a living, whereas the other twenty-six seem to be possible life modes but are not known to have ever been exploited.

Plotting the inferred life modes for fossil taxa within this grid for successive intervals of time shows the growth in ecological structure in metazoan communities (fig. 7.2) (Bambach, Bush, and Erwin 2007; Bush, Bambach, and Erwin 2011). Two modes appear to have been present in the Avalon assemblage of rangeomorphs, increasing to ten in the White Sea–Ediacaran assemblage with the appearance of surficial and erect nonmotile suspension feeders, possible osmotrophic feeders, and the mat-grazing *Kimberella.* Five modes are represented among the Nama assemblage, including the unusual *Pteridinium* and the earliest skeletal forms. In all, a total of twelve different modes are represented in these three assemblages. By contrast, at least thirty different modes were occupied during the first half of the Cambrian, a considerable increase over the Ediacaran but still less than a third of the modern total. Of these thirty modes, nineteen are represented by fossils with durable hard parts, while the remainder were life modes inferred for soft-bodied fossils from the

Chengjiang and Burgess Shale deposits. This analysis highlights the differences between Ediacaran and Cambrian organisms based on extensive ecological innovations that are first found in the explosion faunas. Among the common ecological types in the explosion faunas that are missing from the Ediacaran fossil faunas are predators, rapidly motile animals, deep burrowers, and pelagic forms (Bush, Bambach, and Erwin 2011).

Uniting all the members of communities into common networks or webs of interaction are the network of energy flow among populations and its associated pyramid of biomass. In such a network, nodes are composed of different species, and the connections between the nodes are the ecological interactions involving energy transfer, such as predation, parasitism, herbivory, and so forth. Ecologists have examined the operations of food webs in considerable detail, both empirically and from theoretical perspectives. The objective of food web studies is to develop a comparative analysis of network structures to understand the mechanisms by which such webs are built and operate. For example, in living communities, ecologists describe the patterns of connections between species, the distribution of links among predators and prey, and the mean trophic level of species in the web, but the Ediacaran and the Cambrian present a special and crucial challenge involving the construction of the earliest animal food webs. Animals did not simply replace various microbes within existing food webs; rather, they built new webs that overlay, expanded, and interacted with the microbial ones. Studies of recent marine ecosystems have shown strong synergistic effects of within-trophic-group species diversity on abundance or biomass. For example, positive interactions among species seem able to produce higher species diversity than would otherwise be possible (Cardinale et al. 2006; Worm et al. 2006; Emerson and Kolm 2005), but this result remains controversial. Such features are not easily captured by food web studies unless the connections between nodes can be shown to differ in strength as a result of differences in degrees of interactions between species. We are unlikely to recover the details required to examine such effects in Edia-

Figure 7.2 (right) A comparison of the occupation of ecological space by modes of life between (A) the Ediacaran Avalon assemblage; (B) the entire Ediacaran assemblage; and (C) the early to mid-Cambrian entire biota, including the Chengjiang and Burgess Shale faunas; and for (D) the early to mid-Cambrian skeletal biota. The purple boxes are modes of life used by the designated fauna. Green boxes are modes of life not documented in the Ediacaran-Cambrian but used in the Recent by the taxa with readily preserved hard parts. Blue boxes are modes of life not documented in the Ediacaran-Cambrian but used in the Recent by the taxa with a diverse fossil record. From Bambach, Bush, and Erwin (2007).

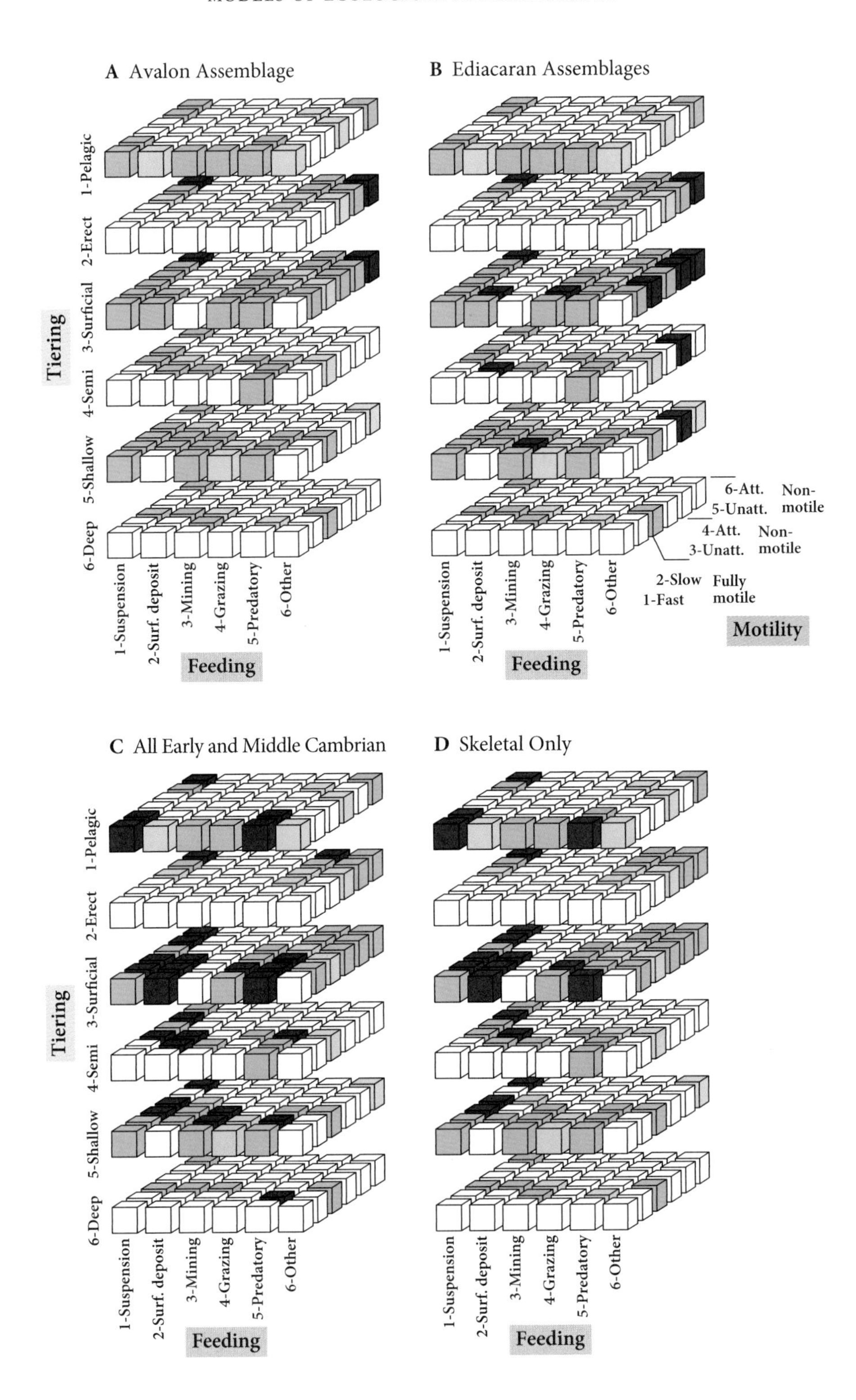
A Avalon Assemblage
B Ediacaran Assemblages
C All Early and Middle Cambrian
D Skeletal Only
Tiering
1-Pelagic
2-Erect
3-Surficial
4-Semi
5-Shallow
6-Deep
Feeding
1-Suspension
2-Surf. deposit
3-Mining
4-Grazing
5-Predatory
6-Other
Motility
6-Att.
5-Unatt.
Non-motile
4-Att.
3-Unatt.
Non-motile
2-Slow
1-Fast
Fully motile

caran and Cambrian food webs, but we can certainly infer something of the topologies of the fossil webs from comparison with living examples.

When the topologies are compared, it turns out that the distribution of trophic types within Cambrian communities was remarkably similar to that of modern communities. The first evidence of a possible similarity came from descriptions of communities from the Burgess Shale fauna in which predators were far more common than many paleontologists had expected (Conway Morris 1979a, 1986; see also Butterfield 2002). Ideally, we would like to be able to reconstruct the web of interactions with some confidence, but this task is challenging even in modern ecosystems where, at least in principle, one can observe feeding behaviors or examine gut contents to determine trophic relationships. The reality, however, is that reconstructions of most modern food webs rely less on such direct evidence and more on the nature of feeding structures and the known feeding habits of related species. Thus, we can use a similarly practical approach to begin an exploration of the earliest metazoan food webs.

Using morphological data, limited direct information on feeding habits from gut contents, and other assumptions, the most likely food webs for the Chengjiang and Burgess Shale faunas have been reconstructed (Dunne et al. 2008). As the best-preserved paleoecological assemblages of the Cambrian, these two faunas provide a basis for exploring early ecological relationships in explosion faunas. The study included eighty-five species from the Chengjiang fauna and 142 from the Burgess Shale. With the data available, it appears that many species preyed on the same taxa or were preyed upon by the same taxa, so they were assumed to be ecologically equivalent. These species were agglomerated into "trophic species" following an approach also used for modern food webs. This simplifying assumption produced thirty-three and forty-eight trophically distinct species for the Chengjiang and Burgess Shale faunas, respectively. At the outset of this research, significant differences between the architectures of these Cambrian food webs and modern marine webs were expected. Surprisingly, the diversity, complexity, resolution, and network structure are remarkably similar to what is described in modern marine food webs (fig. 7.3). There was understandable concern about the uncertainties associated with describing some of the trophic relationships, but testing the results by choosing alternative assignments revealed that the network's basic architecture was remarkably refractory to those problems. The fundamental structure of metazoan marine food webs appears to have been established during the Cambrian radiation and has persisted into modern oceans despite the vicissitudes of mass extinctions and the waxing and waning of different animal groups. Thus, the network structure of species feeding interactions seems to be

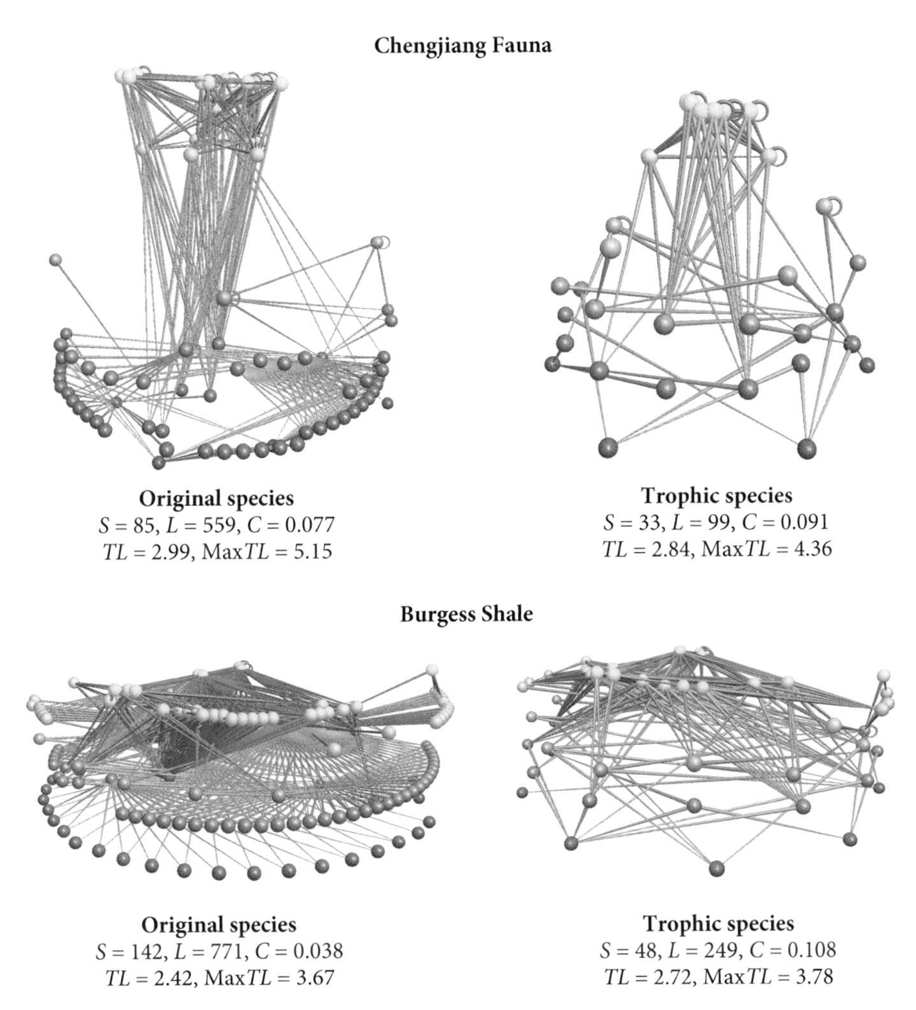

Figure 7.3 Food webs for the Cambrian Chengjiang and Burgess Shale faunas. The spheres represent taxa, species in the original species web, and trophically indistinguishable groups of species in the trophic species web, and lines represent feeding links. The vertical position of the taxa corresponds to their trophic level as determined by the food web analysis. Basal taxa (primary producers and detritus feeders) are shown at the bottom of the network (red); the highest trophic level is shown at the top in yellow. S: number of taxa (nodes) in the webs. L: number of trophic links. C: connectance. TL: mean trophic level in the web. MaxTL: maximum trophic level of a species in the web. Images produced with Network3D software written by R. J. Williams. From Dunne et al. (2008).

largely independent of the species that make up the web or even of the clades that they represent. This fascinating result suggests the promise of establishing more general notions of the operation of such networks independent of the organisms and clades that fill various roles. We must be sensitive, however, to the possibility that this approach may not be sufficiently robust—may be too coarse—to identify very real differences in the architecture of Cambrian food webs relative to those of today. More importantly, although this work provides a snapshot of the likely structure of these two webs, it provides little insight into the evolutionary mechanisms responsible for constructing webs of trophic interaction.

Positive feedback in a network amplifies a signal and under certain circumstances can allow the network to expand. Ecosystems also exhibit positive feedback, either through specific adaptations that expand opportunities or through more diffuse interactions, generating their impact through network effects rather than specific adaptations. Stanley presciently described the Cambrian radiation as "[a series of] events [that] automatically triggered the formation of a series of self-propagating feedback systems of diversification between adjacent trophic levels" (Stanley 1973, 1486). At issue, however, are the specific ecological mechanisms that generated positive feedback systems in the Cambrian and the links between trophic levels. Stanley explicitly linked this process to the invasion of vacant adaptive zones rather than network effects, but network studies can provide a rigorous approach to these problems.

Up to this point, and indeed throughout this chapter, we have focused on diversity—the number of taxa present—rather than on factors associated with the number of individuals of different species (abundance) or of biomass. Adequately characterizing differences in abundance or biomass is a great challenge for paleontologists because there are few localities at which relative abundance information has been collected; in addition, the collections are often composed of many generations of individuals, assembled from community associations that existed over hundreds to thousands of years (J. B. C. Jackson and Erwin 2006). Such time averaging makes it very difficult to acquire and analyze critical ecological information. For example, evenness is an ecological measure of the distribution of relative abundances within a community. Studies have shown that communities that are highly uneven (with one or a few species being much more common than other species) are much less resistant to environmental stress than those that are more even (Wittebolle et al. 2009). Although Walcott did not tabulate accurate relative abundance data during his collection of the Burgess Shale fossils, his collections seem to include the most specimens and thus are likely to provide at least a rough guide to relative abundance (Conway Morris 1986). Walcott's assemblages appear to have been highly uneven, with some highly abundant species, such as the arthropod *Marrella,* but few specimens of many others. To the extent that the Burgess Shale community is indicative of the Cambrian, its species abundance pattern thus indicates a lack of ecological robustness.

MECHANISMS PROMOTING DIVERSIFICATION

Several companies have sought to capitalize on concerns about global warming by fertilizing nutrient-limited areas of the ocean to spur primary productivity. The burst of productivity would eventually lead (in theory) to an increase in the organic "rain" to the ocean floor, increasing the sequestration of carbon in the deep sea. Applying a similar logic, "bottom-up" models of diversification generate diversity by injecting new resources at the base of the food chain. Nutrient limitation in marine systems is a common bottom-up phenomenon and leads to a logical scenario: if primary productivity is limited by the lack of nutrients (phosphorus, nitrogen, iron, and silica), the injection of more nutrients should enhance productivity. More broadly, increased ecological opportunity could reflect a change in environmental conditions or primary productivity, including an increase in oxygen levels or nutrient availability. In contrast, "top-down" models of diversification hypothesize that the primary control comes from consumers (herbivores or predators) higher up in a food chain who reduce the sizes of the populations of their prey species, thus freeing resources that may then support additional species at lower levels of the food web. Both bottom-up and top-down models have been proposed as ecological explanations for the Cambrian explosion (see Butterfield 2001, 2009, 2011).

Prominent among the bottom-up models are scenarios in which nutrient supply to the oceans is increased through changes in the physical environment. The geological and isotopic evidence for late Neoproterozoic tectonic activity, as described in chapter 3, led several workers to invoke increased nutrient flux from the land as permitting an expansion of primary productivity and the evolution of new trophic levels (Brasier 1990, 1992; P. J. Cook and Shergold 1984; Elser et al. 2006). The tectonic activity largely occurred tens to hundreds of million years before the increase in trophic complexity that is recorded during the early Cambrian, however. Moreover, those tectonic events coincide with the aftereffects of the Marinoan and Gaskiers glaciations, making it difficult to disentangle the relative contributions of tectonics and climate. The widespread late Ediacaran and earliest Cambrian phosphate deposits have led to another nutrient-based hypothesis in which phosphates, silicon, trace elements, and perhaps other nutrients built up in deeper waters of stratified oceans during the Ediacaran and then were delivered by upwelling into shallow waters during the early Cambrian (Brasier 1991, 1992).

The most frequently invoked bottom-up model for Cambrian diversification is an increase in oceanic oxygen levels, discussed at length in chapter 3. The bulk of geochemical evidence suggests that deep-water oxygen levels increased substantially, probably largely in the late Neoproterozoic, although the exact timing and rate of increase of oxygen levels remains a subject of intense investigation. It is generally assumed that the signatures from deep-water geochemical proxies are indicative of a concurrent increase in shallow-water oxygen levels, which was a necessary precondition to the development of benthic marine ecosystems of metazoans.

The primary difficulty with all bottom-up scenarios is that even though expansion of primary productivity might have created larger populations, there is no necessary reason it should generate a diversification of new lineages. In the present oceans, high nutrient flux correlates with high productivity but not as well with high diversity. Given a stable trophic source, marine communities are quite diverse in regions of low productivity, suggesting that it is the relative stabilization of trophic supplies rather than their amount that was the major key to higher Cambrian diversities. Increased nutrient flux is not a sufficient explanation for increased Cambrian diversity.

A model of diversification that involves predator-prey dynamics across more than one trophic level was suggested by Stanley (1973, 1976). He proposed that cropping by herbivores would drive the diversification of algae, thus creating more feeding options and more niches for herbivores and allowing them to diversify in turn. The subsequent increase in diversity would increase the number of prey types available to predators and eventually facilitate the establishment of new trophic levels. In the absence of significant herbivory, competition among algae would result in a limited number of more successful species, which would have large populations but which would support only a limited number of herbivore species. In this scenario, the initial changes were imposed by generalist herbivores, creating empty ecospace that permitted algal diversification, a top-down system, which led to an increase in specialist herbivores and their predators, a bottom-up system.

Studies of modern ecosystems suggest that most top-down trophic cascades do not generally have network-wide effects. Ecologists have studied trophic cascades in considerable detail, establishing that although they are widespread, they are also context dependent (Schmitz, Krivan, and Ovadia 2004; Sergio et al. 2009; Wootton 1994). It is difficult to establish whether diversity within any particular community has arisen from a trophic cascade because one cannot assume that top predators necessarily structured the ecosystem or that their effect had influenced the entire community rather than a particular species or chains of species. Trophic cascades require not only that the species at the top of the cascade limit the population size of prey but also that the reduction in prey influences the abundance of the next-lower trophic level. In the absence of either condition, a trophic cascade will not develop, and there will be no effect on overall diversity. Beyond its hypothesized role in generating a trophic cascade, the onset of predation has long been invoked as a trigger for the acquisition of skeletons during the Cambrian and indeed is sometimes hypothesized to be responsible for the explosion. Skeletonization evidently occurred nearly simultaneously across a variety of taxonomic lineages (Bengtson 2002; Bengtson and Conway Morris 1992; Stanley 1976; Vermeij 1989). At all trophic levels that can be distinguished, however, a majority of Cambrian organisms lacked a durable skeleton (e.g., cnidarians, ctenophores, some chordates, chaetognaths, priapulids, annelids, and many arthropods), and the intensification of predation may simply reflect a general expansion of trophic relationships during the early Cambrian. Although the appearance of skeletons was the marker by which the Cambrian was first recognized, the Cambrian radiation is clearly not simply about the acquisition of skeletons. It is certainly plausible that protection from predation was an adaptively important evolutionary innovation during the early Cambrian. Skeletons are also involved in systems of structural support and locomotion, however, and such systems may have played an important role in skeletonization of some clades.

Trophic cascades are ecological phenomena, but their relationship to the development of trophic structure in evolutionary time is unclear (Post and Palkovacs 2009). There are a number of cases of eco-evolutionary linkages during the Ediacaran and Cambrian, including the advent of bioturbation, the evolution of zooplankton, and the effects of filter feeding. Together, these cases illustrate the need to understand the role of organisms themselves in causing system-wide ecological and evolutionary change.

Ecosystem engineering describes activities of organisms that directly or indirectly cause physical and chemical changes resulting in direct effects on resource availability for other organisms (Cuddington et al. 2007; Gutierrez and Jones 2008; C. G. Jones, Lawton, and Shachak 1997; Wright and Jones 2006). Ecosystem engineering is concerned primarily with habitat creation or modification and does not include direct trophic interactions such as predation. Physical ecosystem engineering encompasses both changes in the environment owing to physical structures produced by organisms, such as reefs (autogenic engineers), and activities that change the physical state of materials, such as burrowing (allogenic engineers). Of course, some organisms are both autogenic and allogenic engineers. Burrowing provides perhaps the paradigmatic example of allogenic ecosystem engineering: its net effect is a wholesale change in the chemical gradients within the substrate, oxygenating the sediment and enhancing the productivity of many microbial systems and of the organisms that subsist on them. Ecosystem engineering may have both positive and negative effects; when it increases physical stress for other species, it tends to have a negative effect on species diversity, but when it reduces physical stress, it can have a positive effect (Badano, Marquet, and Cavieres

2010). At a landscape scale, ecosystem engineering creates new habitats and thus increases species richness. Because the students of ecosystem engineering have been ecologists, there has been relatively little concern with its evolutionary implications, particularly over long timescales (Erwin 2008; Post and Palkovacs 2009).

Evolutionary outcomes have been more clearly addressed in discussions of niche construction, which involves the activities of organisms that influence their environment in ways that affect the fitness of the population (Laland and Sterelny 2006; Odling-Smee, Laland, and Feldman 2003). Both niche construction and ecosystem engineering can cause ecological inheritance, that is, the persistence of biologically produced environmental modifications over multiple generations, which influences the evolution of species (fig 7.4). Structures such as beaver dams and reefs as well as resources such as oxygen are used by multiple generations and thus form examples of ecological inheritances. Although advocates of the importance of ecosystem engineering initially neglected evolutionary issues, students of niche construction have emphasized the resulting evolutionary effects on the niche-constructing organism, although they have not been concerned with the spillover effects on other species. Although these concepts are closely related, there are good reasons to keep the two distinct. Many ecosystem engineering activities have no apparent selective effect upon the species that engage in them or the effects are sufficiently diffuse that they are difficult to attribute to a single species (as in the case of burrowing). Moreover, it is very difficult for paleontologists to assess the fitness effects of niche construction in the past. So, although the terms *niche construction* and *ecosystem engineering* identify aspects of a larger spectrum of organismal interactions with their physical, chemical, and biological environments, it seems operationally useful to distinguish ecosystem engineering and niche construction (Erwin 2008). (As an aside, it is worth noting that there are echoes of well-established ecological concepts such as keystone species, foundation species, and ecological facilitators in these notions.)

Erwin and Tweedt (2012) examined the rise of ecosystem engineering during the Cambrian radiation via the database of first occurrences of higher taxa, introduced in chapter 6, together with a compendium of the generic ranges of fossils from the early Cambrian of south China, including the Chengjiang biota (G. X. Li et al. 2007). The Chinese data include 876 named genera, of which 582 were considered to be valid. For this analysis, the stratigraphy and stages were updated to those described in chapter 2. Because activities can have both physical and chemical effects, the classification of ecosystem engineering activities is not straightforward. Burrowing through sediment is

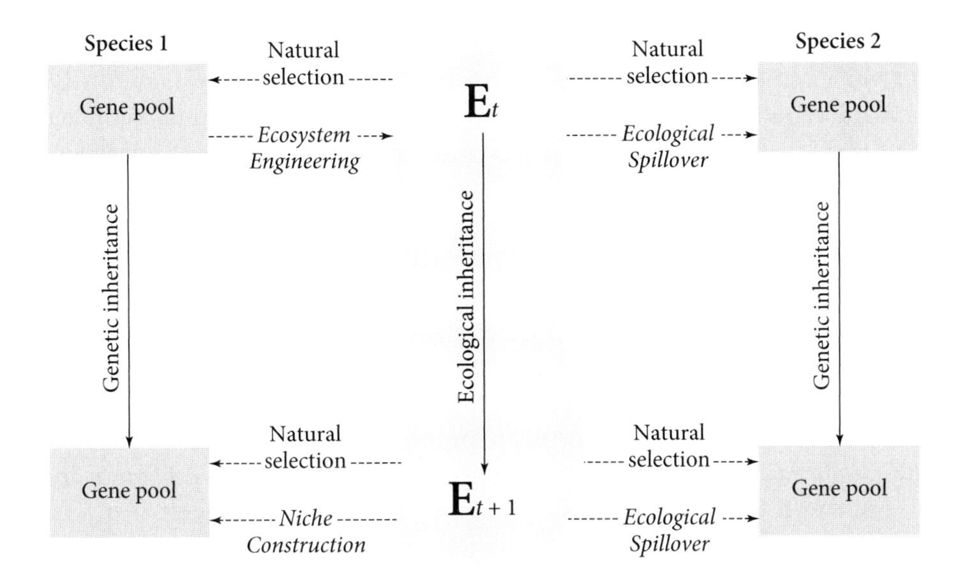

Figure 7.4 A schematic of the relationships between genetic inheritance and ecological inheritance during ecosystem engineering. Niche construction involves the production of external phenotypes that may persist in the environment as an ecological inheritance and affect selection on subsequent generations of the species. The consequences of niche construction thus modify the selective environment of the species. The effects of ecosystem engineering activities can range much further, however. In this example, the activities of species 1 modify the environment of the species, generating immediate ecological spillover for species 2 as well as ecological inheritance that may modify selection for species 1 (niche construction sensu stricto) and ecological spillover to affect, in a future generation, selection on species 2.

a form of physical ecosystem engineering, but it can have substantial effects on the redox gradient within the sediment (Lohrer, Thrush, and Gibbs 2004; Mermillod-Blondin and Rosenberg 2006; Nogaro et al. 2009), producing chemical effects as well. Similarly, filter feeding is a trophic behavior that often has the effect of changing both physical and chemical characteristics of the water, with follow-on effects on the composition of the biota. Other sources of biogenic mixing have been identified. For example, when large numbers of jellyfish swim together, they can produce substantial viscosity-enhanced mixing (Katija and Dabiri 2009). In each of these cases—burrowing, filter feeding, and biogenic mixing by swimming organisms—the consequences are highly dependent on abundance within a population.

To tabulate the different types of ecosystem engineering, Erwin and Tweedt followed the suggestion of C. G. Jones, Lawton, and Shachak (1994) in distinguishing between physical and chemical ecosystem engineering. Relevant activities during the Ediacaran-Cambrian include structural or architectural activities (building structures), including sediment stabilization, sediment bioturbation, and chemical engineering. Structural modification during the Ediacaran and Cambrian, for example, includes the formation of reefs by archaeocyathids. Sediment stabilization by microbial mats is widespread during the Ediacaran but declines during the early stages of the Cambrian. Chemical engineering includes nutrient and carbon transfer between different components of an ecosystem, ventilation or oxygenation of the water column, ventilation of the sediments, and the creation of biogeochemical gradients. Significantly, there is no indication of either active suspension feeding or organisms capable of viscous transport of fluids during the Ediacaran. Although this classification was developed for the events of the Ediacaran and Cambrian, it closely matches the functional classes independently identified by Berke (2010): structural engineers (encompassing both reef constructors and sediment stabilization), bioturbators, light engineers, and chemical engineers. This classification is not exhaustive, but it does provide a basis for a preliminary examination of the extent of ecosystem engineering.

The first appearances of engineering activities (table 7.1) indicate that ecosystem engineering during the Ediacaran was limited to minor structural engineering, microbial grazing by *Kimberella,* and sediment stabilization by microbial mats and possibly sessile forms like the rangeomorphs. Sponges appear to have provided the only active ventilation of the water column. This situation shifted considerably during the Cambrian, with the addition of a number of structural and bulldozing engineers and the disappearance by the mid-Cambrian of sediment stabilization (likely as a consequence of the increased bioturbation). A large number of clades arose whose activities resulted in ventilation of either the sediment or the water column. The generic data from south China provide a more detailed look into the ecological shift of the first four stages of the Cambrian (table 7.2). Diversity was low during Stage 1, increased significantly with the pulse of first appearances of small shelly fossils during Stage 2, and then increased greatly during Stage 3, coincident with the introduction of a variety of different types of ecosystem engineers.

The importance of ecosystem engineering for the growth of biodiversity may best be explained by analogies to the sources of economic growth. Both economic growth and the growth of biodiversity involve the allocation and investment of scarce resources, and as such the economic models provide an interesting perspective on the factors that may promote the construction of biodiversity. Many early economic models identified technological innovation as the driver of almost all economic growth (rather than, say, increased popu-

Table 7.1 Physical and chemical ecosystem engineering types and effects for the principle Ediacaran and Cambrian engineering clades

Clade	Physical Ecosystem Engineering			Chemical Ecosystem Engineering	
	STRUCTURAL	BULLDOZING	SEDIMENT STABILIZATION	VENTILATION	NUTRIENT TRANSFER
Ediacaran					
Microbial mats			X		
Organic buildups	X				
Rangeomorphs			X		
Kimberella		X		X	
Sponges	X				X
Cambrian				X	X
Sponges	X			X	
Archaeocyathids	X			X	
Chancelloriids	X			X	
Brachiopods				X	X
Phoronids			X	X	
Priapulids and annellids			X	X	
Trilobites		X		X	

Source: From Erwin and Tweedt (2012).

Table 7.2 Summary data with the diversity of different Cambrian clades, by stage

Clade	Stage 1 eMeis	Stage 2 m/l Meis	Stage 3 Qiong/Catl	Stage 4 Longw
Sponges (nonarchaeocyath)		5	24	
Archaeocyathids			33	
Cnidarians	3	8	8	3
Ctenophora			4	
Annelids			2	
Hyolithids	5	23	27	2
Helcionellids		34	10	1
Miscellaneous Mollusca		35	3	1
Chancelloriids		3	3	1
Brachiopods		12	18	3
Phoronids			2	
Sipunculans			2	
Priapulids and other worms			20	
Chaetognath			1	
Protoconodonts	1	4	4	
Lobopodians			8	
Anomalocarids			3	
Bradoriids			34	
Trilobites			113	54
Other arthropods			51	
Echinoderms			2	1
Vetulicolians/chordates			12	
Enigmatic Chengjiang metazoans			19	
Enigmatic tubular fossils	1	18	7	
Enigmatic sclerites	1	27	11	0
TOTAL	**11**	**169**	**421**	**66**
Small shelly fossils	11	164	65	11
As % total fauna	100	97	15	17
% trilobites	0	0	27	82
No. of ecosystem engineers		24	76	4
% ecosystem engineers		14	18	7

From G. X. Li et al. (2007).

Li and colleagues compiled the data based on Chinese stages, but for easier comparison to Table 7.1, the data were converted to International Cambrian Stages (Stages 1 to 4). The primary ecosystem engineers are shown in bold. Many trilobites also disturbed the sediment, as did some of the "other arthropods," which underestimates the actual diversity of ecosystem engineers. The figure summarizes the numbers and percent of small shelly fossils, percentage of trilobites per stage, and the number and percent of ecosyswtem engineers.

Abbreviations: eMeis, Early Meishucunian; m/l Meis, middle to late Meishucunian; Qiong/Catl, Qiongzhusian and Canglangpuan; Longw, Longwangmiaoan. From Erwin and Tweedt (2012).

lation size, education, or other such variables) but included it as an exogenous variable, that is, a variable external to the model (a free parameter) that could be cranked up or down at will to achieve a desired result. This exercise was unsatisfactory in that it provided little insight into the conditions under which technological innovation occurred or into the relationships between technological innovation and other variables. In fact, this practice is formally identical in ecology to simply changing the carrying capacity without addressing the processes that are involved. More informative work involves models in which growth is an endogenous rather than an external variable (see Warsh 2006 for a review and C. I. Jones 2002 for more technical detail).

The ideas economists developed about the sources of technological innovation turn out to have considerable resonance for biology and particularly for events such as the Cambrian explosion. Economists recognize that technological inventions vary in whether they can be used by multiple users at one time. A TV signal or calculus can be used simultaneously by many people without degrading performance; these goods are called nonrivalrous goods. Economists contrast them with rivalrous goods such as a bicycle or an iPod that can only be used by a single person at a time. A closely related concept is that of excludability, that is, whether or not one can easily exclude another from using the good. Although it is easy to exclude another from using a cell phone or a car (you put the cell phone in your pocket and lock the car), excluding someone from catching a fish in the ocean is more difficult. A matrix of the combinations of rivalry and exclusion is shown in figure 7.5. Remarkably, the most significant inventions or innovations for economic growth have involved the quadrant of nonrivalrous, nonexcludable goods (Romer 1990). Some of the novelties are ideas—for example, the calculus—whereas others—such as concrete—are, well, more concrete. Ideas are often the most difficult to earn a profit from (hence the development of patent law), but have actually had the greatest positive feedback effects in generating continuing economic growth. Intermediate in nature are economic goods such as highways or railroads from which the public derives great benefit. Both highways and railroads are excludable goods, but they create great spillover effects (just think about all the gas stations and fast-food establishments at highway interchanges).

Characterizing adaptations by their degree of excludability and rivalrousness provides a way of considering evolutionary innovations, and they can be mapped into this framework to illustrate the critical importance of ecosystem engineering to evolutionary innovation (Erwin 2007, 2008). Figure 7.6 illustrates some biological equivalents to figure 7.5. Competition for space in the intertidal zone is always intense, and living space is a classic example of a highly excludable, rivalrous good. At the opposite extreme, the production of oxygen by oxygenic photosynthesizers is perhaps the paradigmatic biological

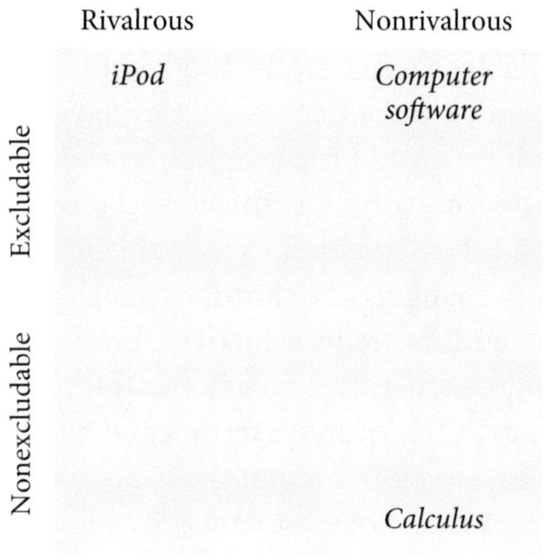

	Rivalrous	Nonrivalrous
Excludable	*iPod*	*Computer software*
Nonexcludable		*Calculus*

Figure 7.5 The division of economic goods into rivalrous versus nonrivalrous and excludable versus nonexcludable has proven to be a powerful means of understanding the relationship between types of goods and their effect on economic growth. Economic theory has demonstrated that nonrivalrous, low-excludability goods have the greatest spillover effects and are the most likely to generate economic growth. They are often known as public goods because many of them are best provided by governments and there are few ways for a private company to get economic returns from these goods. Rivalrous goods are goods that can have only one user at a given time, whereas a non-rivalrous good can be used by multiple users simultaneously without degrading the good. Excludable goods encompass goods for which it is easy to prevent others from using them.

example of a nonrivalrous, low-excludability good. It is true that in particular settings oxygen can become a rivalrous good, but on a global scale it is not limiting today. Such biologically significant goods are often the product of ecosystem engineering and produce spillover effects that may affect much of an ecological network.

The profound ecological changes between the Ediacaran and Cambrian are surely critical drivers of the Cambrian explosion (fig. 7.7). We have argued that both the ecological changes and the attendant changes in the physical environment were constructed by the activities of early organisms. Whatever the causes and rates of oxygenation of the oceans during the Ediacaran, the change in redox conditions opened up opportunities for benthic organisms. Exploiting these organisms fully required the delivery of labile, nutrient-rich organic matter to the seafloor to serve as food, however. Bioturbation was likely an important factor in the positive feedback process because, like many aspects of ecosystem engineering, it is nonrivalrous and nonexcludable. By

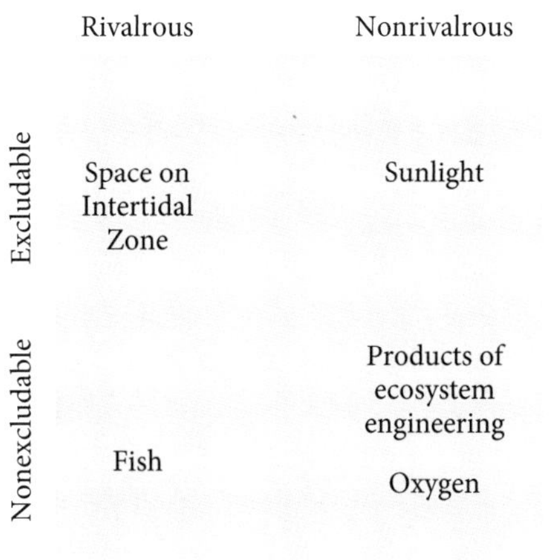

	Rivalrous	Nonrivalrous
Excludable	Space on Intertidal Zone	Sunlight
Nonexcludable	Fish	Products of ecosystem engineering Oxygen

Figure 7.6 Biological "goods" can also be distinguished, as in figure 7.5. Nonrivalrous, low-excludability goods are often the consequence of ecosystem engineering and produce spillover effects that reverberate throughout an ecological network.

recycling nutrients more rapidly, bioturbation greatly increased microbial biomass and produced a more complex geochemical structure within the sediments (a greater oxygen supply, shifting redox gradients, etc.) (McIlroy and Logan 1999). The enhanced microbial biomass in turn provided additional resources for burrowing animals, further increasing bioturbation and expanding the depth of the sediment available for bioturbation, which in turn added more labile organic matter to the reservoir. All this activity must have had a strong positive feedback effect on metazoan population sizes and, in some circumstances, on biodiversity (Meysman, Middelburg, and Heip 2006).

Species innovations in the nonrivalrous, nonexcludable quadrant generate the sort of consequences that spill over to affect the abundance or diversity of other species. Because they open up new opportunities rather than subdividing existing resources, these sorts of change seem most likely to develop in settings in which competition is low. This notion matches the expectation of paleontologists and evolutionary biologists that innovation is most successful in empty ecospace. Thus, applying Romer's (1990) insights into the dynamics of economic growth produces a simple model for growth of biological diversity that requires neither an equilibrial carrying capacity nor an invocation of empty niches. It does not mean that resource limitation, competition, or even empty niches (in a restricted sense) do not occur; rather, to understand the generation of biological diversity, there is no need to assume that they do. More importantly, the approach allows us to differentiate adaptations by their generative effects on other species and on the network of ecological interactions.

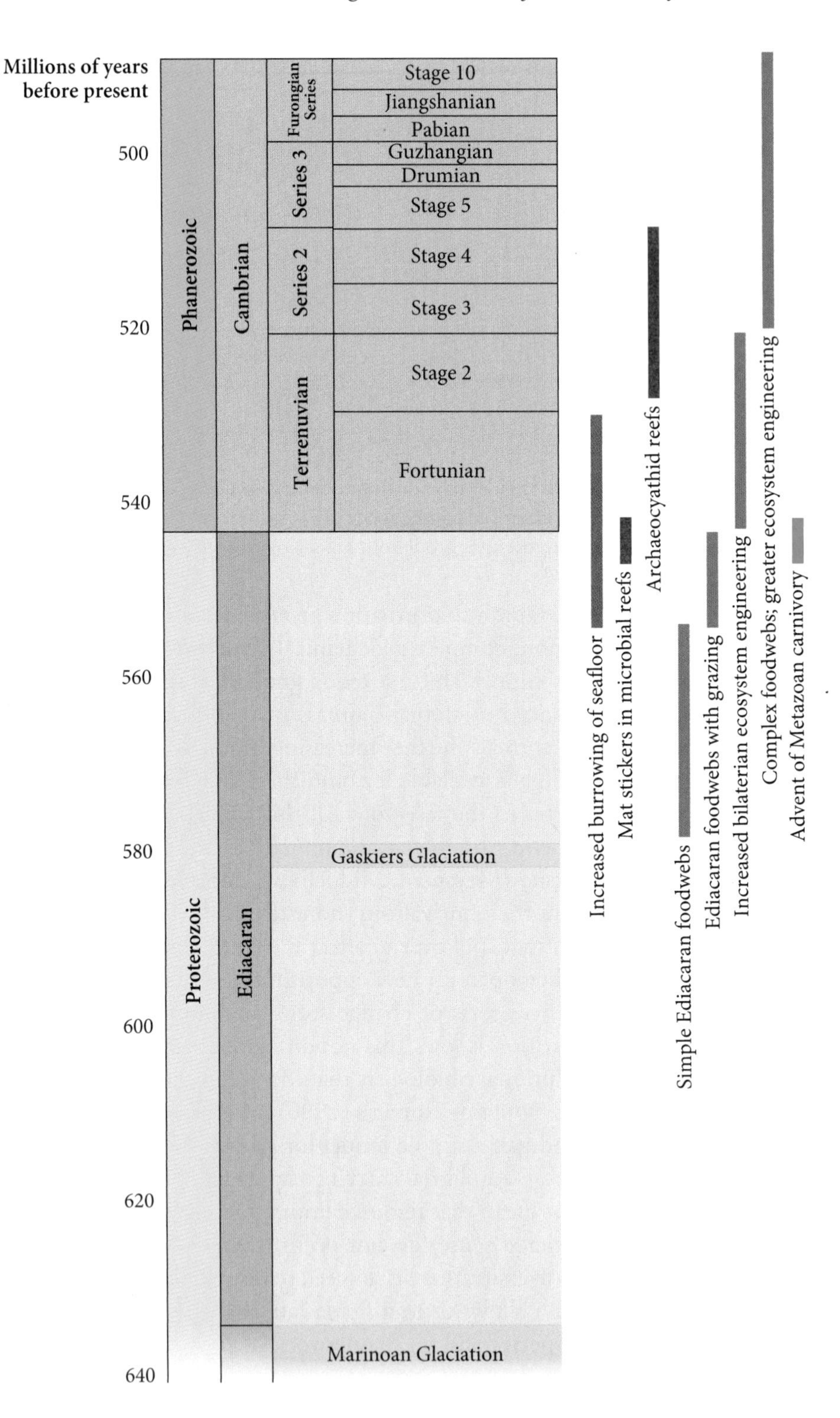
Millions of years before present
500
520
540
560
580
600
620
640
Phanerozoic
Cambrian
Furongian Series
Series 3
Series 2
Terrenuvian
Stage 10
Jiangshanian
Pabian
Guzhangian
Drumian
Stage 5
Stage 4
Stage 3
Stage 2
Fortunian
Proterozoic
Ediacaran
Gaskiers Glaciation
Marinoan Glaciation
Increased burrowing of seafloor
Mat stickers in microbial reefs
Archaeocyathid reefs
Simple Ediacaran foodwebs
Ediacaran foodwebs with grazing
Increased bilaterian ecosystem engineering
Complex foodwebs; greater ecosystem engineering
Advent of Metazoan carnivory

Figure 7.7 **(left)** Time line of major ecological changes through the Ediacaran and Cambrian periods.

The success of the novel morphologies of the Ediacaran and Cambrian reflected the ecological opportunities generated during this span of time by changes in both the physical and biologic environment, but these novel morphologies reflect an equally important expansion in the developmental capacity of early metazoans. The formation of new cell types, the spatial and temporal deployment of these cell types, their coordination in a developing embryo, and the evolution of entirely new levels of interaction such as tissues and organs were also critical components of the Cambrian explosion. In the next two chapters, we take the story forward to consider the developmental component of the macroevolutionary triad; in particular, we will look at the new networks of regulatory interaction that were established and exploited by early Metazoa.

CHAPTER EIGHT

The Evolution of the Metazoan Genome and the Cambrian Explosion

As the sequencing of the human genome came to a close in 2000, an informal betting pool developed among molecular biologists as they wagered on the likely number of protein-coding genes. Most estimates clustered around 50,000 genes, but they ranged from 27,000 to as high as 160,000. When the genome sequence was substantially completed in 2003, the total number of genes found were fewer than 25,000. No one in the pool had picked a number that low. Estimates had been based partly on the 14,601 genes of the fly *Drosophila melanogaster,* which had been recently sequenced (Adams et al. 2000), and most biologists, knowing that the human genome contained multiple copies of many *Drosophila* genes, reasonably concluded that the total number of human genes would be some multiple of that number. After all, humans seem to be far more complex than flies.

Continued genome sequencing of other metazoans has not done much for humanity's pride. Even the morphologically simple cnidarians have genomes that contain between about 18,000 (the sea anemone *Nematostella*; Putnam et al. 2007) and 20,000 (*Hydra*; Chapman et al. 2010) protein-coding genes, significantly more than flies, and the small-bodied "water flea" *Daphnia pulex,* a freshwater crustacean, has at least 30,907 genes, more than twice as many as *Drosophila* and nearly 30% more than humans (Colbourne et al. 2011). As far as we know, those genes are all useful to *Daphnia.* By way of comparison, the number of genes in unicellular eukaryotic genomes is commonly less than half that of metazoans; choanoflagellates have only about 9200 genes.

The relatively limited size of the human genome is especially surprising when considering that the human body is composed of trillions of cells and more than four hundred different somatic cell morphologic types, including epidermal cells, nerve cells, muscle cells, and blood cells (Vickaryous and Hall 2006). This tally is still incomplete. These cell types are organized into a wide variety of tissues, organs, and organ systems, reflecting our morphological complexity and sophisticated nervous systems. Cnidarians commonly have ten to fifteen cell types and few tissues and organs, whereas arthropods, which are much more complex, may have as few as thirty-seven cell types and seem never to exceed one hundred (table 8.1). Evidently, what counts most is not the number of genes, but how they function.

When it comes to understanding the genetic bases of the morphological richness and disparity of the Cambrian explosion, our primary interest is in the genes involved in specifying the development of body plans and other morphological features. That is to say, it is not the genes that control basic cellular functions (so-called housekeeping genes) that are of interest, but the genes that regulate the development of the morphology from egg to adult. Comparative studies of living organisms allow us to reconstruct the genomes and learn something of the developmental processes shared between both the organisms that participated in the explosion and others further back on the metazoan tree. In this way, it is possible to study the genes that regulate morphological development in their living representatives and then to trace them back to their LCAs. We shall refer to those developmental regulatory genes as composing the "developmental genome."

In this chapter, we first discuss the genetic basis of development and then the evolutionary processes involved in generating changes in development. One of the most surprising discoveries in studies of the comparative developmental evolution of various living animal groups has been the widespread conservation of genes and gene networks that generate eyes, appendages, and other important morphologic innovations. This unexpected evolutionary conservation allows us to compare the development of living groups and infer the development of their ancestors in the Neoproterozoic and Cambrian, the topic of chapter 9.

THE REGULATION OF METAZOAN DEVELOPMENT

Estimates of the number of genes in whole genome sequences are based on the number of protein-coding, or structural, genes. But genomes also contain a variety of noncoding regulatory sequences and many noncoding regions whose function is not known. Regulatory sequences trigger gene transcription, which produces an RNA sequence complementary to the DNA template. The RNA sequence is then processed to become messenger RNA (mRNA) (fig. 8.1). In

Table 8.1 Variation in genome size and the complexity of regulatory elements across a range of modal organisms for which whole genomes have been studied.

	Monosiga brevicolis	*Amphimedon queenslandica*	*Trichoplax adherens*	*Nematostella vectensis*	*Drosophila melanogaster*
Genome size (Mb)	41.6	?	98	450	180
No. of genes	9100	?	11,514	18,000	14,601
No. of cell types	1	12	4	20	50
No. of miRNA	0	7	7	3	49
No of metazoan transcription factors/families	?/5	57/?6	35/9	Min 87/10	Min 87/10
No. of bHLH genes	10	16	27	68	59

Note: Amphimedon is from Fahey et al. (2008). The number of cell types is from Valentine, Collins, and Meyer (1994), except for *Trichoplax,* which is from Srivastava et al. (2008); and the number of microRNAs (miRNAs) comes from Sempere et al. (2006), Gimson et al. (2008), and Wheeler et al. (2009). Notice that, according to Wheeler et al. (2009), the seven miRNAs in demosponges are not homologous to any eumetazoan miRNAs. The ten transcription-factor families are the homeodomian, forkhead, p53, Myc, Sox/TCF, ETS, HOX, NHR, POU, and T-box, with the data for *Monosiga* from King et al. (2008) and other taxa from Larroux et al. (2008); true *Hox* class genes are missing from *Monosiga* and *Amphimedon.* The number of basic helix-loop-helix (bHLH) genes is from Degnan et al. (2009).

most eukaryotes, including metazoans, the transcribed genes include coding sequences, or exons, are interrupted by "noncoding" sequences, called introns. A eukaryotic gene is thus rather complicated. First, it contains DNA that does not code for its protein product. Second, many gene sequences can be spliced and processed in more than one way, with different starting positions and different combinations of exons included in the processed transcript, and sometimes introns may be processed as well. Such alternative splicing allows the generation of a diversity of alternate transcripts (isoforms) from a single gene (Keren, Lev-Maor, and Ast 2010; Nilsen and Graveley 2010), an issue we will return to later in this chapter. Developmental patterning involves signaling and regulatory architectures that regulate the combinatorial expression of genes within cells throughout development in a gene regulatory network (GRN).

Unicellular forms have no development in the metazoan sense, that is, a development involving the differentiation and deployment of cell types. Uni-

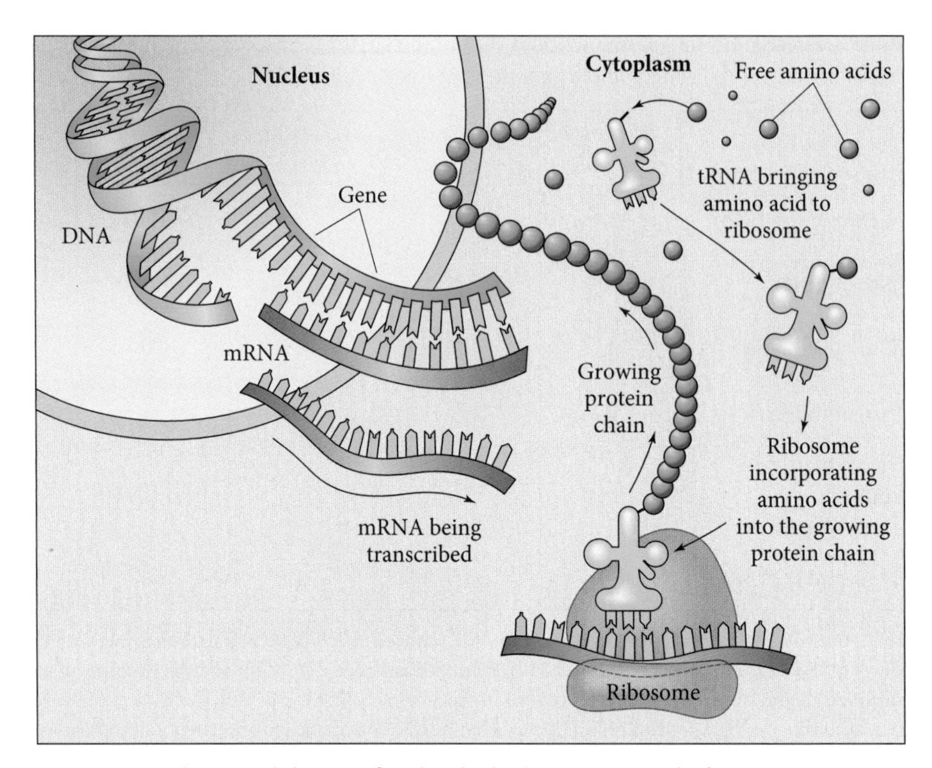

Figure 8.1 In the central dogma of molecular biology, one strand of DNA is transcribed to produce a complementary RNA sequence that is then processed to produce a messenger RNA (mRNA). In most eukaryotes, including metazoans, the transcribed genes include coding sequences, or exons, interrupted by "noncoding" sequences, called introns, that are transcribed into RNA but excised before translation. The connected exon transcripts are spliced by a protein complex to form mRNA. The mRNA is then translated by an RNA protein complex to form a protein.

cellular organisms must accommodate temporal and spatial variations in their biological and physical environments. From early in their history, unicellular eukaryotes evolved methods to cope with environmental variability and biotic interactions via molecular regulatory systems that sensed ambient conditions and responded appropriately. Their regulatory systems concerned behavioral and physiological, but not developmental, regulation. It is quite likely that such unicellular systems provided a basis for the mechanics of the developmental regulatory system in metazoans. Indeed, the genome of the putative metazoan sister clade, the unicellular Choanoflagellata, contains genes deployed in four of the major regulatory signaling pathways used in metazoan development (see chapter 9) (King 2004; King, Hittinger, and Carroll 2003). Some of those genes have been found in the genomes of other related unicellular groups present before metazoans arose, although the relatively sophisticated signaling pathways of metazoan development had not yet taken shape.

Protein-Based cis-*Regulation*

The process of development leading from an egg to the adult involves proliferation and differentiation of a variety of different cell lines, producing a treelike architecture. These cell lines form a nested hierarchy of cells, tissues, organs, and organ systems. The early development of most sponges is somewhat different from eumetazoans, and sponges may have invented a number of unique developmental routines. In crown metazoans, the egg contains maternal gene products, chiefly localized mRNAs that include regulatory elements of the developmental genome. Early development makes use of maternal gene products to increase the speed of early development. At some time during cleavage and continuing throughout development, regulatory genes in the embryo's genome begin to produce molecules that affect their subsequent gene expression patterns and thus replace earlier regulation by maternal gene products. Two of the more important components of this process, at least for our purposes, involve signal transduction and transcription factors (fig. 8.2). In signal transduction, an extracellular signaling molecule activates a receptor in a cell membrane, causing activation of a signaling pathway inside the receiving cell that eventually delivers a transcription factor protein to the nuclear DNA. The transcription factor includes a sequence that binds to a complementary sequence of the DNA, usually quite near the target gene, and changes its activity by triggering transcription or repressing activity. Although there are many transcription factor genes, there are surprisingly few signaling pathways in metazoans: the most common pathways involve the Wnt, Notch, TGF-b (transforming growth factor b), and Hedgehog genes. Signaling pathways make it possible for genes expressed in one cell to cause the transcription of genes in the nuclei of other cells. The entire intercellular signaling pathways containing these genes are highly conserved and are used many times during development; they have been called signaling cassettes.

Gene expression is generally achieved by specific combinations of transcription factors whose binding sites lie adjacent to or within a protein-coding sequence. These regulatory sites are organized into modular blocks known as enhancers; they can be quite complicated, containing binding sites for several transcription factors and often with multiple binding sites for each transcription factor (fig. 8.3). Furthermore, multiple enhancer modules are often associated with a target gene. Activation of the enhancer modules signals a sequence adjacent to the coding region, known as a promoter, to initiate tran-

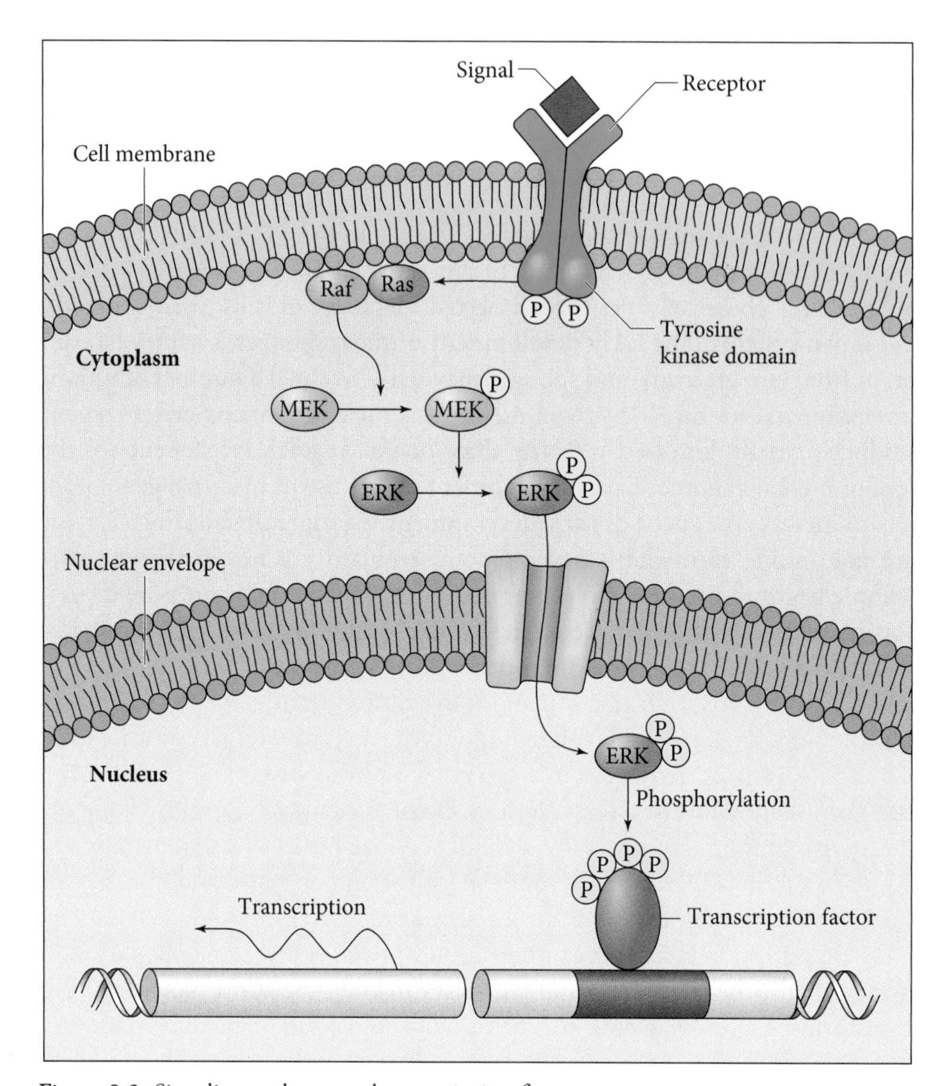

Figure 8.2 Signaling pathway and transcription factors.

scription of the target gene. Many enhancers lie upstream from the target gene (downstream is the general direction of transcription) and therefore are not transcribed, while other enhancers are found in introns and, although transcribed, are excised from the gene's mRNA transcript before its translation. This method of gene regulation via signaling to enhancers is called *cis*-regulation; *cis* is Latin for "on this side" and refers to the location of regulatory elements near the gene that is to be expressed (or, in some cases, to be silenced). These developmental control systems influence cell size, shape, patterns of differentiation, and even the orientation of cell divisions.

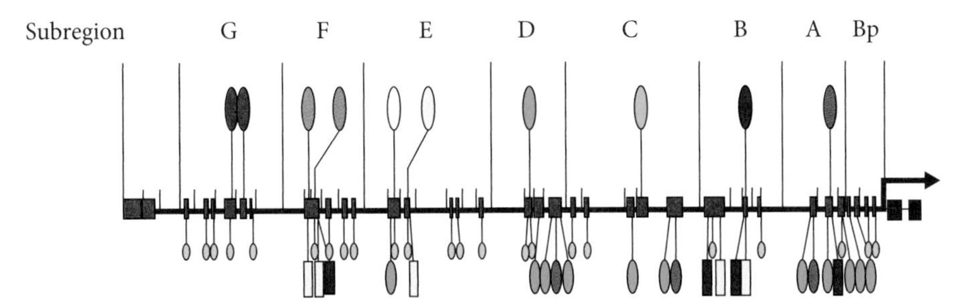

Figure 8.3 Structure of the *Endo16 cis*-regulatory region in a sea urchin and the accompanying protein interactions. The horizontal line is a 2300 basepair DNA sequence necessary and sufficient to produce gene expression experimentally. Transcription factors that bind at unique locations are shown above the line, and transcription factors that bind at several locations are shown below the line; different colors indicate distinct proteins. G-A indicate functional modules of the gene. From Davidson (2006).

Target genes are commonly regulatory genes themselves, and may be their targets in turn, forming a hierarchically structured network of regulatory interactions. Developmental signals tend to proceed from general ones, such as signals that specify body axes, to increasingly more specific ones, such as signals that designate the location of organs within the body; these developmental genes pattern the body plan. As the pattern unfolds, genes are expressed even more locally to specify the numbers, locations, and functions of specific cell types in each tissue. Depending on the combination, dosages, and timing of the delivery of the transcription factors that bind to its enhancer modules, the patterns of gene transcription vary greatly among cell types. Not surprisingly, genes subject to considerable regulation possess larger regulatory regions than other genes. Thus, although there are genes whose products contribute only to specialized cell types, most cell types owe much of their individual characters to a particular combination of genes, the majority which are also expressed in other cell types, albeit in different combinations. As cell lineages proliferate during growth and development, their patterns of gene expression change because the daughter cells can receive new signals.

RNA-Based Regulation

In addition to genes that express regulatory proteins, the developmental genome includes genes that produce various RNA molecules that are not involved in translation but have regulatory activities of their own. Genes that express regulatory RNAs are widespread in eukaryotes, and their expression is controlled by transcription factors. One type of these RNA molecules, microRNAs (miRNAs), is implicated in gene regulation by processing the RNA transcripts of protein-coding DNA before they are translated. These miRNAs are formed by processing an RNA transcript to produce a short, single-stranded RNA molecule that binds to a specific messenger RNA (mRNA), preventing its translation. For brevity, we refer to the protein-coding genes whose mRNAs are affected as the target genes. The effect of miRNA binding is to reduce the expression of the target gene and thus the effective dosage of the target gene product. The production of miRNAs tends to be restricted to one or a few tissues or cell types, so miRNAs thus fine tune the pattern of expression of genes. In some cases, miRNAs act to stabilize development during environmental perturbations. For example, the miRNA miR-7, which is widespread across bilaterians, is known to buffer the development of sensory organs in *Drosophila* against environmental perturbations that would otherwise affect dosages of miR-7 target genes and thus alter their developmental outcomes (X. Li et al. 2009). Overall, it seems that a primary function of miRNAs may be to stabilize the expression of particular cell types.

Developmental Gene Regulatory Networks

The interactions among signaling molecules and transcription factors form developmental gene regulatory networks (GRNs). Relatively few of these networks have yet been experimentally elucidated in detail, but figure 8.4 illustrates one example, the complexity of patterns of gene expression and the associated *cis*-regulatory activation of genes within the early sea urchin embryo at thirty hours after fertilization. At this early stage, the embryo is still under the influence of maternally produced signaling molecules. Notice the expression of the gene *Endo16* in the endoderm compartment; later blastula and postgastrulation patterns of *Endo16* expression are also shown. Despite their complexity, the basic architectures of GRNs can be described in terms of some common and distinctive elements. GRNs are actually quasi-hierarchical because the upper level controls the initial stages of development followed by establishment of morphological patterning and finally by the regulation of cell differentiation and morphogenesis (Davidson 2006; Davidson and Levine 2008). Furthermore, some elements of GRNs can be redeployed in toto as modules and can function in new positions within the network (Erwin and Davidson 2009).

Modularity

Some genes influence, to a greater or lesser degree, many different phenotypic traits and are described as pleiotropic. Pleiotropy is one way to increase

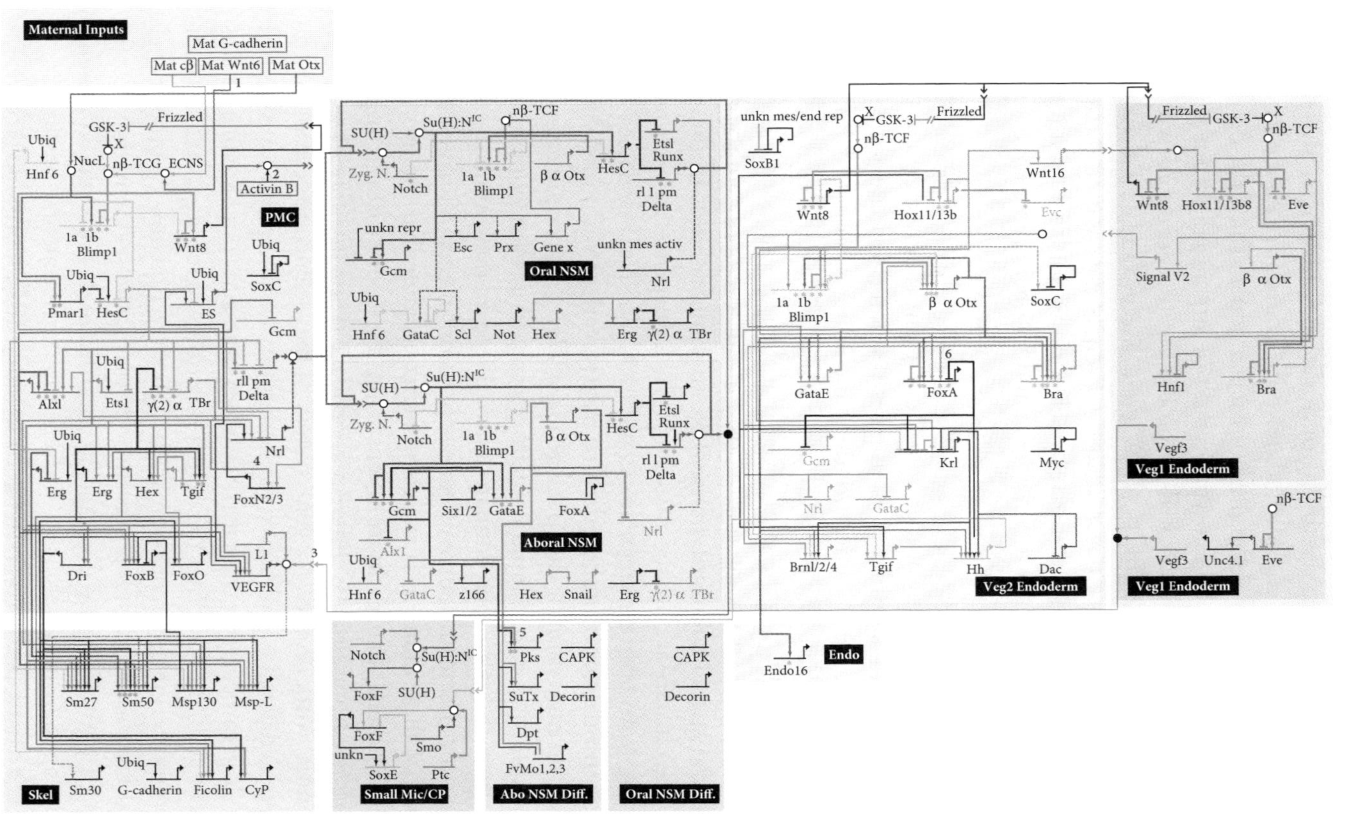

Figure 8.4 The network of gene interactions for sea urchin embryo endomesoderm, through the cleavage and blastula stages, up to about 30 hours following fertilization. The color-coded background boxes represent the domains of the embryo, identified by the black boxes. The horizontal lines represent gene sequences; colored lines depict the regulatory interactions, with the arrows representing a positive interaction, while a bar represents repression. The interactions shown here have been experimentally verified by perturbation and gene expression studies. The network depicted is that of 21 November 2011. The network is described in Davidson (2006) and subsequent papers; the most recent depiction of the network is available at www.biotapestry.org.

gene usage without increasing gene number, but a gene whose product works well during signaling in one cell type may be disadvantageous if expressed in a similar pattern in another. In such a case, selection would disconnect the expression events, adjusting the timing or dosage as appropriate. This selective bias favors the establishment of modules with coordinated gene expression patterns involved in the development of each module. Thus, the domains of gene expression within cell genotypes would be organized within the GRN as developmental modules devoted to producing the organs to which they contribute (Klingenberg 2008; G. P. Wagner, Mezey, and Calabretta 2005; G. P. Wagner, Pavlicev, and Cheverud 2007). Even if a gene is usefully expressed in multiple modules, it is scrutinized by selection within each module. Early models of the relationship between a gene and its phenotypic effect assumed that there was a relatively simple mapping between the two, but the mapping between genotype and phenotype has been subjected to considerable evolution during the deployment of metazoan body plans (Hansen 2006).

Another form of modularity is apparent among regulatory genes that mediate anterior-posterior differentiation, including a family of homeobox genes called *Hox* genes. *Hox* genes are often clustered together and are usually aligned along the DNA in the same order that they are expressed, in temporal and spatial sequence, from anterior to posterior in developing embryos (called temporal and spatial colinearity) (fig. 8.5). In some cases (as in the fly *Drosophila*), the clusters are broken into parts. The *Hox* genes are each regulated by signaling from upstream regulators, and each gene has a restricted domain of expression that sometimes overlaps or partly overlaps with other Hox gene domains. The expression of each *Hox* gene (or *Hox* gene combination) mediates the expression of a unique regulatory cascade in each domain. Consequently, even though other domains may have extremely similar expressed genotypes, the upstream source of gene expression in each domain is unique. Thus, each of these domains is a developmental module that can evolve independently through changes in its downstream regulatory cascade to produce distinctive structures, such as different appendage types, in different segments.

The initial work on the *Hox* gene system of flies seemed to reveal that *Hox* genes were dedicated to producing segmentation; as more phyla were studied, however, it became apparent that these genes were active in all eumetazoans, and, at least in bilaterians, that their chief function was more general: to mediate antero-posterior differentiation (although they function in other developmental tasks as well). In segmented body plans, the segmentation became coordinated with *Hox* gene domains, as in arthropods (also true in annelids and, as in fig. 8.6, in vertebrates). Indeed, the developmental possibilities of the *Hox* system are partly responsible for the incredible morphologic diversity

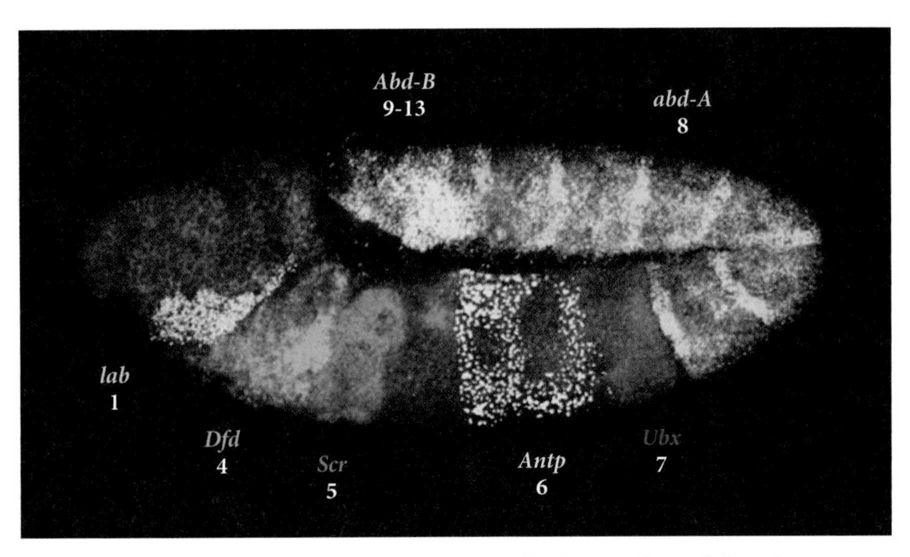

Figure 8.5 Image of gene expression patterns in a developing *Drosophila* embryo, displaying the spatial expression patterns of *Hox* gene transcripts. Anterior to left, with staining for labial (*lab*), Deformed (*Dfd*), Sex combs reduced (*Scr*), Antennapedia (*Antp*), Ultrabithorax (*Ubx*), Abdominal-A (*abd-A*), Abdominal-B (*Abd-B*). Their orthologous relationships to vertebrate *Hox* homology groups are indicated below each gene. From Lemons and McGinnis (2006).

of arthropods. In nonsegmented bilaterians, *Hox* genes simply specify different cell and tissue types along the main body axis, or they may mediate the development of organs at particular locations. No *Hox* genes are known in sponges, and they may never have been present; they are also absent in placozoans, although they may have been lost from such a reduced genome.

There are many gene clusters within the metazoan genomes, each of which has a history of growth and change. The *Hox* cluster is among the best studied and serves as an example of the sorts of evolutionary changes that have occurred within transcription factor families that were evolving during the late Neoproterozoic and early Cambrian as the body-plan characteristics that we identify as phyla were evolving. The expansion of the founding *Hox* gene into an entire family of transcription factors occurred via a series of tandem gene-duplication events (Balavoine, de Rosa, and Adoutte 2002; Finnerty and Martindale 1999; Gehring, Kloter, and Suga 2009; Pearson, Lemons, and McGinnis 2005; Ryan et al. 2007; G. P. Wagner, Amemiya, and Ruddle 2003) within the stem of eumetazoans and stems of their phylum-level clades.[1] *Hox* genes have also been lost from some genomes, perhaps in lineages undergoing morphological simplification or losing some features regulated by a

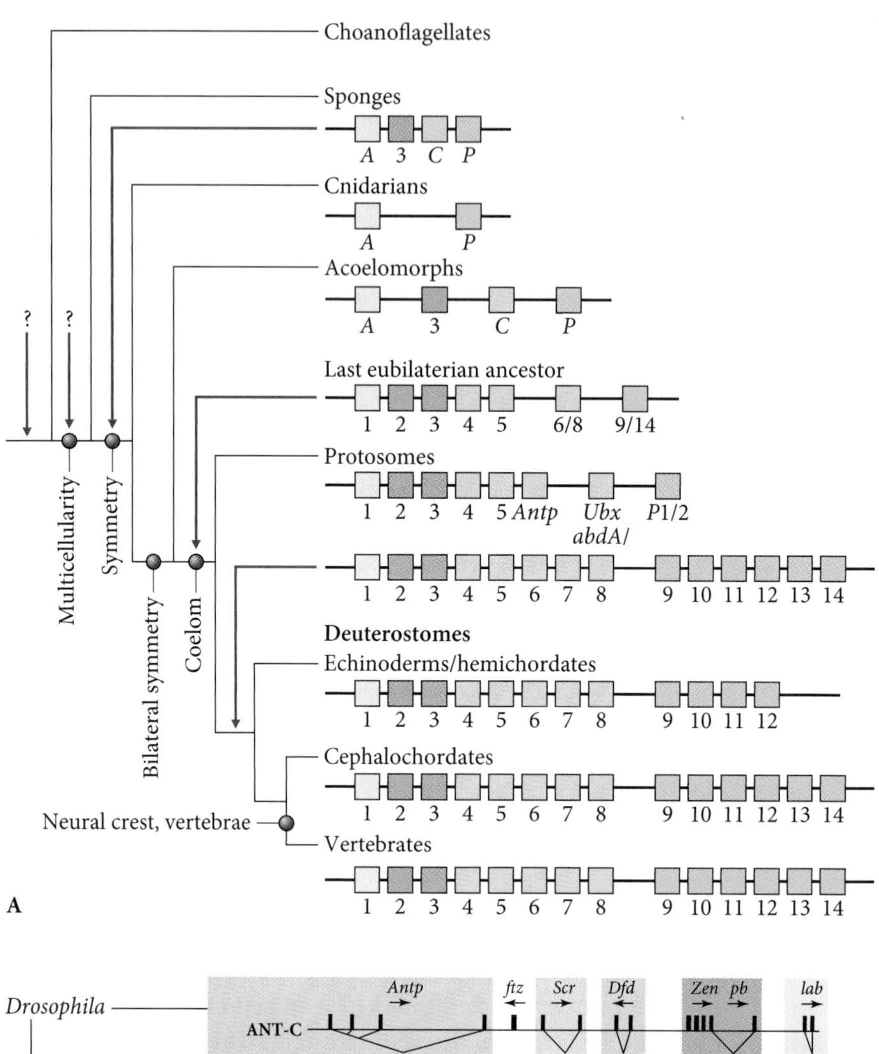
Choanoflagellates
Sponges
A 3 C P
Cnidarians
A P
Acoelomorphs
A 3 C P
Last eubilaterian ancestor
1 2 3 4 5 6/8 9/14
Protosomes
1 2 3 4 5 Antp Ubx abdA/ P1/2
1 2 3 4 5 6 7 8 9 10 11 12 13 14
Deuterostomes
Echinoderms/hemichordates
1 2 3 4 5 6 7 8 9 10 11 12
Cephalochordates
1 2 3 4 5 6 7 8 9 10 11 12 13 14
Vertebrates
1 2 3 4 5 6 7 8 9 10 11 12 13 14
Multicellularity
Symmetry
Bilateral symmetry
Coelom
Neural crest, vertebrae
A

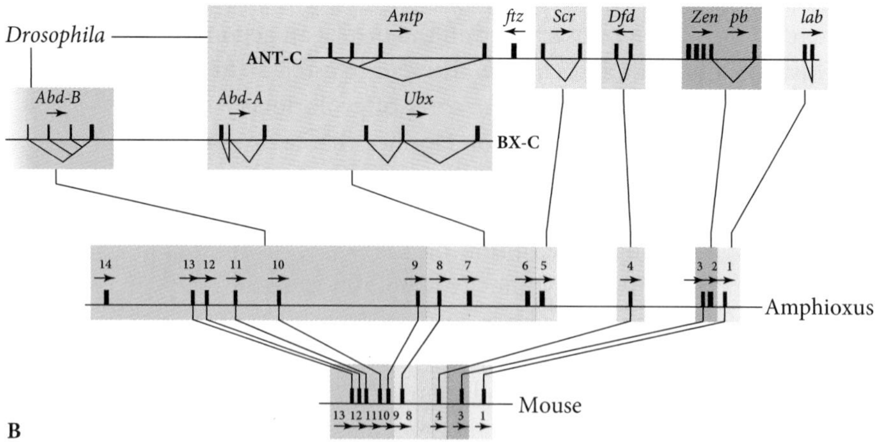
Drosophila
ANT-C
Antp
ftz
Scr
Dfd
Zen pb
lab
Abd-B
Abd-A
Ubx
BX-C
Amphioxus
Mouse
B

Figure 8.6 (left) Diversification of *Hox* genes across Metazoa. (A) Diversification of *Hox* genes across a simplified metazoan phylogeny, including the inferred composition of several last common ancestors (red arrows). (B) Structural representation of organization of genes for *Drosophila, Amphioxus,* and mouse (*Mus*), showing the relative distances between genes. ANT-C, *Antennapedia* complex. BX-C, *Bithorax* complex. Colored boxes correspond to the various paralogy groups within the *Hox* family, representing genes that are derived from a common ancestral *Hox* gene. (B) from Duboule (2007).

specific *Hox* gene. As a result, no two phyla have precisely the same *Hox* gene complement; in some cases, the *Hox* clusters even differ among classes within phyla (Hejnol and Martindale 2009). Duboule (2007) identified four different types of *Hox* clusters: organized, with tight spatial colinearity; disorganized, in which the genes cover a much larger region; split, as in *Drosophila*, with elements of the cluster on different chromosomes; and atomized, with the genes scattered and no cluster evident (fig. 8.7). Sufficient phylogenetic evidence is now available to indicate that an atomized cluster was likely present in basal eumetazoans, whereas the most tightly organized cluster occurs in vertebrates. In general, temporal colinearity of *Hox* gene expression appears to be associated with intact *Hox* clusters (Chiori et al. 2009). The similarity of arrangement between arthropods and vertebrates suggests that their last common ancestor (the last common protostome/deuterostome ancestor, or PDA), must have had a single, disorganized cluster of about eight genes, but subsequently evolved into atomized, split, and organized structures in

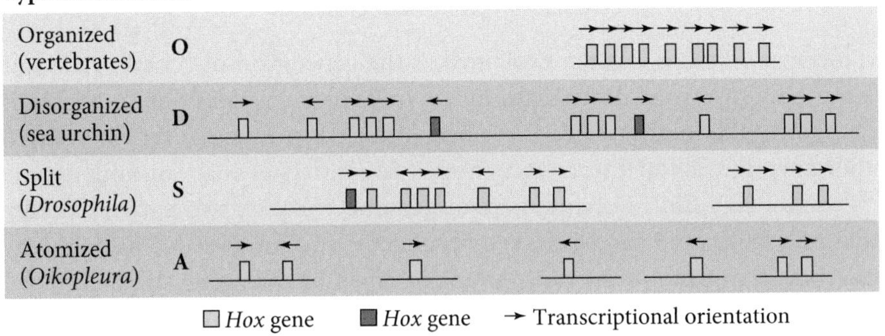

Figure 8.7 The structure of *Hox* gene complexes differs among various metazoan clades. Not all *Hox* genes are tightly clustered, and although it is not evident in this figure, the scale of the genes and the number of intervening genes differ considerably. The figure notes the structure of the genes based on the analysis of Duboule (2007): A, atomized; D, disorganized; O, organized; S, split.

various bilaterian clades, with the most tightly organized structure in vertebrates (prior to the two genome-duplication events that produced four clusters). This pattern of *Hox* gene evolution, if true, implies that global regulation must have been operating to drive the consolidation of the cluster in chordates and the preservation of clusters in numbers of other lineages (Duboule 2007).

EVOLUTION OF METAZOAN GENE REGULATION AND DEVELOPMENT

Through comparative studies of the developmental processes between very different animal clades, developmental biologists can essentially look back through time to the genomes of animals of the Ediacaran and Cambrian, when the LCAs of many of these clades lived. These studies demonstrate that the generation of the incredible diversity of animal forms is largely a consequence of the formation of new networks of gene interaction and new ways of regulating these interactions (Carroll 2008; Davidson 2006; Levine and Tjian 2003). Most gene families basic to the metazoan developmental toolkit evolved very early in the history of animals (and some within their ancestors) and have been highly conserved during their subsequent evolution. Simulation studies using a simple computational model of gene regulation, but with sufficient complexity to be pertinent, show that it is the combinatoric possibilities of regulatory networks that yield novelty (Lobo and Vico 2010). Despite a limited number of structural genes, the simulation generated symmetries, segments, and modularity among a diverse array of body plans.

Sources of Genomic Change

Evolutionary change is often explained as the generation of genetic variation via mutation followed by changes in the frequency of genes due to selection or random genetic drift. There have been great controversies over the relative importance of each of these mechanisms, controversies that continue today. With some exceptions, each metazoan genome contains two copies of most genes, one from each parent, and any change in nucleotide sequence (a mutation) within a gene produces a variant or allele of that gene. If alleles have different selective values, those that are selectively inferior may be eliminated from the gene pool of a population, whereas those that are selectively advantageous will, on average, tend to increase in frequency. The frequencies of different alleles at a given locus will change as the environment changes, and although there are many complications to this simple picture, evolution has sometimes been minimally defined as a change in gene (allele) frequencies.

A wealth of genomic data has emphasized the importance of selection in driving most evolutionary change. Indeed, for proteins, there is abundant evidence that their evolutionary trajectories through time are chiefly due to natural selection rather than neutral evolution (Endler 1986; Hurst 2009; A. Wagner 2008; G. C. Williams 1966). Most organisms harbor large reservoirs of genetic variation, and mutations can steadily add to this reserve; thus, organisms are able to respond to environmental challenges or opportunities by changes in gene expressions that are based on fitness differentials among individuals in populations. As the environment challenges the morphology and physiology of lineages of organisms—their phenotypes—populations respond through selective changes in their genotypes through differential reproduction.

Population size plays a critical role in determining the efficacy of selection versus genetic drift—stochastic changes in the frequencies of different alleles. Evolutionary biologists have long known about the importance of population size, but population genetic models often assume large population sizes. When effective population sizes are above 10^9 or so, as they are in many single-celled eukaryotes, selection is a very efficient evolutionary force. (The effective population is essentially the number of individuals that influence genetic variability and is often far less than the total population size.) As population sizes decrease, however—and effective population sizes in vertebrates may often be much less than 10,000 individuals—the efficacy of selection begins to drop and the importance of genetic drift increases. Genetic drift is the tendency of allele frequencies to change owing to random sampling of the gene pool from generation to generation. Because there is a negative correlation between body size and population size, a complicated evolutionary relationship exists in higher eukaryotes, with increasing developmental complexity, increasing body size, and decreasing population size. Lynch has forcefully advocated that various aspects of eukaryote genome architecture, including many features of development, may reflect the decreasing efficacy of selection as a consequence of this complex trade-off (M. L. Lynch and Conery 2003; M. L. Lynch 2006, 2007). We have no way of assessing population sizes during the Cryogenian diversification of basal metazoans, although invertebrates tend to have much larger populations than vertebrates and most bilaterian body sizes appear to have been small during the Ediacaran, suggesting large population sizes. Mean body sizes increased substantially during the Cambrian radiation, but the population sizes of those Cambrian species cannot yet be estimated. Consequently, whether the genomic architectures of diverging clades in the Ediacaran and Cambrian were heavily influenced by drift is not clear. Although most of the developmental genetic toolkit was already in place by that time, drift may have played a role in the diversification of elements of the toolkit, such as expansion of transcription factor families, and in other changes in regulatory control.

The importance of drift relative to selection at small effective population sizes is an example of nearly neutral models of evolutionary change in which mutations have only small effects on fitness. Nearly neutral theory grew out of the now discredited neutral theory that posited that many mutations had no differences in fitness. The predictions of change under nearly neutral models and selection can be very difficult to distinguish, however, and it remains unclear how much of the genome of most species has been shaped by each factor (Hurst 2009). Mutation is the other major driver of evolutionary change, but it has traditionally been seen as an input to the filtering of selection or to the fluctuations due to genetic drift. Nei, however, has suggested that mutation may have a role in shaping evolutionary change equal to that of selection (Nei 2005, 2007). Certainly, mutation pressure may be a source of non-random introduction of variation, biasing the direction of evolutionary change (Stoltzfus and Yampolsky 2009). Because there are specific sorts of developmental regulatory mutations that could have played a direct role in the early diversification of Metazoa, we will return to this subject later in this chapter. In addition, although these questions can be addressed empirically in modern groups, it is difficult to assess whether mutation pressure has changed over time within the genomes of early basal and later metazoans during the Ediacaran and Cambrian.

Sources of Metazoan Genes

As metazoan body plans evolved along many diverging pathways to produce the rich morphological disparity observed both in fossil and living faunas, it seemed reasonable to expect correspondingly large divergences in their genomes. After all, different phenotypic characters are underpinned by different genes or alleles; it seemed likely that arthropods and chordates, for example, shared few genes. When genes began to be sequenced from a wide variety of organisms, however, the results indicated much more genetic similarity among even distantly related groups than had been suspected. That housekeeping genes—genes that make eukaryotic cells work—are quite similar across the metazoan phyla was not so surprising. When the entire genomes of species in any two different phyla are compared, however, on average, two-thirds or more of the genes are closely related, and most of them are distributed so widely among metazoans that they must have originated before the Cambrian explosion. Therefore, those genes have had many hundreds of millions of years to diverge, and although some differences have emerged, they can still be classed into families of closely related genes, such as *Hox* genes. Some regulatory genes have changed so little that, when substituted for a native gene in the genome of a different phylum, they will function to mediate the development of the same features that the native gene mediated. A classic case is the gene *Pax6*, which is expressed high in the cascade that regulates eye development in many metazoans. The mouse version of this gene can be expressed in *Drosophila* and eyes will still develop, although these lineages have been separated since well before the Cambrian (Gehring 2004; Halder, Callaerts, and Gehring 1995).

GENE DUPLICATION Most new genes are believed to result from gene duplication, with subsequent divergence of the gene sequence and functions. Gene duplications have long been seen as a major source of evolutionary innovation. Indeed, some evolutionary biologists have argued that they are the primary factor behind the evolution of complex organisms (Nei 1969, 2007; Ohno 1970). Following gene duplication, daughter genes may subsequently specialize for different subsets of the parent gene function (subfunctionalization) or originate a function different from the parent gene (neofunctionalization) (Force et al. 2005; Innan and Kondrashov 2010; M. Lynch and Katju 2004; Van de Peer, Maere, and Meyer 2009). This process of gene duplication and divergence produces families of genes with DNA sequences similar enough, especially in their binding regions, that they can be recognized as descending from the same ancestral gene. The true frequency of gene duplication is probably underestimated because some duplicated genes may have diverged so much that their common origins are no longer evident. Duplications may involve just one gene, a few genes, or even a whole genome. In many gene families, the duplicated genes have divergent expression patterns under different environmental conditions (Colbourne et al. 2011); presumably, they underpin a variety of adaptive physiological responses to life in highly heterogeneous and variable environments.

Whole-genome duplications have been common in plants, but have by contrast played a much smaller role in animal evolution. Two whole-genome duplications occurred near the base of vertebrates and probably another in bony fish. There are other whole-genome duplications known among insects and amphibians, but most of them seem to be relatively recent (they occur near the top of their phylogenetic trees), and evidence suggests that metazoan lineages with duplicated genomes tend to become extinct (Van de Peer, Maere, and Meyer 2009). The increased gene numbers resulting from whole-genome duplications are soon reduced (e.g., although humans have such two duplications in their ancestry, they do not have especially large genomes). Genes that function to regulate development, however, tend to be preferentially retained following whole-genome duplications, which can obviously lead to the potential for increased regulatory complexity.

OTHER SOURCES OF GENES Other metazoan genes have been created by domain shuffling. A protein domain is a sequence that folds into a compact three-dimensional structure that can function and evolve independently of other domains within the protein. Many proteins comprise multiple domains, and domain shuffling—that is, the bringing together of domains that have previously functioned in separate genes—is an important source of new genes. Another source of new genes among single-celled organisms is "lateral" gene transfer (as opposed to "vertical" gene inheritance within a lineage), where genes move between different lineages. Lateral gene transfer, however, is rare among metazoans, where the best-established examples occur in the more basal groups (such as the genus *Hydra* among cnidarians; Chapman et al. 2010). Lateral gene transfer has also been implicated in the acquisition of the ability to biomineralize in a demosponge (D. J. Jackson et al. 2011). There is little evidence that lateral gene transfer contributed significantly to genome growth associated with the Cambrian explosion, however.

ALTERNATIVE SPLICING An important caveat to the poor correlation between protein gene number and morphological complexity is that many gene sequences can be manipulated to produce more than one product. Recall that metazoan genes are complicated and can be processed into mRNA in more than one way: different exons from a transcribed gene can be either excised or included by the splicing machinery and thus combine to form different proteins. Such alternative splicing allows the generation of a diversity of alternate transcripts, called isoforms, from a single transcribed sequence (Keren, Lev-Maor, and Ast 2010; Nilsen and Graveley 2010). When different isoforms are translated, they produce different gene products that are commonly expressed in different cell types, although they usually have somewhat similar functions. The 24,000 human genes can in theory produce more than 100,000 different proteins, and there is even one *Drosophila* gene that could in theory generate 38,016 different isoforms, which is more than the total number of gene sequences in the organism itself (Schmucker et al. 2000). Much alternative splicing involves the shuffling of sequences that are adjacent on a chromosome, but recent studies have shown that some isoform-like proteins involve the splicing of sequences from different, and sometimes distant, DNA sequences (Gingeras 2009). Preliminary data suggest that complex animals have a much higher percentage of genes that can produce isoforms than do simpler ones (perhaps 20% in flies and nematodes and perhaps as many as 94% in humans, although it is not certain how many of those isoforms are actually functional; Pan et al. 2008; E. T. Wang et al. 2008). So, one way the ability of relatively few genes to code for many outcomes can be increased is through the evolution of isoforms.

The ubiquity of alternative slicing provides yet another level of complexity to the diversity of molecular sequences and raises interesting issues about just what is a "gene." The production of proteins from DNA sequences has been the traditional view of a gene, but DNA sequences may have a wide variety of functions beyond translation and transcription into proteins. The concept of a gene has been evolving from a physical entity to a logical one, assembled from various pieces of the genome. Many of the evolutionary implications of this emerging view have yet to be explored.

Whole-genome sequences are now available for a number of phyla, and sequencing efforts are under way for many others, but the task of comparing the genes among phyla and evaluating their significance is far from complete. The genomes of several phyla that branched early in the metazoan tree have nevertheless been sequenced, and they provide a preliminary view of the early phases of metazoan genome evolution, leading up to the Cambrian explosion. By comparing the genomes of unicellular eukaryotes allied to Metazoa with those of sponges and by comparing the genomes of sponges to those of other early metazoan branches, it is possible to determine the latest possible branch among crown phyla at which a given metazoan gene, or a least the gene family to which it belongs, appeared in the metazoan genome. A sponge (the demosponge *Amphimedon*), representing the earliest crown metazoan phylum, has been sequenced (Srivastava et al. 2010). The resulting comparisons with gene families known in other metazoans at that time produced a much-needed and conservative look at gene sources. Of 4670 sponge gene families that were identified as present in some metazoans, about 73% are represented in nonmetazoans and thus represent a substantial metazoan inheritance from unicellular eukaryotes; many of these genes are housekeeping genes. Of the 1286 gene families known only in metazoans, about three-fourths arose along the metazoan stem, between the LCA of crown unicellular forms and crown sponges, by gene duplications. Slightly more than 1000 new protein domains or domain combinations also appeared along this stem, for about 15% of sponge gene families have no domains that are known outside of animals. The functions permitted by the sponge genome include not only developmental signaling and gene regulation, but also programmed cell death, cell-cell and cell-matrix adhesion, and, of course, cell type differentiation and specialization, among others (see Srivastava et al. 2010). Sponges invented many basic functional elements of metazoan development, expanding the genome from their unicellular inheritance. As this evolutionary accomplishment probably took close to 100 million years, it provides some perspective on the evolutionary abruptness of the rise of the disparate bilaterians that composed the bulk of Cambrian explosion fauna, which evidently took just a few tens of millions of years. So far as is known, neither comparably novel genomic innovations nor comparably novel body plans have evolved since the explosion.

Putnam and colleagues (2007) analyzed the genome of the cnidarian sea anemone *Nematostella,* providing much information on genome evolution between the sponges and the eumetazoans. Some 7766 gene families have members present in a sea anemone and in either vertebrates or arthropods; thus, each of these families must have been represented in the stem eumetazoan. Of those families, nearly 80% (6182) have gene members known to be present in clades other than Metazoa and thus must have been present in the last common unicellular ancestor of Metazoa; they were certainly present in the very first metazoans. Another 3% of those gene families contain domains that are also present in premetazoan groups but not in the combinations in which they are found in Metazoa. In other words, these genes have been assembled from ancient gene parts. Another 2% contain an ancient domain to which a metazoan-unique domain has been added. This leaves about 15% of cnidarian/PDA gene families that have no domains known outside of animals, most of which must have evolved after animals appeared but before the origin of the LCA of cnidarians and bilaterians (Putnam et al. 2007).

Expansion of the Eumetazoan Genome

Sponges deserve praise for their inventiveness because much of their genomic contribution was in the form of new genes, virtually doubling the numbers found in their unicellular forerunners. Sponges have made relatively modest use of their abilities for cellular differentiation, however, and although some sponges are beautiful, they are certainly not morphologically elaborate. These facts lead to the suspicion that sponge developmental regulatory systems are significantly less sophisticated than eumetazoan ones and that their *cis*-regulatory abilities were relatively primitive. But three of the four most common eumetazoan signaling pathways—*Wnt, Notch,* and *TGF-b*—evolved along the sponge stem, presumably as mediators of sponge-cell differentiation. Some elements of the fourth major pathway, *Hh* (Hedgehog), are also present in sponges, although this full pathway was finally fully assembled along the eumetazoan stem (Richards and Degnan 2009).

With recent data from the unicellular *Capsaspora* and from the choanoflagellate *Monosiga,* it is clear that many transcription factors were present in some unicellular eukaryotes and that their numbers expanded through domain shuffling, divergence, and gene co-option along the metazoan stem (Sebe-Pedros et al. 2011). Transcription factors underwent a further impressive expansion along the eumetazoan stem (fig. 8.8) (Larroux et al. 2008). The class of transcription factor genes with the homeobox domain is particularly diverse. Sponges have thirty-one such genes, and the number doubled to sixty-one or sixty-two along the eumetazoan stem, rising to eighty-two along the bilaterian stem. All told, the transcription factor classes Larroux and colleagues studied show fifty-eight genes in sponges, between eighty-seven and ninety-two in the eumetazoan LCA, and 115 in the bilaterian LCA. These numbers will certainly change as more genomes are included and explored, but their relative proportions are likely to hold. Although, as we have seen, the total number of protein-coding genes in metazoan genomes is not a good indicator of the branching order or the morphological complexity of their body plans, this growth of the developmental genome, occurring over the critical period of the late Neoproterozoic, must surely indicate an enhanced potential for regulating the development of novel morphologies and of enhancing body-plan complexity.

The rough sequence of gene introductions is being recovered as we reconstruct the genomes of LCAs. Genes that have become extinct within the lineage leading to a given genome cannot be studied in quite this way, but such evidence as we have indicates a history of gene origination and expansion that supports the account we have given above. The genome of the fly *Drosophila melanogaster* has been extensively studied, and through comparison to the genomes of other species, allows a broad understanding of the sequence of genomic growth throughout its particular ancestry (Domazet-Loso, Brajkovic, and Tautz 2007). Those data can be recast into a graph that indicates the times of origin of genes still found in living fly genomes in terms of the major nodes on the entire tree of life (fig. 8.9) (Marshall and Valentine 2010). Notice that after the establishment of the eukaryotes (which came to employ in the neighborhood of 10,000 genes in their genomes), there is nearly a doubling in new gene introductions along the metazoan stem; an even more outstanding phase of gene introduction clearly occurs next, along the eumetazoan stem, more than redoubling the number introduced during the origin and early history of sponges. Before the Cambrian explosion, gene expansions recorded

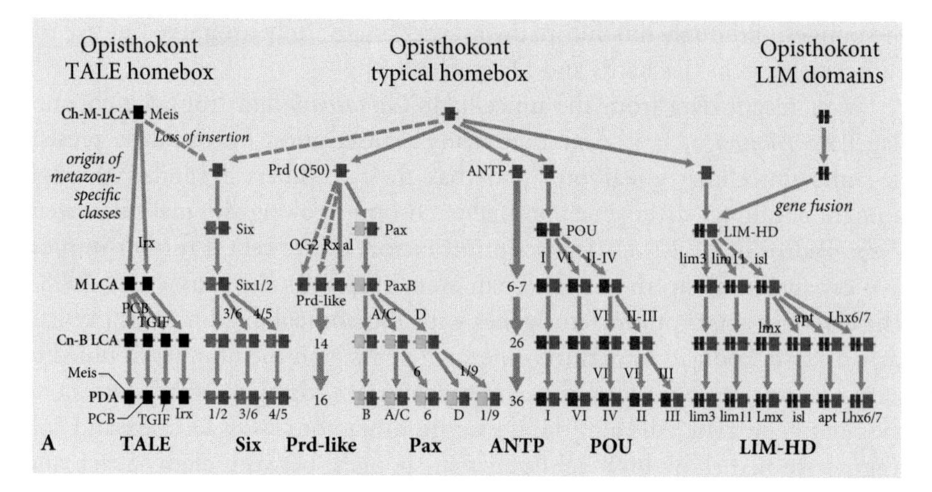

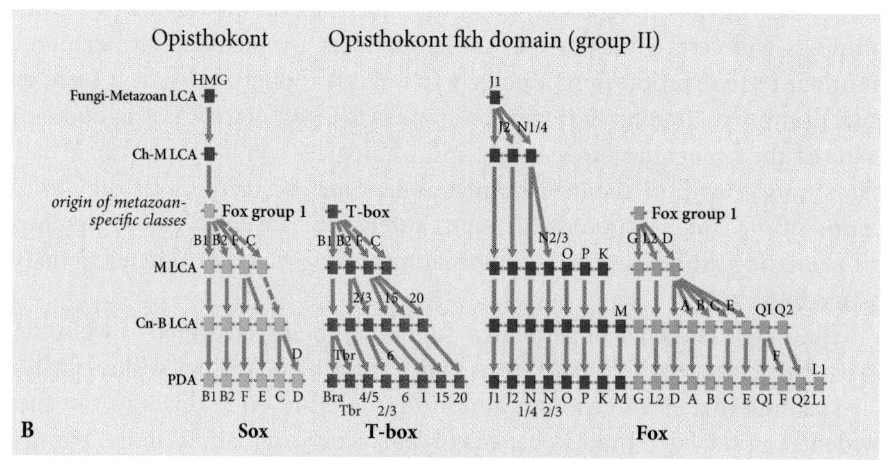

Figure 8.8 Reconstruction of the expansion of transcription factor families along a series of key nodes. (A) Homeobox gene classes. (B) Sox, T-box and Fox gene classes. Dashed lines indicate poorly supported inferences. Ch-M: Choanoflagellate-metazoan last common ancestor (LCA). M: Metazoan. Cn-B: Cnidarian-Bilaterian. PDA: Protostome-deuterostome ancestor. From Larroux et al. (2008).

in fly genomes seem related to periods of change in grades of organization, which makes sense because they involved increasing morphological and developmental complexity associated with the assumption of new modes of life. During the origin of the bilaterian, protostomian, and even the arthropodan clades, few new genes were added to, or at least retained in, the modern *Drosophila* genome. The origin of the group that includes crustaceans and insects (the Pancrustacea) certainly involved an important rise in gene introductions,

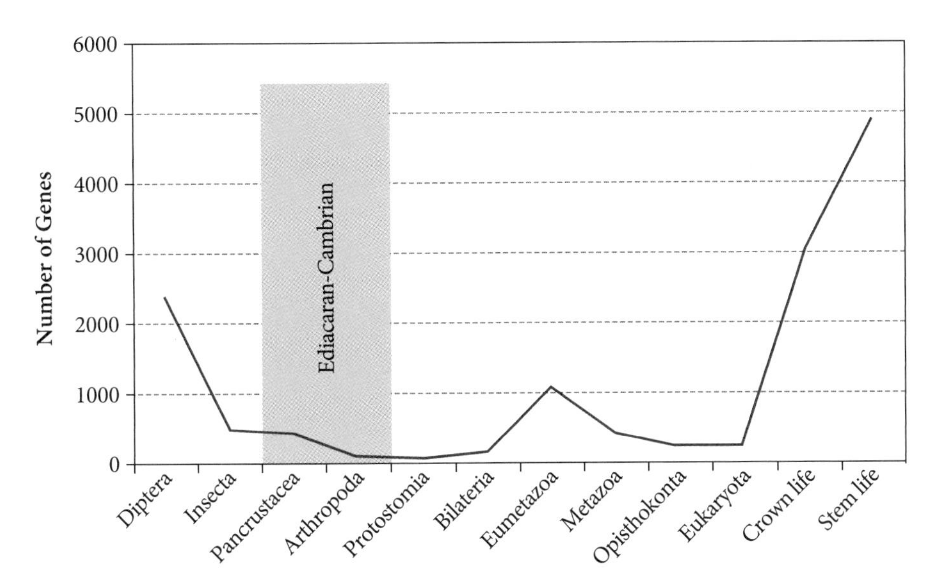

Figure 8.9 Times of origins of 13,382 genes within the genome of *Drosophila melanogaster.* Genes had their origin within the clade indicated, but before the time of origin of the clade to the left. Tan bar shows the likely position of the Cambrian explosion relative to the lineage divergence, with the origin of arthropods and pancrustacea predating the Cambrian explosion. From Marshall and Valentine (2010), based on data from Domazet-Loso et al. (2007).

however, and, finally, *Drosophila* retains within its genome nearly a couple of thousand genes that originated during the evolution of flies themselves (Diptera). The appearance of new genes associated with rise of insects (before 400 Ma) and the later origin of flies may be linked to the invasion of land and the acquisition of nonaquatic adaptations—again, to adaptations to new modes of life—but these latter genes, the youngest large cohort of genes in the genome, include many that have appeared during the origin of the familial, generic, and specific morphologies that characterize *D. melanogaster*.

Usually, the genomes of species belonging to separate bilaterian phyla do differ in a fair number of genes—as many as 30% but commonly fewer. It is plausible that, on average, late-appearing genes are involved in the fine tuning of morphology within a narrow adaptive zone or niche, and that they turn over more rapidly than the genomic elements associated with the basic features of a metazoan or eumetazoan body plan. Thus, the genomes of the largely stem organisms of the Cambrian explosion doubtless contained reasonable percentage of genes that were associated with the particular adaptive zones and niches they occupied but that are now beyond our ability to discover.

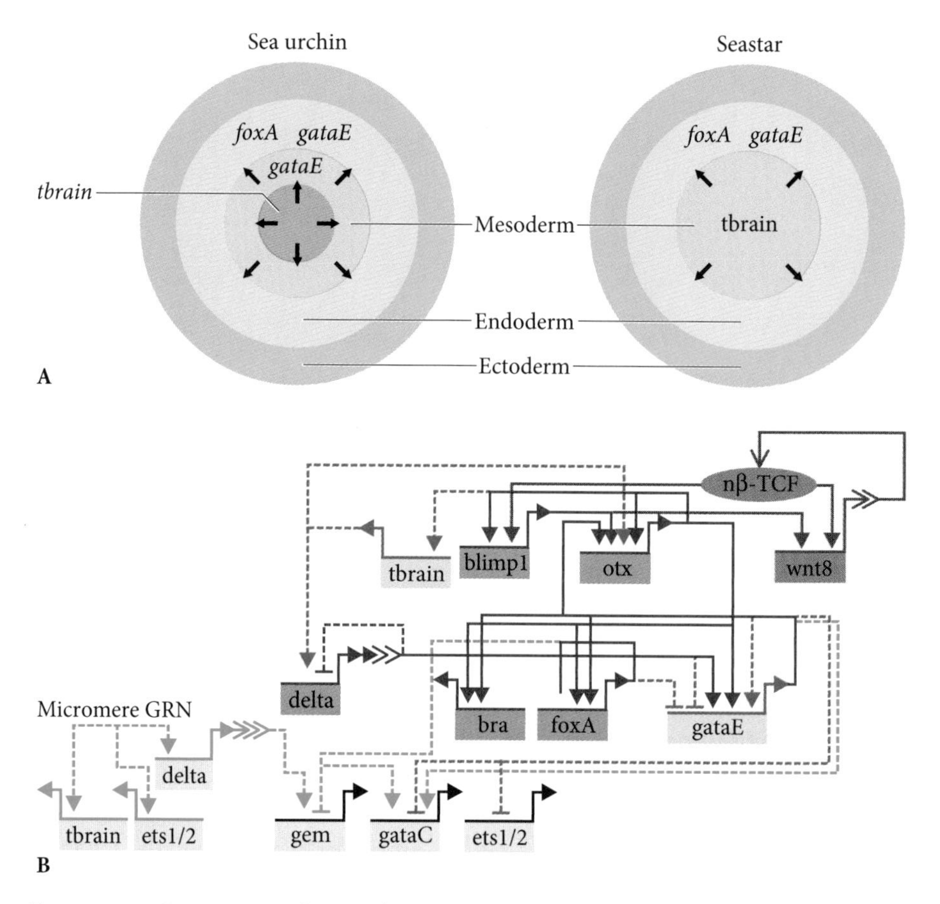

Figure 8.10 Comparison of sea urchin and starfish kernels for endomesoderm development. (A) Schematic represesntation of sea urchin and starfish blastulae viewed from the vegetal pole. In both clades the mesoderm (blue disk) is derived from the central vegetal plate region and the surrounding cells will give rise to endoderm (tan ring). In modern sea urchins, an additional mesodermal lineage (the micromere lineage indicated by the pink disc) forms at the centralmost vegetal plate and forms the blastocoel during gastrulation. Some gene expression domains are shown: e.g., *tbrain* is expressed in the micromere cell lineage in sea urchin and in the mesodermal and endodermal territories in sea star, and gataE is expressed within the endoderm and mesoderm in sea urchin but only the endoderm in sea star. The arrows represent *Delta-Notch* signaling from one cell territory to another. (B) The GRN depicting endomesoderm specification in sea urchins and sea stars at blastula stage. Each horizontal line above the gene name represents a *cis*-regulatory module. The regulatory interactions found in common in both taxa are shown in red, while those occurring in only sea urchin are shown dashed in green, and those only occurring in sea star are shown dashed in blue. In sea urchins, the nuclearization of β-catenin is critical for the establishment of endomesoderm and (*continued*)

EXPANSION OF REGULATORY INTERACTIONS The various kinds of gene regulatory network elements differ in their evolutionary lability. At one extreme, comparison of the developmental GRN for endomesoderm development in sea urchins with that in starfish reveals a recursively wired, highly conserved module, a subcircuit of regulatory gene interactions that compose the core of the network (fig. 8.10). This type of module is called a kernel. Although many of the regulatory interactions upstream and downstream of this kernel have been modified in the 500 million years since these two clades shared a common ancestor, the kernel itself has been preserved (Hinman, Nguyen, and Davidson 2007; Hinman, Yankura, and McCauley 2009; Hinman et al. 2003). This kernel has also been identified within zebrafish, establishing that it has been conserved across deuterostomes (Tseng et al. 2011). Obviously, this conserved regulatory module has been highly refractory to modification. In general, such kernels are genes linked by mutually reinforcing feedbacks so that simply silencing any of the component genes would disable the system (Davidson 2006, 2009; Davidson and Erwin 2006, 2010a; Erwin and Davidson 2009; Hinman and Davidson 2007; I. S. Peter and Davidson 2009, 2010). Although it is possible to imagine circumstances under which a gene could be substituted within a kernel—for example, if the morphology of the group has been greatly simplified—a number of kernels have been identified that had their origins early in the bilaterian radiations.

The formation of a kernel has several evolutionary consequences. First, these network subcircuits of regulatory genes define the spatial pattern of part of the developing embryo. In the example from echinoderms, the genes involved in the kernel define the endomesoderm. Second, because the genes are recursively wired, multiple regulatory interactions are required for the function of any single gene. Third, because kernels are so refractory to modification once they form, selection can act on them as a unit rather than on specific interactions within the kernel. And fourth, although there is no evidence of any unusual evolutionary processes in the original construction of kernels, once they are formed they change the opportunities for successful evolutionary innovations in their region of the developmental network. Unless a kernel

---

Figure 8.10 (*cont.*) forms a positive feedback loop with *blimp1* (shown in brown). The role of nuclear β-catenin has not been examined in sea stars, but is likely to be conserved. Many of the initial molecular events acting downstream of the nuclearization of β-catenin are conserved in these two echinoderms (i.e., all of the interactions shown in red), and are defined as a kernel. Based on Hinman, Yankura, and McCauley (2009).

is modified (perhaps owing to morphologic simplification), the kernels establish limits to the pathways of morphologic variation for the region of the body plan for which they are responsible (Davidson and Erwin 2010b). Because kernels are the most refractory elements of developmental gene regulatory networks, their formation can force subsequent evolution to other components of a network.

Developmental GRNs also contain modular elements called plug-ins (Davidson and Erwin 2006), as epitomized by signaling cassettes. Because the targets of these plug-ins are enhancer modules to which their transcription factors bind (see above), target genes may be added or subtracted from a signaling cassette, usually through sequence changes at binding sites within the enhancers or by replacement of the transcription factor at the end of the cassette. Changes in transcriptional regulation usually involve both the number and distribution of the transcription factor binding sites, but they may be caused by changes in the sequence, and thus the specificity, of the transcription factors themselves (V. J. Lynch and Wagner 2008; Wray et al. 2003). Some of the genetic elements used in metazoan cassettes are found in unicellular eukaryotes, although their functions there have not been determined. Still another network feature consists of regulatory switches that operate upstream of the various modules to determine when and where they are expressed within the developmental system. Although conserved as switches, they are not necessarily dedicated to particular developmental functions. Such switches were present in the stem eumetazoan ancestor. At the periphery of the GRN and downstream of the various types of regulatory modules are the batteries of structural genes and their low-level regulators that produce the differentiated cell and tissue types. Indeed, most classical studies of microevolution are performed on mutations of these peripheral structural genes and evaluated for their consequences by processes of selection. Alterations in those differentiation gene batteries have little effect on the structure or function of the network architecture, yet they provide the physical components and their physiological and behavioral features that permit multicellular organisms to function.

One of the most impressive aspects of this hierarchical view of the structure of developmental GRNs is how well it tracks the dynamic of evolutionary changes across the Linnaean hierarchy (fig. 8.11). A reasonable hypothesis given existing data is that the formation and stabilization of regional patterning within the developing embryo was achieved largely, and perhaps exclusively, through the formation of kernels. At the opposite extreme, differences between species within genera, and probably between closely related genera, often involve differences in structural gene batteries at the distal tips of regulatory networks. Intermediate differences in morphology, often recognized in

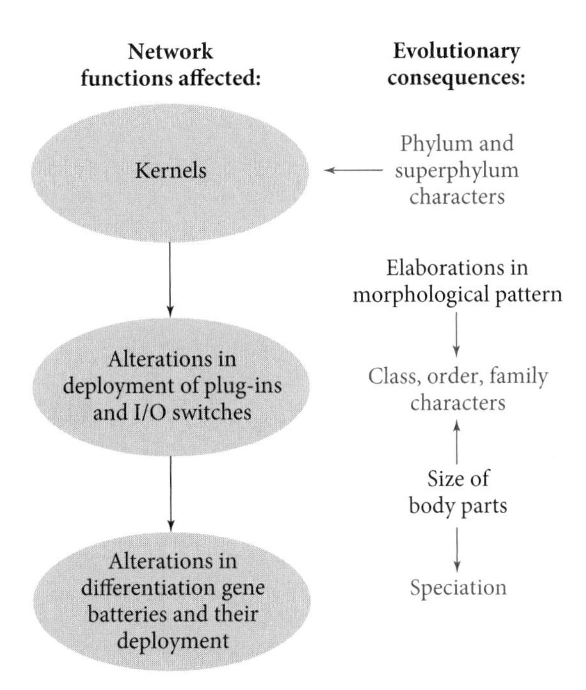

Figure 8.11 Differences in evolutionary lability depending on where in a developmental GRN evolutionary changes occur. The left column shows changes in dGRN components; the right column shows the evolutionary consequences, and, in red, how they might be interpreted within a Linnaean taxonomic system.

a Linnaean scheme as orders and families, may commonly be produced by changes in regulatory patterns around kernels, particularly by plug-ins and switches.

EVOLUTION OF MICRORNAS In general, the number of microRNA (miRNA) genes in a genome correlates with the complexity of the body plan, and it follows that their numbers increased before and during the Cambrian explosion within the genomes of lineages that display increasing complexity. Many miRNAs seem to have been conserved once they evolved, although there are indications of turnover within some miRNA complements (fig. 8.12). About forty or so miRNAs have been identified in a sea anemone genome, and several of them were conserved in later-branching lineages (Grimson et al. 2008). There are 706 miRNAs known in humans (Kosik 2009). The most reasonable interpretation of this pattern is that the progressive addition of miRNAs has been important in enhancing the ability to stabilize specific cell and tissue types in different lineages and in buffering their development

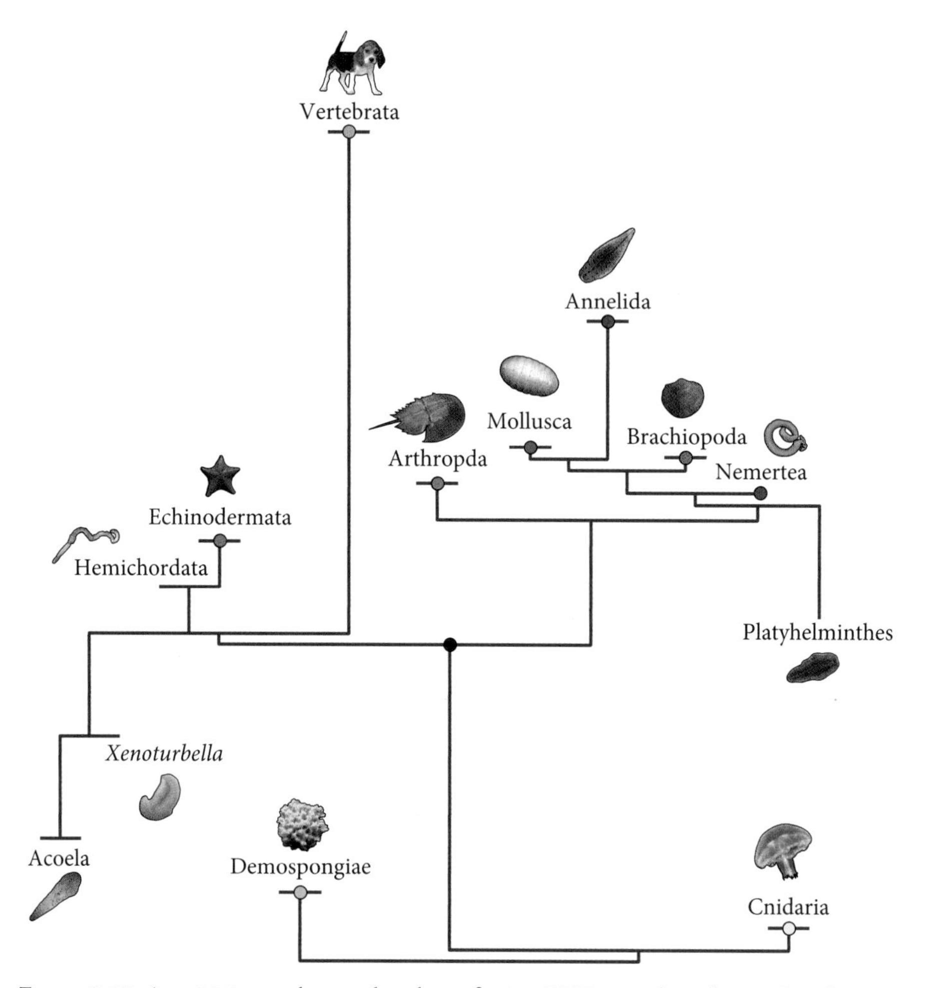

Figure 8.12 Acquisition and secondary loss of microRNAs in selected taxa. One hundred and thirty nine miRNA families from twenty-two metazoan species were coded, with the branch lengths proportional to the total number of miRNA genes gained at that point minus those that were secondarily lost (scale bar represents ten genes total). Increases in morphological complexity are correlated with increases to the miRNA toolkit, and secondary simplifications in morphology correlate with relatively high levels of secondary miRNA loss. From Erwin et al. (2011).

against environmental perturbations, although they play other regulatory roles as well (Christodoulou et al. 2010; Grimson et al. 2008; Sempere et al. 2006; Wheeler et al. 2009). A single miRNA can downregulate (i.e., lower the effective dosages of) hundreds of genes, thus overseeing major differences in cell lineages. The strong correlation between the richness of miRNA repertoires and the complexity of body plans suggests that the function of these molecules contributes significantly to that complexity. One challenge for future work is to understand the relationships between miRNA regulation and other systems that regulate development programs. In addition to miRNAs, other classes of small RNAs are essential for development, some of which (named Piwi-interacting RNAs after the gene family they bind with) act in the nucleus to regulate the activity of some mobile genetic elements (Grimson et al. 2008). Other, larger RNAs that do not code for proteins seem also to be used in regulating genomic machinery, but little is known about their functions (Jacquier 2009; Mattick, Taft, and Faulkner 2010).

Expansion of Cell Type Numbers and Functions

Cell types, like individual genes, have family trees and in many cases appear to have evolved from multifunctional ancestral cell types, becoming specialized for a narrower range of functions or even a single function (Arendt 2008); this trend recalls the subfunctionalization of duplicate genes. Arendt, Hausen, and Purschke (2009) argue that this process of "cell-type functional segregation" was common in the early history of animals and that cells may have thus become *less* complex rather than more complex. A similar argument for reduction in cell complexity has been made by McShea (2002) with regard to the evolution of multicellular forms from unicellular ancestors because eukaryotic unicells must each perform the whole range of metabolic and behavioral functions necessary for life, including reproduction, whereas cells in metazoans can be relieved of some of these tasks by specialized cell types and can become simpler. The evolution of photoreceptors has provided a model of functional segregation within eumetazoans (Arendt, Hausen, and Purschke 2009). The earliest eye may have sensed only light intensity, but it required shading of light-sensitive parts of the cell. Later, cilia were incorporated to enable cells to respond to light cues by movements, that is, by phototaxis. Arendt and colleagues consider the possibility that all three of these functions (photoreception itself, light shading via pigment granules, and steering via locomotory cilia) were originally found within a single cell type. As eyes evolved, three distinct functional cell types eventually appeared, separating these functions. Furthermore, many cell lineages have certainly evolved functions that were not within the range of their ancestral cells, recalling the neofunctionalization of duplicated genes. A likely example is that rhabdomeres (microvilli of cells that are arrayed at the receptive surface of many protostome eyes) do not have a role in photoreception in living vertebrates but are found in the ganglion cells (neurons) that form the inner layer of the vertebrate eye.

Because differentiation of specialized cell types accompanied the evolution of tissues and organs, cells became more specialized and more complex and, presumably, contributed to improved adaptation. Distinctive evolutionary trajectories have underlain the integration of the morphological, physiological, and behavioral systems of each metazoan body plan. In chapter 9, we attempt a brief look at some features of the nodes and stems that may lie in the history of the major clades of living metazoans.

Lessons from the Evolution of the Developmental Genome

One clear theme in the evolution of the developmental genome has been the co-option of regulatory genes to serve new functions. The protein products of structural genes tend to have dedicated functions that are defined by their structures and chemical properties, and their biochemical jobs remain about the same during evolution. When structural genes are reassigned to new addresses and activated in new contexts and combinations through the evolution of regulatory pathways within the developmental genome, profoundly novel morphologies may result.

We have speculated that developmental regulators evolved from physiological regulators in unicellular eukaryotes so that genes used to regulate physiological or behavioral functions were used for cell differentiation in multicellular ones. In support of this notion, genetic motifs with orthologs in the developmental genome are found in genes of choanoflagellates and their allies, although commonly there has been some sequence evolution or even domain shuffling within the genes along the metazoan stem. In addition, within the genomes of crown representatives of each early phylum-level branch of the metazoan tree are genes that came to serve functions that had not yet evolved when the genes originated; these genes were clearly recruited later to regulate features newly arising along the stem of the next major branch. If there is an important exception to this generality, it is probably found primarily among the sponges. Their evolution involved the origin of a significantly larger genome than their unicellular ancestors; it is as if they did not have all the *cis*-regulatory abilities that are so well documented among crown eumetazoans and had to rely on inventing new genes to meet adaptive challenges. This relative deficiency was evidently overcome along the eumetazoan stem, eventually permitting the explosive radiation of many animal body plans.

The average crown eumetazoan (with, say, about 20,000 protein-coding genes) has a suite of many thousands of housekeeping genes, largely inherited from sponges, that permit its cells to function. In addition, this hypothetical animal contains many signaling and transcriptional genes in its developmental genome that lie above the modules of structural gene batteries and their low-level regulators. Finally, there are a few thousand genes devoted to coping with the taxon-specific requirements and demands of physical and biological aspects of its niche and relevant features of its broader adaptive zone. Although these latter genes may differ quite a bit among taxa—they may well make the greatest contribution to the genetic distances among metazoan genomes, given the rich variation within the environmental mosaic—they should be relatively young on average. They have turned over as lineages invaded new regions of ecospace or as environmental conditions shifted, more or less gradually eliminating the need for one suite of such genes but requiring their replacement by a new suite, perhaps even roughly similar in number.

Although according to gene counts in sequenced genomes the number of protein-coding genes that are addressed by transcriptional regulators has not grown excessively in eumetazoans, the increasing numbers of isoforms during the evolution of the more complex body plans make it impossible as yet to establish the relative numbers of distinctive gene functions in their genomes. Earlier, we mentioned that the genome of *Daphnia*, a small crustacean, harbored many more than 30,000 protein-coding genes; they may be required for adaptation to the extraordinarily broad range of environmental conditions the species encounters. That high gene count resulted largely from gene duplications, presumably with subsequent divergence in function among the duplicates, but *Daphnia* exhibits little gene splicing. Other organisms seem to increase their functional diversities by evolving isoforms, which may not get counted as separate genes. Why the *Daphnia* genome evolved by duplicating genes rather than by creating isoforms is not understood. At any rate, portions of the GRN that are devoted to dealing with the expression patterns of all such genetic elements are likely to have become increasingly complex as the numbers of gene functions increased and must be interwoven with and adapted to the network of expression patterns of the nominal protein-coding genes.

Developmental Patterning of Body Architectures

The developmental genetic system—the developmental toolkit—that we have described forms the foundation for the basic architecture of eumetazoan body plans, but how are they assembled to construct such complex architectures? Expression studies of the activities of transcription factors and signaling systems revealed extensive similarities in members of widely divergent clades, initially through comparisons of gene expression patterns in vertebrates and *Drosophila*. The similarity in gene expression patterns among major body axes of these groups as well as in other developmental systems reflects conservation of transcription and signaling factors. In some cases, genetic interactions appear to be conserved as well, as shown in the studies of the role of *Pax6* in eye formation in different phyla.

Developmental biologists had little reason to expect similarities in the genes or developmental networks responsible for generation of body axes, appendage formation, segmentation, eye formation, the developing gut, or the nervous system in clades with such very disparate body plans, but that is what early gene expression studies began to suggest. As comparative developmental studies expanded, the extent of highly conserved developmental genes became a focus of considerable attention (see Amundson 2005; Carroll, Grenier, and Weatherbee 2001; R. A. Raff 1996; Wilkins 2002). The similarity of patterns among distinctive, complex body-plan patterns led to a number of papers suggesting that the LCA of protostomes and deuterostomes had shared a large suite of regulatory elements (Balavoine and Adoutte 2003; Carroll, Grenier, and Weatherbee 2001; Davidson and Erwin 2006; Erwin 2006a, 2009; Hejnol and Martindale 2008; Marshall and Valentine 2010; Martindale, Finnerty, and Henry 2002; Martindale and Hejnol 2009; W. E. Muller et al. 2004; Peterson and Butterfield 2005; R. A. Raff 2008), which has led in turn to two alternative views of the nature of the PDA. In one view, the PDA was a fairly complex animal, possessing antero-posterior and dorsal-ventral differentiation; a differentiated head with a tripartite brain and nerve cord; sophisticated sensory systems, including a primitive eye with both ciliary and rhabdomeric photoreceptors; a differentiated gut; and possibly segmented mesodermal blocks, a heart, and appendages (e.g., Baguña et al. 2001; Carroll, Grenier, and Weatherbee 2001; De Robertis and Sasai 1996; Kimmel 1996; W. E. Muller et al. 2004; Ohno 1996; Schierwater and Kuhn 1998; M. P. Scott 1994; Shenk and Steel 1994). The alternative view is that although the PDA was triploblastic and was probably an elongate vermiform animal with differentiated body axes, the genes that regulate development of derived features in crown protostomes and deuterostomes may have been devoted to simpler and more generalized tasks, such as specifying the position of cell or tissue types along the axes. During the explosion, they were thus co-opted to their present roles as more complex, semihierarchical network patterns evolved to mediate the development of complicated organs and structures. In the following sections, we describe some of those highly conserved developmental tools, focusing on the ones shared among *Drosophila* and vertebrates as a bridge to the discussion of the metazoan LCAs in chapter 9.

Main Body

Bilaterians are characterized by antero-posterior and dorsal-ventral axes and in many cases possess appendages and segmentation. Gene expression studies suggest considerable conservation of the underlying developmental mechanisms that produce these patterns.

BODY AXES The *Hox* genes described previously have a primary role in establishing antero-posterior patterning within developing bilaterian embryos. In bilaterians, the *Hox* genes are expressed in a colinear fashion so that the anterior part of the developing larvae is controlled by genes at the upstream end of the gene cluster, with more posterior portions controlled by progressively more downstream genes (McGinnis and Krumlauf 1992). As noted, the *Hox* genes experienced a series of duplications during the early history of Metazoa, so the PDA had a suite of at least eight genes (Balavoine, de Rosa, and Adoutte 2002; de Rosa et al. 1999). These eight *Hox* genes expanded to fourteen in early chordates, and most vertebrates have four distinct clusters, representing a twofold duplication of the entire cluster, followed by considerable gene loss (Duboule 2007; Ferrier and Minguillon 2003). *Hox* gene domains establish the basic antero-posterior axes in the developing embryo, although some of these genes are used in other contexts as well (e.g., proximo-distal differentiation). Across bilaterians, the most anterior and posterior sections of the body axes are controlled by non-*Hox* genes: *Otx/Otd* in the anterior (Hirth et al. 2003) and caudal (*Cad*) in the posterior (Copf, Schroder, and Averof 2004; de Rosa, Prud'homme, and Balavoine 2005).

Dorsal-ventral axis specification is mediated by a gradient in activity in *Dpp/BMP-4* that is antagonized by *Sog/Chordin*. The vertebrate homolog, *BMP-4*, is expressed at high levels in the ventral side of vertebrates and is retarded dorsally by *Chordin*. In *Drosophila* and other ecdysozoans, *Dpp* is expressed dorsally and retarded ventrally by *Sog*. The mirror symmetry in expression patterns suggests that the vertebrate and arthropod dorsal-ventral axis have been inverted relative to each other (De Robertis and Sasai 1996), and gene expression studies indicate that this pattern has been conserved since the PDA.

APPENDAGES The astute observer will have noted a conundrum posed by the evolutionary relationships of Cambrian lobopods, the anomalocarids and related organisms, and the arthropods in figure 6.29: the first and last have legs, but *Opabinia* and the other dinocarids make due with lateral, gill-bearing lobes or flaps (X. L. Zhang and Briggs 2007). *Anomalocaris* has the jointed, preoral appendages that make it such a fierce predator (by Cambrian standards, at least) but no legs or other jointed appendages. If the phylogenetic tree in figure 6.29 is at least roughly correct, either appendages evolved independently in lobopods (including Onychophora and Tardigrada) and Arthropoda or the lateral flaps of the anomalocarids are developmentally homologous with lobopod appendages and in turn developed into arthropod limbs (Janssen et al. 2010).

This problem is but a small part of a larger question about the similarities in appendages across the protostomes and deuterostomes. The *Drosophila* transcription factor distalless (*Dll*) and its vertebrate ortholog (*Dlx*) exhibit conserved gene expression patterns (G. Panganiban and Rubenstein 2002; G. E. F. Panganiban et al. 1997). *Dll/Dlx* is also involved in development of other parts of the embryo, particularly in the peripheral nervous system. The similarities between arthropod and vertebrate development are even more extensive: limb primordia in *Drosophila* are induced by *Wnt*, where they express *Dll*; in vertebrates, the induction of neural crest begins with *Wingless* (a *Wnt* homolog) followed by *Dlx* expression. The posterior compartment of the antero-posterior limb axis is controlled by hedgehog (*Hh*) in *Drosophila* and by its vertebrate ortholog sonic hedgehog (*Shh*) in mice. The boundary between the anterior and posterior compartments in arthropods is formed as *Hh* induces the expression of decapentaplegic (*dpp*); *Bmp-2*, a *dpp* ortholog, serves the same role in vertebrates. There are other similarities in the formation of the proximo-distal and dorsal-ventral axes between *Drosophila* and vertebrates. Despite these considerable similarities, Shubin, Tabin, and Carroll (1997) argued that the similarities were due to independent co-option of signals from a morphologically more primitive common ancestor. An alternative explanation was offered by Minelli (2003), who noted that segmented appendages are only found in clades in which the main body is segmented (it depends a bit on one's definition of segmentation; see below). He suggested that this coordinated pattern of gene expression was initially associated with formation of the central body axis and was co-opted independently.

A recent study of the development of the simple and unsegmented legs of onychophorans provides useful insight into the problem of appendage formation. The onychophoran limb is regionalized by *Dll* as well as the other gap genes required by arthropods (homothorax, extradenticle, and dachshund). This recognition supports earlier suggestions that the genetic cascade for regionalization preceded segmentation and was later co-opted in the more complex appendages of arthropods (Janssen et al. 2010). Thus, appendages per se are likely homologous across the Cambrian lobopods, *Anomalocaris* and its relatives, and other stem panarthropods, while articulated, segmented appendages are unique to arthropods.

SEGMENTATION Vertebrates, annelids, and arthropods are all segmented, and this modular construction has been used to explain why each of these clades is both morphologically and taxonomically diverse. The type of segmentation is distinct in each group, however. Indeed, even the nature of segmentation is contentious. Serial repetition of organs is generally not considered segmentation, despite Budd's (2001) claim that organs rather than the entire body can be segmented. In *Drosophila*, the pair-rule and segmentation genes progressively subdivide the embryo into units via top-down segmentation, but other arthropods have a very different regulation of segmentation. Spiders, for example, use *Delta-Notch* signaling in a manner very similar to vertebrates (Stollenwerk, Schoppmeier, and Damen 2003). Vertebrates use bottom-up segmentation via a *Delta-Notch* signaling cascade to produce somites; in annelids new segments bud from a growth zone (Chipman 2010; Tautz 2004).

Thus, there was no reason to expect similarities in the developmental patterns underlying segmentation across the various bilaterian clades. Engrailed (*en*) and its vertebrate homologs, however, are expressed in the posterior compartment of *Drosophila* segments and in the posterior portion of vertebrate somites, although the pattern of segmentation in *Drosophila* is much different from other arthropods. The *Delta-Notch* system used in vertebrates is also responsible for spider segmentation, and caudal and its vertebrate ortholog *Cdx* are also involved in segmentation (Copf, Schroder, and Averof 2004). Together, the similarity in these gene expression patterns led to suggestions that these genes were operating to produce segments in the PDA (Balavoine and Adoutte 2003; De Robertis 2008; Kimmel 1996; Seaver 2003). As suggested by the morphological differences in the pattern of segmentation, this similarity now appears to be an overinterpretation of the gene expression patterns. Neither comparative anatomy nor the fossil record indicates that the earliest representatives of lophotrochozoans, ecdysozoans, or deuterostomes were segmented, although the ancestral ecdysozoan may have had some metamery, and the earliest deuterostome may well have had segmentally repeated gill slits in the head (Chipman 2010). As in previous cases, it appears that gene networks initially established for axis determination and differentiation in bilaterians were subsequently and independently co-opted for segmentation in different lineages (Chipman 2010).

Organs and Organ Systems

The apparent conservation of developmental pathways among disparate bilaterian groups has also been observed within specific organs and organ systems.

GUT The different origins of the mouth and anus in protostomes and deuterostomes is fundamental to the distinction between the two groups and provides little basis to expect developmental similarities in gut and endoderm formation, yet expression studies of GATA transcription factors and *forkhead* reveal surprising similarities. GATA transcription factors are expressed in early endoderm formation, as are *brachyury* and *goosecoid* in the developing foregut (Arendt, Technau, and Willbrodt 2001; Zaret 1999). Early in animal evolution, duplication of a cluster of four *Hox* genes established a second cluster that is involved in gut patterning in many Bilateria (Brooke, Garcia-Fernandez, and Holland 1998; Ferrier and Minguillon 2003). Known as the Parahox genes, this cluster also has similar expression patterns across bilaterians, but not in cnidarians (Chiori et al. 2009).

HEART AND CIRCULATORY SYSTEMS Heart formation in both *Drosophila* and vertebrates involves the gene tinman and its vertebrate homolog *Nkx2.5* and, with the co-conservation of other genetic elements, suggested a common source for this process (Bodmer and Venkatesh 1998; R. P. Harvey 1996; Olson 2006). Rescue experiments in which the mouse *Nkx2.5* gene was expressed in *Drosophila* generated visceral mesoderm but not heart development, however. Further, nematodes do not have a circulatory system, but the nematode ortholog of these genes, *ceh-22*, is involved in pharyngeal muscle development. Thus, the ancestral function of this gene was likely visceral mesoderm cell development, probably of a contractile muscular tube rather than of a heart (R. P. Harvey 1996; Ranganayakulu et al. 1998; Tanaka et al. 1998).

CENTRAL NERVOUS SYSTEM The development of the anterior regions of a developing bilaterian embryo is closely linked to the formation of the brain and central nervous system. Vertebrates have a tripartite brain, composed of fore-, mid-, and hindbrains followed by a dorsal spinal cord. In *Drosophila,* the cerebral ganglion is followed by a subesophageal ganglion and then the ventral nerve cord. *Otx* (in vertebrates) and its protostome ortholog *Otd* are responsible for patterning the most anterior components, followed by expression of *Pax2/5/8* transcription factors near the midbrain/hindbrain and subesophageal domain and then the anteriormost *Hox* genes. The similarities in gene expression patterns led to lengthy discussions about whether a tripartite brain was the ancestral condition of the PDA (Arendt and Nubler-Jung 1999; Hirth et al. 2003; Lichtneckert and Reichert 2005). Much of the uncertainty stems from debates over the presence of a central nervous system (CNS) in hemichordates and chordates. This controversy continues (Arendt et al. 2008; Lowe 2008; Nomaksteinsky et al. 2009), but a recent study found signaling centers homologous to three vertebrate signaling centers in the hemichordate *Saccoglossus kowalevskii* (Pani et al. 2012). This suggests that ancestral deuterostomes possessed a very complex developmental system for the CNS, which degenerated in cephalochordates but has been preserved in hemichordates. There are strong similarities in gene expression patterns of nervous system development between some protostomes and deuterostomes, which suggest to some that a similar CNS must have been present in the PDA, yet it is not clear that they could not represent independent co-option of gene expression patterns for a regionalized nerve network.

The pattern of staggered gene expression of *Hox* genes in protostome and deuterostome central nervous systems is also found in the staggered *Hox* gene expression of acoels, which lack a CNS, suggesting that the apparent homology of CNS structures in protostomes and deuterostomes could reflect co-option of a more ancient axial pattering system (recall that molecular phylogenetic studies now place acoels as an early diverging branch of deuterostomes). Studies of *Hox* gene expression in acoels suggest that the ancestral role of these genes is regionalization of the CNS and that *Hox* gene expression was subsequently co-opted for more complex patterning of other tissues along the antero-posterior axis (Hejnol and Martindale 2009).

We have already noted that similar expression patterns of *Pax6* in eyes across a variety of metazoans and ectopic expression studies of the vertebrate *Pax6* gene in *Drosophila* led to claims that eyes might be conserved from the PDA. There is considerable difference between a simple photoreceptive cell and a complicated eye with an image-forming lens and associated structures. Sadly, this distinction has been lost in some discussions of the evolution of eyes. The occurrence of both ciliary and rhapdomeric photoreceptors in cnidarians and across bilaterians indicates that a photoreceptive cell occurred in basal eumetazoans, as did at least two types of opsins (light-sensitive proteins) (Plachetzki and Oakley 2007; Vopalensky and Kozmik 2009). The lens-forming eyes of bilaterians required the co-option of lens cystallin genes from a variety of different sources (Piatigorsky 2007), and the repertoire of opsins expanded. Coordinating the development of these eyes involves an interactive network of transcription factors in addition to *Pax6*: other *Pax* genes, sine oculis (*Six*), eyes absent (*eya*), and dachshund (*dac*).

Complexity of the Urbilateria

The proposals that the PDA was a complex animal were not unreasonable when they were made given the limited comparative developmental data then available. If the urbilaterian was quite complex and was greater than about 1 cm long, the trace fossil record suggests that it could not have existed before at least 555 Ma, the age of *Kimberella*, and possibly not before the early Cambrian (Valentine, Jablonski, and Erwin 1999). Claims for a complex urbilaterian assume that highly conserved developmental elements also conserve their accompanying suite of network interactions, however. As we discuss in

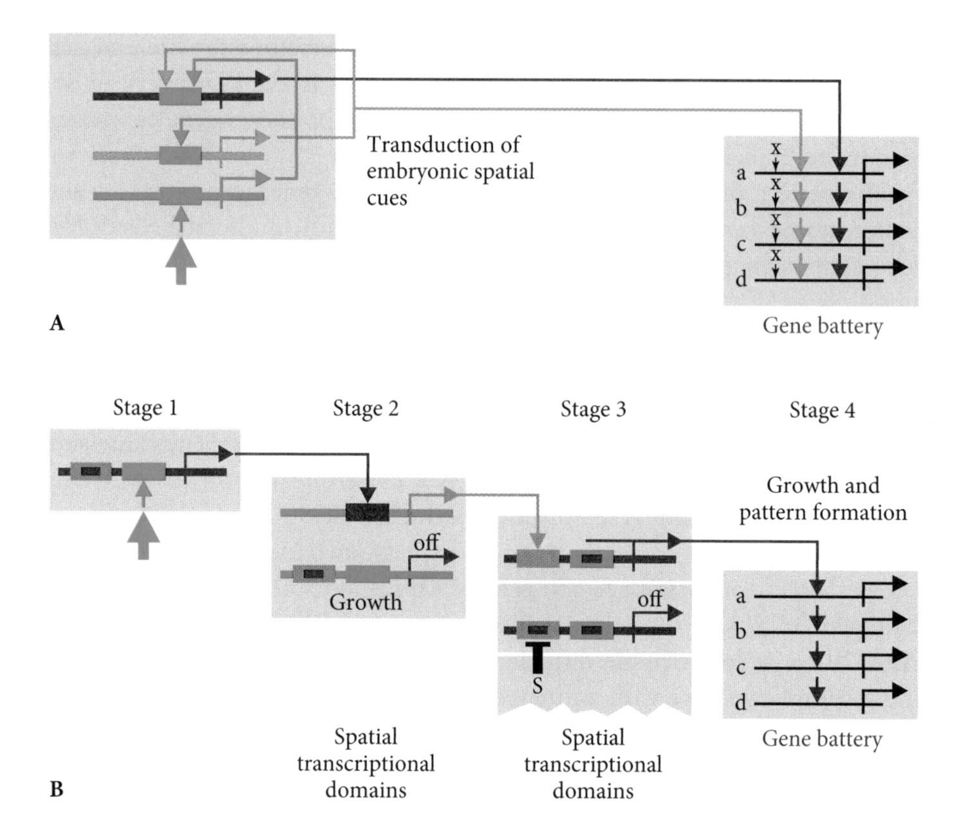

Figure 8.13 Gene intercalation in the developmental gene regulatory networks during evolution. The boxes represent transcriptional domains where the state of the domain is dependent upon the presence of a gene product. (A) The genes in the box to the left transduce spatial embryonic cues (thick green arrow) and acrivate an initial gene (green), which in turn activates two additional genes (orange and red), all of which produce transcription factors. The orange gene also cross-regulates the red gene. These genes in turn regulate the gene battery to the right. These genes encode proteins used for a differentiated cell type. Each gene has at lease two *cis*-regulatory inputs, indicated by the orange and red arrows. Each gene may also have other *cis*-regulatory inputs, indicated by the "x." (B) A gene regulatory network at a later state of evolution following intercalation of two new spatial transcriptional domains (stage 2 and stage 3). The early cell differentiation gene battery is now incorporated in a pattern formation system that controls the differentiation gene pattern. Only the red gene from (A) is shown in this figure, activated at its initial embryonic address through the green gene. A new regulatory linkage has appeared so that the transcriptional activator from the red gene now controls the purple gene (purple arrow). A second *cis*-regulatory module has been added to the red gene, allowing it to be activated by the purple gene product, or repressed by a signal (S) from an underlying spatial domain (stage 3). The result at stage 4 is to mount the differentiation gene battery on the morphological structures developed from stages 2 and 3. From Erwin and Davidson (2002), redrawn from fig. 5.7 of Davidson (2001).

more detail in chapter 9, the discovery that many of these genes, although not necessarily their developmental gene networks, were also present in cnidarians and sponges poses some problems for the complex urbilaterian hypothesis. For example, eight genes that play significant roles in mesoderm formation among bilaterians have been found in cnidarians (Martindale, Pang, and Finnerty 2004). Because cnidarians do not possess mesoderm, the presence of these genes is clearly not diagnostic for mesoderm. We conclude from this data that the original role for many of the highly conserved developmental genes across Metazoa was not associated with the development of complex morphological features (eyes, appendages, guts, etc.) but rather with cell-type specification.

We favor an alternative to the complex urbilaterian scenario whereby the ancestral role of many of these highly conserved developmental genes may have been the formation of simple gradients or specification of particular cell types. These developmental control genes were ideally suited to being co-opted because subsequent evolution led to the formation of more complex regulatory networks. Specifically, as shown in figure 8.13, we believe that spatial and temporal regulatory control regions were intercalated into simpler regulatory networks to produce the greater regulatory specificity required in most protostomes and deuterostomes (Erwin and Davidson 2002). Such intercalary evolution was described previously for *Pax6* in eyes (Gehring and Ikeo 1999) and explains the presence of these highly conserved genes near the apex of a regulatory cascade but does not require a conservation of the remaining developmental cascade. The similarities between the developmental processes of various bilaterian clades are often in differentiated cell types rather than in complicated morphogenetic processes. Thus, neurons and their associated ganglia and the slow contractile muscles of hearts, ciliary and rhabdomeric photoreceptors, and so on are conserved, but not the developmental processes that produce hearts, eyes, or segmentation. Morphogenetic processes begin by establishing a general pattern in the appropriate region of the developing embryo, followed by more specific developmental cascades.

NOTE

1. A word about types of homology among related genes: If the same gene is found in two different species, it is described as orthologous. Such orthologous genes are in contrast to a gene that undergoes a gene-duplication event, producing two related genes within the same organism; these genes are described as paralogous.

PART FOUR

Evolutionary Dynamics of the Cambrian Explosion

CHAPTER NINE

Ghostly Ancestors

Many ancient books are known to have existed only because they are referred to in other works, even though no copies are known to have survived into today's world. The extinction of books is not limited to ancient tomes; some novels that we know that Darwin read seem to have vanished. Although a text may be lost forever, its existence and even something of its contents can often be gleaned from references in surviving works. The same holds true for many ancient organisms that have left no fossil record. Their former existence is revealed, and indeed required, by other organisms, living or fossil. These missing forms are ghost lineages. The Ediacaran is haunted by a diverse and disparate population of such ghosts, which to paleontologists seem to be clamoring for recognition.

Comparative morphology has long provided scientists an avenue with which to reconstruct these evolutionary ghosts across the 500 million to 600 million years or more since they lived. Decades of molluscan specialists, for example, have discussed the putative characteristics of the "hypothetical ancestral mollusk," better known as HAM. Based on characters that seemed to be shared by the major classes of mollusks, HAM served as a foundation for discussions of molluscan evolution in the days before phylogenetic analysis. Phylogenetic analysis provides a more powerful and rigorous approach, however, because placing known organisms in the tree of life allows something of the character of their ancestral morphologies to be inferred. In addition, although the genomes of Neoproterozoic and Cambrian groups are irretrievably lost, we now have whole genome sequences and comparative developmental studies for living representatives of a growing number of metazoan clades. This comparative developmental information, as discussed in chapter 8, provides an entirely new source of data for assessing patterns of metazoan evolution.

Such comparative developmental evidence reveals something of the nature of vanished ancestral lineages. In this chapter, we resurrect a few of those ghosts, ones representing critical nodes or biological advances that proved to be of special significance in the unfolding of metazoan history. By comparing the developmental genes—and, even more importantly, their deployment within developmental gene networks in the developing embryos of modern organisms—biologists have identified highly conserved developmental patterns, which can be referred to the nodes in the tree where they first appear. Although fossils, particularly representatives of stem lineages, often provide unique character combinations that are important for understanding developmental and morphologic evolution, few of the nodes we discuss here are themselves represented by fossils, so we discuss fossil representatives only where particularly relevant. Chapter 10 integrates chapter 8 and this chapter on comparative developmental evolution with the fossil record discussed in chapters 5 through 7 to present our synthesis of the early evolution of metazoans and the Cambrian explosion. First, though, some ghost stories.

NAMING CLADES AND NAMING TAXA

As we have seen, most of the animals that appeared before and during the Cambrian explosion for which we have a fossil record do not share a last common ancestor with crown taxa below the level of phylum. Therefore, they represent stem members of living phyla or, as in the case of the Cambrian lobopods, of even larger clades. Although the daughter branches of the LCA of two phyla will evolve independently, those daughters do not possess the morphological features that will eventually characterize and define these taxa (fig. 9.1). In fact, the founding members of each daughter branch, being daughter species of the LCA, differed morphologically from the LCA and from one another only in the ways that closely allied species differ from one another, and they would usually be classed in the same genus on Linnaean-style morphological grounds. Thus, in general, the earliest members of large clades will resemble members of the group from which they branched far more closely than later members of the clade they have founded. From a phylogenetic perspective, these ancestors and their immediate descendants are nondiagnosable as stem members of their phyla. Only when new morphological features associated with new body plans begin to be assembled do stem lineages become identifiable in the fossil record (see fig. 9.1). This situation produces an interesting problem for the paleontologist, who must rely entirely on morphology to place fossils within the phylogenetic tree. Clearly, the greatest difficulties occur during the successive appearance of higher taxa and thus particularly just before and during the Cambrian explosion. The node occupied by the

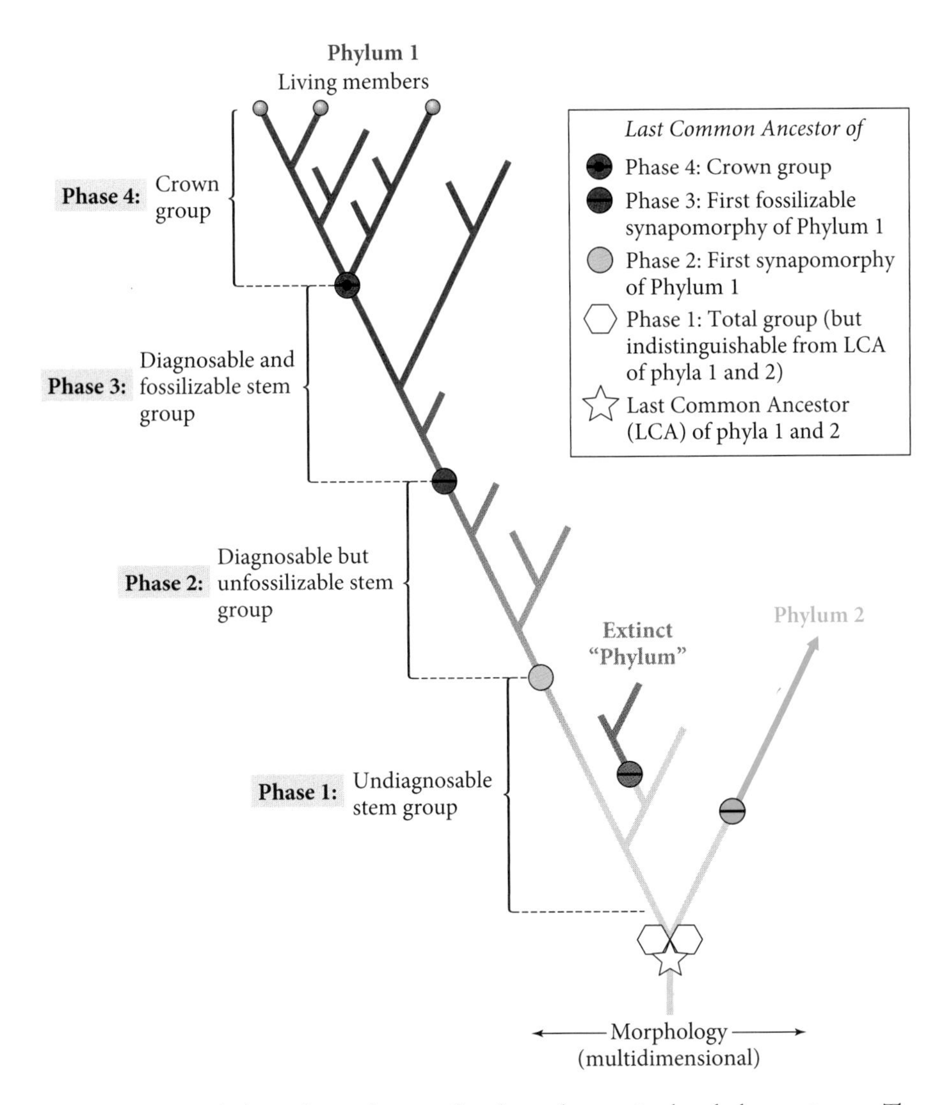

Figure 9.1 Morphological significance of nodes and stems in the phylogenetic tree. The stem group of a clade that evolves a phylum-level body plan is founded by a species that closely resembles its ancestor (star) and its sister (hexagon) and that cannot be diagnosed morphologically as a member of the phylum (phase 1). Subsequent anagenetic divergence, however, produces features that permit such diagnosis should fossils be preserved (phase 2). When members of this clade become fossilizable (phase 3), they are diagnosed as members of the stem group. Finally, members of living clades appear (phase 4; dark red circle with the dot is the crown ancestor), and it is such crown groups that have provided the basis for defining phyla. Modified from Marshall and Valentine (2010).

LCA of the living clades of a phlyum represents their crown ancestor; crown clades are distinguished because members are still present, not because they necessarily possess any special features.

Thus, the major nodes on the tree of life that separate large clades (such as those ranked as phyla or other higher taxa) do not indicate the origin of the body plans that characterize them today. The nodes indicate the presence of a speciation event, but the genomic and morphological changes involved in the speciation may have nothing at all to do with the changes that led to the eventual body plans of the derived taxa. The clades that we recognize as phyla earned their status through tens of millions to hundreds of millions of years of evolution. On the other hand, their distinctive architectures, once evolved, lasted much more than half a billion years and certainly merit recognition as the physiological and biomechanical wonders that they are. If we were secure in our knowledge of all the phylogenetic patterns and their relations to specific body plans, the nomenclature of major clades could be revised, but as we write, we must continue to use the familiar Linnaean names for phyla, which can be placed within the phylogenetic hypothesis presented in figure 4.5.

THE ANCESTRAL METAZOAN (EARLIER THAN 780 MA)

Multicellularity has arisen perhaps twenty times or more in different eukaryotic lineages, but only three of those lineages—metazoans, fungi, and embryophytes—have gone on to produce bodies with differentiated cell types, which raises the question of whether some clades possessed more of the features required for multicellularity than others, and of this subset of lineages, whether cellular differentiation was favored in only a few of them (Buss 1987; Kaiser 2001; King 2004). Because choanoflagellates are the best living candidates for the metazoan sister group (fig. 4.5), it is not surprising that their genome contains some genes otherwise unknown outside of Metazoa. For example, some of those genes have homologs that are components of signaling pathways commonly used in cell-type differentiation and patterning in metazoans, whereas other genes code for proteins—cadherins—used for cell adhesion in metazoan tissues, yet in both cases they are found in single-celled or colonial organisms. In all, more than seventy protein domains known to be shared by choanoflagellates and metazoans are unknown elsewhere (King 2004; King and Carroll 2001; King, Hittinger, and Carroll 2003; King et al. 2008). The genomes of other, earlier-branching protistan groups, however, are also beginning to yield gene domains not previously known in unicellular organisms, although some of their proteins lack domains characteristic of their presumed metazoan homologs. Thus, it appears that many of these genes had not yet been assembled into the form they take in metazoans at the point that choanoflagellates diverged. Elements of several metazoan transcription factors and signaling families are also present in choanoflagellates, although most are less diverse than in metazoans.

An excellent example of the early appearance of transcription factors is the LSF/Grainyhead (GRH) family (Traylor-Knowles et al. 2010), which has two subfamilies, LSF and GRH. The LSF subfamily is known in choanoflagellates and even fungi, but it is much more diverse in metazoans, where it is involved in a wide variety of functions. The GRH subfamily is only known to be involved in the development and maintenance of epithelia, an important metazoan novelty (see below). Indeed, as Traylor-Knowles and colleagues note, the appearance of GRH subfamily genes seems to coincide with the appearance of epithelia.

As for signaling families, there are seven major signaling pathways in metazoans. Elements of four of them are known in choanoflagellates, although what becomes the JAK/STAT signaling pathway is represented by only a single gene (Larroux et al. 2008). We can nevertheless see in the choanoflagellate genome the rudiments of what will become sophisticated molecular developmental programs in metazoans.

THE METAZOAN STEM (EARLIER THAN 780 MA)

The whole-genome sequence of the demosponge *Amphimedon* reveals that sponges possess large numbers of genes otherwise found only in eumetazoans, where some are involved in functions that do not occur in sponges. A variety of families of transcription factors have been identified, and some exhibit tissue-specific activity, indicating a degree of regional gene expression. Furthermore, the genome of a homoscleromorph sponge has components of six of the seven major signaling pathways that operate in metazoans (Nichols et al. 2006). There are, however, relatively few members within each family of transcription factors, which is consistent with the view that the sponge regulatory machinery lacked the complexity and probably the developmental control found in more-derived metazoan clades; in addition, some of the signaling pathways, although much more eumetazoan-like than those in choanoflagellates, lack elements found in eumetazoans. Furthermore, complex developmental gene regulatory networks have not been identified in sponges, which may limit the complexity of morphologic structures that can be formed (Adamska et al. 2007; Degnan, Leys, and Larroux 2005; Fahey et al. 2008;

Larroux et al. 2006; W. E. G. Muller 2003). The last common ancestor of sponges and eumetazoans was nevertheless able to specify multiple cell types, establish coherent body architectures, and array the differentiated cell types to produce multicellular structures (Adamska et al. 2011).

If the molecular phylogenies indicating that sponges are paraphyletic are correct, the eumetazoan clade ancestor would necessarily have had a sponge-like architecture. Some eumetazoan-type features that are unusual in sponges are found in various living sponge groups, including gastrulation (Boury-Esnault et al. 1999), intercellular junctions, and basal membranes (Boury-Esnault et al. 2003). Some sponges even feed as predators (Vacelet and Boury-Esnault 1995). Demosponges also have many of the genes that contribute to epithelial structures among eumetazoans, although additional novel domains appeared in the lineage leading to eumetazoans (Fahey and Degnan 2010). Although these features are functionally similar to those in eumetazoans, it is uncertain whether any of them are metazoan homologs; their presence does confirm that each can evolve within sponges, however. The early radiation of sponges probably gave rise to some groups that are now extinct, including a diversity of archaeocyathans and possibly chancelloriids, in addition to producing crown sponge groups (chap. 6).

These extinct spongiform clades provide unique character combinations missing from living clades, and, in combination with morphological and developmental data from living species, they aid in generating a scenario for their early evolution (fig. 9.2). The history of the metazoan stem involves the evolution of cell differentiation, which was accompanied by the origin of many regulatory genes and the enlargement of gene families, thereby establishing a fairly extensive regulatory toolkit to support specialized cell types that could be organized into tissues. Judging from molecular clock dates, the processes of invention of the stem sponge morphologies and their supporting genomes may have required, or at least lasted for, nearly 100 million years or more before the eumetazoan LCA appeared. Arising from the sponges, the eumetazoan stem evolved a genomic system that permitted increasing morphological complexity without requiring the invention of a novel gene or gene family for each new function.

THE EPITHELIOZOAN LAST COMMON ANCESTOR

The genus *Trichoplax* of the phylum Placozoa (chap. 4) may represent the most basal living epitheliozoan. Like sponges, placozoans lack muscles and nerves, but their tissue-like construction, enabled by the presence of cell junctions, is unlike any living sponge, and their epithelial cells exhibit cellular polarity. The

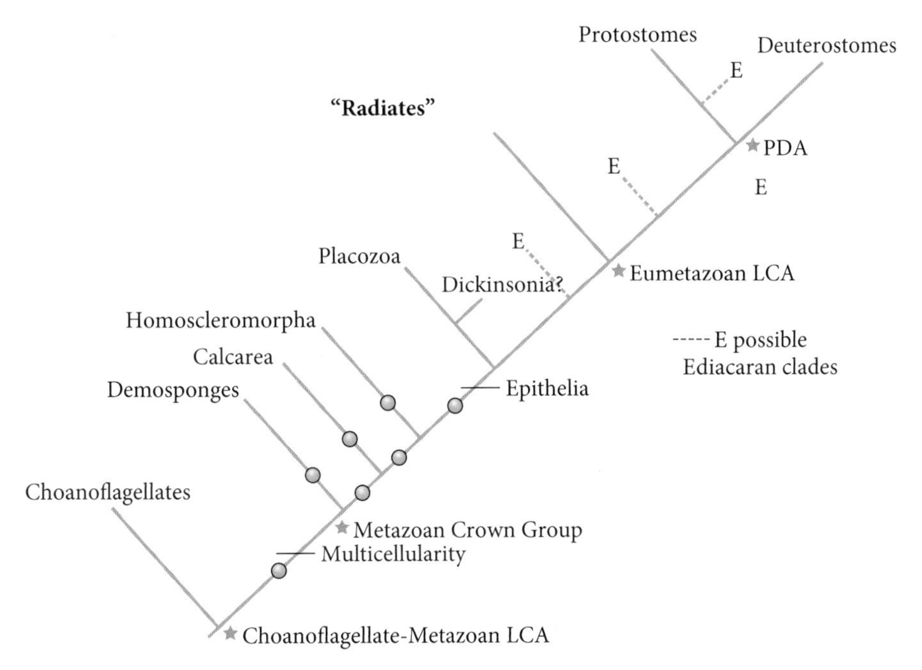

Figure 9.2 A hypothesis for the branching order among sponge, epithelial, and major eumetazoan-grade clades; notice that the tree includes Ediacara macrofossils. Stars identify last common ancestors, and closed circles are on stems and denote a division between undiagnosable and diagnosable stems. This tree is extremely spare, but the many stems indicate the difficulties in interpreting the placement of the fossils that happen to be found but that lack, as they commonly do, successions of fossils that link them to ancestral or descendant forms. Many divergences from both undiagnosable and diagnosable stems surely occurred within many clades.

lack of any clear antero-posterior orientation in placazoans is shared with the radiate phyla. Indeed, placozoans show no indication of any of the bilateral features that can be found in cnidarians and ctenophores. Their genome, however, has many genes that are important in eumetazoan development but that are unknown in sponges, including at least thirty-five different transcription factors representing at least four different families (table 8.1). Several signaling pathways are present as well. Although placozoans lack nerve or muscle cells, organs, a through gut (or even a gut lumen), and an antero-posterior axis, they have members of gene families associated in a broad sampling of bilaterians with the specification of neurons and myocytes, with antero-posterior patterning, and with organogenesis. In placozoans, some cell types are multifunctional; the organisms do respond to stimuli, even though no specialized nerve cells are present (Srivastava et al. 2008).

The evolutionary path from sponge to placozoan remains entirely speculative. That there are large missing gaps along this path is indicated by the placozoans having many of the genes that are important in eumetazoan development and that are unknown in sponges. Did placozoans arise from the larvae of one of the extinct branches of sponges by paedomorphosis (i.e., by shifting reproduction into the larva)? Or perhaps placozoans evolved from a very aberrant sponge that had shifted from filter feeding to extracellular digestion? Or did they come from a branch of sponges that evolved adult epithelia? Or have placozoans descended from a more complex eumetazoan that is long extinct? The pathway between sponge and eumetazoan is one of the most puzzling of all.

What was the morphology of the first organisms that could be assigned to Eumetazoa and from which the LCA of cnidarians and bilaterians evolved? The answer is partly a matter of definition. If we define eumetazoans as having epithelial tissues, placozoans would branch later than, or represent a branch of, that stem ancestor. If, however, we require the stem ancestor to have nerves and muscles as well, placozoans branch at an earlier position along the ancestral path, assuming that they have not lost nerve and muscle cells along the way (which is certainly a possibility). In figure 9.2, we arbitrarily require eumetazoans to have epithelia and nerve and muscle cells (it is possible that nerve and muscle cells evolved concurrently to permit responses to sensory inputs). Many of the Ediacara macrofossils (chap. 5) lack any characters clearly associating them with either cnidarians or bilaterians. Like placozoans, these ediacarans do not appear to have organs, which raises the question as to whether they may be related to placozoans. At least three ediacarans (*Dickinsonia*, *Yorgia*, and *Kimberella*) show signs of movement, however, suggesting but not requiring that they had muscle or nerve cells. Sperling and Vinther (2010) argue that *Dickinsonia* may have been a stem placozoan, based on the inference of a common feeding method through external digestion via the sole, the absence of a mouth or a gut, and the ability to move. The size and axial patterning of those evidently mobile ediacarans are not similar to features of the living placozoan *Trichoplax*, but the general argument is plausible.

THE EUMETAZOAN LAST COMMON ANCESTOR (APPROXIMATELY 700 MA)

Proceeding along the phylogenetic tree to cnidarians, we find a situation similar to that of sponges and placozoans in that the range and number of elements of the genetic toolkit have expanded, yet extant cnidarians seem (at least from a bilaterian perspective) not to use much of the inherent potential of these genetic tools. The whole-genome sequences of the anthozoan sea anemone

Nematostella (Putnam et al. 2007) and of the hydrozoan *Hydra* (Chapman et al. 2010) are available, and developmental genes of other cnidarians have been studied. Despite their diploblastic architecture, cnidarians have many genes that mediate the development of tissues and organs in triploblastic forms. Elements of six major metazoan signaling pathways and a wide diversity of transcription factors, including members of at least fifty-six different homeodomain families, are found in cnidarians. Orthologs of many characteristic bilaterian genes are present. Perhaps most remarkable is the presence of seven genes associated with mesoderm formation in bilaterians, which suggests that the tools used to specify that third germ layer were present in the LCA of cnidarians and bilaterians. Despite the unsophisticated appearance of the average sea anemone, cnidarians possess much if not all the developmental machinery for *Hox* gene expression staggered along the antero-posterior axis and differentiation of other morphological elements (Boero, Schierwater, and Piraino 2007; Erwin 2009; Hui, Holland, and Ferrier 2008; P. N. Lee et al. 2006; Martindale 2005; Martindale, Pang, and Finnerty 2004; Matus, Pang, et al. 2006; Matus et al. 2008; D. J. Miller, Ball, and Technau 2005; Ryan et al. 2007; Seipel and Schmid 2006). Recent evidence of a bilateral pattern of gene expression in cnidarians suggests that bilaterality as a character was present, at least to a degree, before the divergence of the cnidarian and bilaterian lineages. Although both *Hydra* and *Nematostella* possess a suite of genes associated with the conserved actin-myosin contractile machinery in bilaterians and have muscle cells, they lack some of the important regulatory elements employed to generate striated or smooth muscle cells of bilaterian types; presumably, cnidarians and bilaterians diverged before those specializations arose (Chapman et al. 2010). Many of the highly conserved transcription factors that had diversified by the origin of the eumetazoan LCA are associated with neurogenesis but appeared after the divergence of the sponges (Galliot et al. 2009).

Because cnidarians have only two well-defined germ layers, systematists have commonly assumed that they branched earlier than the triploblastic Bilateria, thus representing a simpler stage that paved the way to more complex animals with mesoderm. Although this hypothesis is perfectly reasonable, we do not know for certain whether the cnidarian LCA was radially or bilaterally organized; after all, there is no definitive evidence that the radiate-bilaterian LCA was *not* triploblastic, but if it was, the cnidarian stem became somewhat simplified. Some workers (e.g., Martindale, Pang, and Finnerty 2004) have proposed that cnidarians (and presumably ctenophores) might as well be considered triploblastic. Furthermore, notions that these groups might be more developmentally complex than meets the eye are supported by their rich complement of regulatory genes also present in complex bilaterians.

How do we explain this perplexing pattern in which genes that mediate complex developmental functions in bilaterians are found in simpler organisms that do not require those functions? The various scenarios in figure 9.3 emphasize three possibilities. The first is that the Bilateria and the Cnidaria are sister groups, evolving in parallel from a simpler common ancestor (Schierwater et al. 2009) (fig. 9.3A). In this case, their body axes, nervous and sensory systems, and other characteristics reflect independent elaboration of a similar genetic toolkit. A second alternative is that the simpler organisms we see today have evolved from more complex ancestors (now extinct) that required sophisticated genomes, and their genomes are thus relics of more glorified pasts (fig. 9.3B). Reduction from a more complex form seems unlikely for the choanoflagellates and sponges: there is simply no compelling evidence for more complex forms from which they may have descended, although in some studies, molecular evidence has suggested a deeper branch for a more complex organism (i.e., ctenophores preceding sponges as in Dunn et al. 2008). This notion of unknown complex ancestors has also been suggested for placozoans, cnidarians, and other basal metazoan groups, although there is no obvious reason that every one of them should have been simplified. In addition, of course, these genes appear to be functional in living members of the basal groups, which is evidence against them simply being relics. A final possibility is that the genomes in the more basal forms contain regulatory genes that have switched targets or have subdivided their expression patterns among functional modules, coming to mediate functions in the more derived forms that have no analogs in the more basal groups (fig. 9.3C). As is now evident, we see no support for figure 9.3B and general support for figure 9.3C.

These alternative views reflect, in part, different perceptions of the ancestral role of many of these genes. As described in chapter 8, these highly conserved genes or gene networks were first interpreted as being for "segmentation" or "eye formation." In contrast, it might be better to think of many of these developmental regulatory genes as providing axial patterning, or a particular sort of pattern specification, or simply as providing signals that can be employed to express a wide diversity of genes and therefore a broad array of developmental outcomes. The latter position is completely compatible with the view that the increasing regulatory complexity among these basal metazoan clades was deployed in the development of the relatively simple body plans in which they are now found and are still at work. Recruitment and co-option of genes for novel tasks has been common throughout the Metazoa, and we have no a priori means of knowing what kinds of genes might help attain a new grade of organization or complexity (Larroux et al. 2008). Many of the developmental toolkit genes represent expansions of families of genes that are known in more

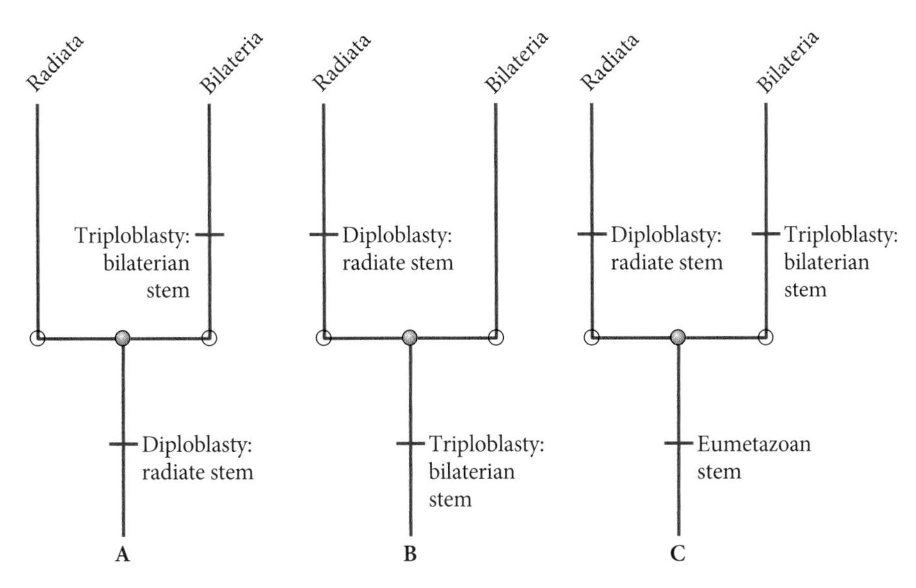

Figure 9.3 Three hypotheses of divergences between Radiata and Bilateria. The last common ancestor might have been a stem bilaterian, with Radiata arising later from within Bilateria, or it might have been a more primitive form. (A) The LCA might have been a stem radiate, with Bilateria arising later from within Radiata, the classical assumption. (B) Clade ancestors of Radiata and Bilateria were triploblastic. (C) A more basal group lacks the body plan of either of those clades within which their ancestral lineages branched, and later those lineages produced body plans that became stem ancestors of the clades; the clade ancestors of Radiata and Bilateria did not have epithelial tissues. See text.

basal groups, and many may have had their functions switched or expanded to mediating novel structures and functions in more derived groups. Developmental data does provide some information to distinguish between the possibilities in figure 9.3, however. Gene expression studies in the demosponge *Amphimedon* by Adamska and colleagues show that both the Wnt and TGF-β signaling pathways help generate the antero-posterior and dorsal-ventral body axes in their larvae, as they do in eumetazoans (Adamska et al. 2007), suggesting that the metazoan LCA was radially symmetrical with patterning by Wnt and TGF-β and with bilaterally patterning originating later, in the eumetazoan LCA (Adamska et al. 2011).

THE STEM BILATERIAN ANCESTOR (700–670 MA)

The crown bilaterian ancestor was by definition triploblastic and bilaterally symmetrical. Because both radiates and bilaterians have rich genomes with many genes in common, it follows that stem bilaterians had similarly rich genomes. Thus, it is quite conceivable that a large fauna of small benthic animals, relatively rich in regulatory genes, evolved before and during the Ediacaran period. Clearly, some, or perhaps many, of those animals reached the structural grade of eumetazoans, and at least one lineage attained the structural grade of the radiate/bilaterian LCA. Of course, significant diversification may have occurred among organisms of that grade. That node produced the founding members of the bilaterians, whose diversification fueled the Cambrian explosion and which dominated the metazoan fauna throughout the Phanerozoic.

There are distinctive differences between the developmental architectures of cnidarians and bilaterians. Martindale and Hejnol (2009) proposed a very speculative hypothesis for these differences, focusing on changes in the sites of gastrulation relative to the animal-vegetal axis of the embryo. They propose that the mouth is homologous in both cnidarians and nonchordate bilaterians, arising in the same hemisphere (animal) and involving many of the same developmental genes (the chordate mouth, however, may be secondarily derived). In anthozoan cnidarians, the site of gastrulation and of the formation of endoderm is at the oral pole, in the animal hemisphere. In most bilaterian embryos, however, the site of gastrulation and endoderm/mesoderm formation is either at the vegetal pole or somewhere in the vegetal hemisphere. Thus, a dissociation of the mouth and gastrulation site occurred along the bilaterian stem, after the LCA of Cnidaria and Bilateria but before the PDA. One consequence of such a change would be that because larval bilaterians tend to develop along an anterior-to-posterior gradient, derivatives from the animal hemisphere can continue with development within the anterior (animal) region even as new tissues and organs are undergoing subterminal growth posteriorly (e.g., see Jacobs et al. 2005). Martindale and Hejnol (2009) suggest that antero-posterior functional specializations, so common in bilaterians, might be promoted by such a developmental pattern, helping produce the hierarchical system of body-patterning genes well known in bilaterians (chap. 8) (fig. 9.4). The somewhat parallel modular patternings, which involve many of the same developmental genes (chap. 8), may owe their genetic similarities to their common derivation from the same network architecture that specified development in early, unsegmented, but elongate and axially differentiated bilaterian forms.

Acoel flatworms have appeared in recent phylogenies as representing the deepest bilaterian branch among living forms, as in figure 9.5A, preceding the branch at the protostome/deuterostome LCA (the PDA) (C. E. Cook et al. 2004; Hejnol and Martindale 2008; Paps, Baguña, and Ruitort 2009;

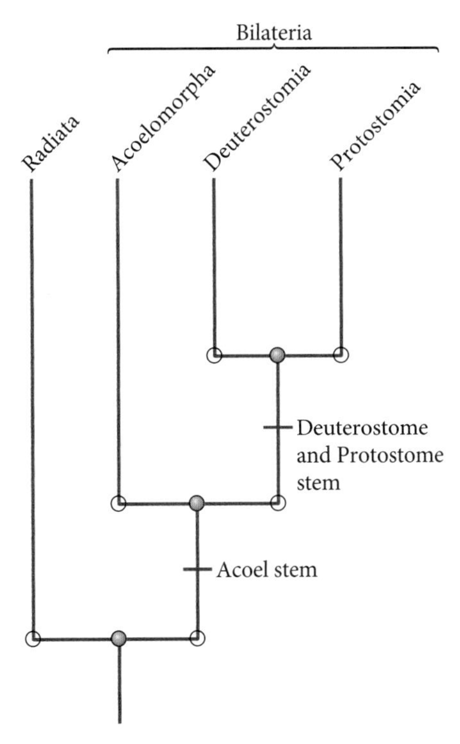

Figure 9.4 A scenario for the evolution of major body-plan features among eumetazoans based on Martindale and Hejnol (2009). In evolving from cnidarians, the basal bilaterians lost an endodermal nervous system, which was partly replaced in bilaterians by nerve nets combined with longitudinal nerve cords and eventually by central nervous systems that varied among clades. If the crown Acoela prove to be deuterostomes, as suggested by recent molecular evidence, the acoelomorpha would be considered an equivalent, now missing, grade of evolution. After Martindale and Hejnol (2009).

Ruiz-Trillo et al. 1999; and other papers). The anatomical simplicity, small body size, habitats, and feeding habits of acoels all supported this assumption. Molecular evidence (Philippe et al. 2011), though, places acoels at the base of the deuterostomes as indicated in figure 4.4 (and see fig. 9.5B). Part of the argument for a deuterostome affinity is that a miRNA acoel gene and a protein-coding acoel gene are found in deuterostomes but not protostomes, and miRNA genes in particular tend to accumulate within lineages, with relatively few losses. Contradictory molecular data sets are not uncommon, and they seem to be most vexing in the basal branches of the metazoan tree. It is worth looking at this sort of problem more closely, and acoels are a good example for illustrating such difficulties.

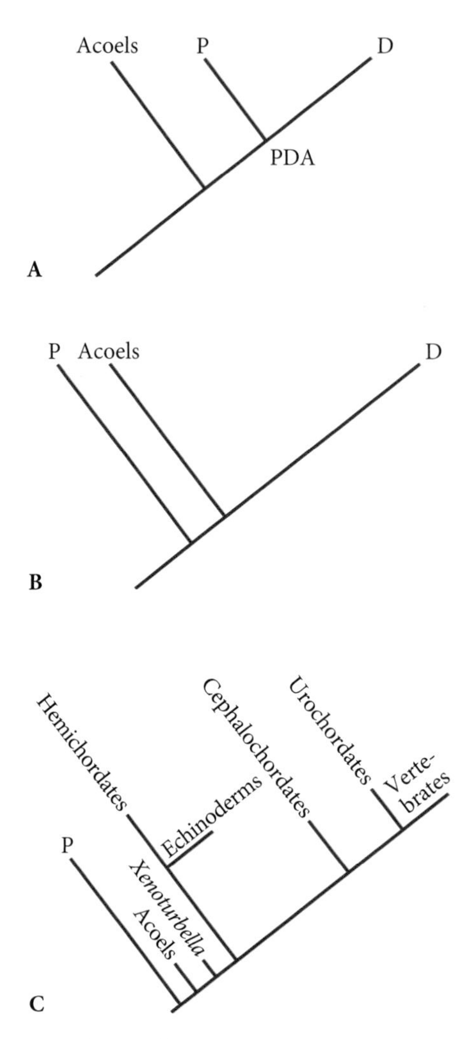

Figure 9.5 Simplified hypotheses of branching among some major eumetazoan clades, with differing positions of acoels. (A) Acoels arose before the protostome/deuterostome ancestor, following the classic view of their phylogenetic position. (B) Acoels are basal to deuterostomes, as suggested by some recent molecular evidence. (C) Acoels are sisters to Ambulacraria.

Recall that acoels are minute (1 mm to 2 mm), are overtly bilateral with antero-posterior and dorsal-ventral differentiation, and are tripoloblastic (fig. 4.13). They lack body cavities within or between any of their tissue layers and, for the most part, even lack a digestive lumen, so their bodies are quite solid. The mouths in cnidarians and ctenophores are also the only openings into their digestive tracts and are homologous with most bilaterian mouths

(Martindale and Hejnol 2009). The acoel nervous system consists of a subepidermal network; coordination may be vested chiefly in the network, recalling the nervous system of cnidarians and ctenophores. In at least some species, the brain is not represented by ganglia, but merely by a thickened anterior cross connection or commissure between nerve cords (Raikova et al. 1998). By contrast, in the PDA, the brain consisted of ganglia, for there is a well-supported molecular developmental and anatomical homology in ganglia between flies and vertebrates (Hirth et al. 2003).

Acoels have long been used as a morphological model for a basal bilaterian. For example, three *Hox* genes are known to be present in acoels, and they are, respectively, similar to genes present in the anterior, central, and posterior *Hox* groups recognized in higher bilaterians (C. E. Cook et al. 2004). These *Hox* genes may have been involved in antero-posterior patterning of the nervous system before becoming co-opted for patterning other tissues (Hejnol and Martindale 2009). Cnidarians have one anterior and one posterior *Hox* gene (Ryan et al. 2007), whereas the reconstructed PDA had at least seven *Hox* genes, and most bilaterians have a cluster of eight to ten or more. The difficulty with acoels as morphological models is that miRNA evidence indicates extensive secondary loss of some of their miRNA complement and thus morphological simplification from a more complex ancestor, limiting their utility as evolutionary analogs.

THE PROTOSTOME/DEUTEROSTOME LAST COMMON ANCESTOR (APPROXIMATELY 670 MA)

Another critical node in the early evolutionary history of the Eumetazoa is the LCA of protostomes and deuterostomes because if we knew both the age and morphology of the PDA, we could answer many questions about the Cambrian explosion. Molecular clock estimates suggest dates of about 670 Ma for this node (Erwin et al. 2011). The fossil record does not contradict this dating (if *Kimberella* represents a group above the PDA, the node must at least predate 555 Ma, implying the presence of a diverse fauna of early bilaterian ghosts). Understanding the morphologic complexity of the PDA is important for two reasons. First, trace fossil evidence has suggested that if the PDA was benthic, morphologically complex, and much greater than a few millimeters in diameter, it could not have existed much before the Cambrian. Because early identifications of some of the small Ediacaran trace fossils have proven to be erroneous (e.g., Sappenfield, Droser, and Gehling 2011 and references therein) (see chap. 5), the ability of the trace fossil record to convincingly inform us of the entire range of Neoproterozoic metazoan activities is in doubt. The burrowing and bioturbation by benthic forms with centimeter-sized cross sections in the very latest Neoproterozoic and earliest Cambrian sediments are quite clear, however, as is the absence of such activity from earlier rocks. Ediacaran structures most likely to be traces are in the millimeter range, and some of them suggest wormlike organisms, although *Kimberella* indicates that some epifaunal forms were larger and slug-like. Second, knowing the morphological complexity of the PDA would tell us the extent of developmental and morphological innovation along the line leading up to it and how much occurred after the separation of different major clades during late Neoproterozoic and Cambrian diversifications. Such knowledge might also shed light on the relationship between the environmental changes of the Ediacaran and the developmental and ecological innovations leading up to the Cambrian explosion. Here we expand on the discussion in chapter 8, emphasizing a morphologically simple PDA.

As discussed in chapter 8, it has been common practice to postulate a rather advanced organism for the PDA (sometimes incorrectly called the Urbilaterian, which literally means the original or primitive bilaterian). As noted, a case for a rather complex PDA can be made from cladistic reasoning based on features in common between protostomes and deuterostomes. For example, lophotrochozoans, ecdysozoans, and deuterostomes each contain clades with blood vascular systems (BVSs), from which one might infer that their LCA had a BVS as well, and coeloms, implying a coelomate ancestor with a BVS. Although it is true that each group also has at or near its base an acoelomate wormlike form that lacks a BVS (the structural grade shown by the wormlike acoels and the xenoturbellids), the microRNA data in chapter 8 indicate that each of these clades has lost miRNAs, a relatively infrequent occurrence. This loss suggests that each of these clades of basal worms became simplified from more complex ancestors; if so, it may provide little information on the morphological and developmental attributes of the PDA. It does, however, raise the intriguing but as yet unanswered question of why these basal forms were preferentially simplified.

The presence of so many of developmental genes shared by cnidarians and bilaterians raises significant problems for interpretations of a sophisticated PDA. The alternative model, which we favor, is that the PDA was not morphologically complex, with eyes, segmentation, and complex morphological differentiation. Rather, most of the shared developmental patterns among the protostomes and deuterostomes were involved in cell-type patterning. If so, it suggests that the PDA was morphologically rather simple, with a variety of cell types, and with antero-posterior and proximo-distal differentiation, but with limited regional patterning or organ formation.

EVOLUTION AMONG GHOSTS OF THE EXPLOSION FAUNA

The transition from the presumably simple worms of early bilaterian faunas to the broad array of body plans in the explosion faunas was accompanied by a significant increase in mean body size, as indicated by trace fossils across the Ediacaran-Cambrian boundary (chaps. 5 and 6). Size increases in clades with mineralized skeletons seem to have unfolded later. For example, the small shelly fossils include the earliest skeletonized mollusks, brachiopods, and other animals as well as sclerites of larger organisms (chap. 6); larger skeletonized representatives of these groups appear later in the Cambrian. Larger body sizes have an array of physiological consequences, many of which require morphological changes that usually involve increased anatomical complexity. Some of the requirements arising from size increases are based on geometry in that the area of an enlarging spherical body increases as the square of its radius, but its volume increases as the cube of its radius. Although these relative increases in area and volume are sensitive to shape, and bilaterians are hardly perfectly spherical, it is nevertheless the case that volume increases faster than surface area when their bodies enlarge (R. H. Peter 1983). Distributing oxygen and nutrients, in addition to removing waste, is also more complicated in larger organisms, and a number of studies suggest that these resource delivery systems scale as length to 0.75, rather than 0.66 as would be expected simply from surface area or volume considerations (J. H. Brown et al. 2004; West and Brown 2005). In either case, increased size creates challenges as well as opportunities.

Morphological features that are affected by size increases are associated with many of the most basic physiological functions. For example, even though bilaterians have three tissue layers, diffusion from the body surface is sufficient to supply the oxygen requirements of even their innermost tissues when body size is sufficiently small, as among acoelomates and pseudocoelomates. As size increases, however, the surface area becomes too small to provide oxygen for the increasing volume of internal tissues and organs, and respiratory surfaces must enlarge, as by the evolution of gills and by increasing aeration by pumping more water across them. There is a similar problem with feeding: food intake must increase to meet the requirements of increasing tissue volume, and the distributional systems must expand as well. Suspension feeders, for example, must evolve additional feeding surfaces, usually by increasing the number of tentacles, filaments, or other feeding organs, by pumping more water over these feeding surfaces, or by both methods. In both of these solutions to feeding in larger animals, the respiratory or feeding structures tend to become more complex, often folded and provided with increasingly elaborate supports. Circulatory systems must be evolved to distribute oxygen and nutrients, and when they appear, they develop at the site of the pseudocoel, the fluid-filled space between germ layers. More (or more powerful) kidney-like excretory organs must be developed. The demand on muscular systems also rises disproportionately with body size, and mechanical supporting elements commonly become increasingly robust. All these functional systems must also be integrated by an increasingly sophisticated nervous system.

The larger body plans that we find during the Cambrian explosion certainly tend to embody these morphological features, although in different forms and proportions among different clades. Why are the various explosion taxa so different from one another? The various eumetazoan clades that participated in the Cambrian explosion had generally separated long before the explosion itself and had adapted to different sets of conditions or solved similar adaptive problems in different ways. We can imagine our primitive bilaterian ghosts diversifying into various marine environments, on and just under the seafloor, adapting to different conditions within the environmental mosaic: to firmer or softer substrates; to low oxygen conditions in, on, or under algal mats; and to feeding on detritus, on other organisms, and on food items suspended in the water column. Although in one sense they are morphologically hypothetical ghosts in that we do not have their fossil remains, in another sense their existences are not hypothetical, for we see many of their descendants during the Cambrian explosion. Those descendants imply that all these life modes and more (such as living in the water column) were occupied by ghosts during the late Neoproterozoic and earliest Cambrian, when the fauna may have been rich in species as well as in body-plan disparity, species that we know, if at all, only from their traces.

CLADE, STEM, AND CROWN ANCESTORS OF THE DEUTEROSTOMES

The clade ancestor of the deuterostomes was descended directly from the PDA and, of course, had the same body plan. We favor a relatively simple form for this ghost as discussed above. The length of time that deuterostomes spent as nondiagnosable stem groups may have been between 50 million and 100 million years, but it seems plausible that diagnosable stems were present by 550 Ma during the late Neoproterozoic. The diagnostic features would have likely included gill slits. Pharyngotremy may have arisen to provide an excurrent pathway for the water in a cilia-powered feeding stream, but it was also employed for respiration, perhaps as body sizes increased. This feature must have been highly useful in establishing a feeding system in swimming forms that when moving through the water can engulf mouth-sized streams charged with suspended food items. At present, it is not possible to be certain whether these feeding and respiratory systems began as benthic or swimming adaptations, but they probably preceded the evolution of the notochord. (Current hypotheses of the branching pattern of deuterostome phyla are given in fig. 6.14.)

Because most living echinoderm clades appear to be highly derived, hemichordates played an important role in scenarios of the nature of early deuterostome evolution before the deuterostome affinities of the xenoturbellids and acoels were indicated by molecular data (Gerhart, Lowe, and Kirschner 2005; Swalla and Smith 2008). Hemichordates include two classes, pterobranchs and enteropneusts, and the latter has been particularly important in identifying morphologic and developmental homologies with chordates. It has been hypothesized on morphologic grounds that the last common deuterostome crown ancestor is likely to have been an enteropneust. Gill slits were probably present in all larger-bodied deuterostome phyla (but are not found in acoels or xenoturbellids) and although absent in living echinoderm groups may have been present in some early echinoderms (Jefferies 1990). The enteropneust nervous system, a subepidermal plexus, could be derived from an ancestor at the grade of acoels and xenoturbellids. The origin of crown chordates involved developing a centralized nervous system, developing the notochord, and probably inverting the dorsal-ventral axis (which involved shifting the mouth) by the time of the Cambrian explosion. The adaptive reason for the evolution of the notochord would seem to be as an antagonist to the muscular contractions associated with swimming by lateral body movements, but the adaptive pathways that led to the many highly derived chordate features remain speculative.

Conserved gene expression patterns have provided another perspective on conserved features of the deuterostome crown LCA, xenoturbellids and acoels aside (Gerhart, Lowe, and Kirschner 2005). The pattern of antero-posterior gene expression is the same in hemichordates and chordates for thirty-two different genes (including *Hox* genes; so, the first gill slit of hemichordates and the first branchial arch of chordates each occur at the same gene expression boundaries, and those of the hemichordate postanal extension correspond with the chordate tail (Lowe et al. 2003, 2006). This gene expression map must be conserved from the LCA of chordates and ambulacraria, which by extension included gill slits and a postanal tail. Although the notochord is a defining feature of the chordates, hemichordates do have a stomatochord with similar gene expression patterns to the notochord. Thus, at the least, the prechordal endomesoderm of chordates and the stomochord of hemichordates appear to be derived from a common structure (Gerhart, Lowe, and Kirschner 2005).

The path to chordates involved multiple gene duplications, which are most obvious in the differing number of genes within the *Hox* gene complex. Cephalochordates have fifteen *Hox* genes (L. Z. Holland et al. 2008) within a single cluster, perhaps the basal chordate cluster number. Jawed vertebrates have four clusters, each of which presumably represents a duplication of the basal chordate cluster. None of the four clusters has an identical *Hox* gene complement, each cluster having lost different orthologs; thirty-nine *Hox* genes remain in these four vertebrate clusters (chap. 8). The duplication of *Hox* clusters is one line of evidence suggesting that jawed vertebrates have probably undergone one or two rounds of duplication of their entire genome. If they had retained all the genes from two duplications, they should have more than 80,000 genes (instead of about 24,000), which has led several workers to argue that although there may have been substantial regional duplications, there were not two whole-genome duplications. Regardless, vertebrates have clearly lost many of the duplicated genes, probably soon after duplication, and it is estimated that about 25% of vertebrate gene families still contain genes that arose during those duplications (Putnam et al. 2007). Presumably, the duplicated genes that have been retained have become subfunctionalized or neofunctionalized, and many of the genes that have been retained have regulatory roles in development.

Gene expression patterns of chordates are inverted with respect to the dorsal-ventral axis of other deuterostomes and indeed of other bilaterians (Arendt and Nubler-Jung 1994; De Robertis and Sasai 1996). This inversion is perhaps most obvious in the expression patterns of bone morphogen protein-4 (*BMP-4*) and chordin. In hemichordates, *BMP-4* is expressed dorsally (Lowe et al. 2006), as in arthropods (Holley et al. 1995), but it is expressed ventrally in vertebrates (Sasai and De Robertis 1997). These expressions entail cascades of gene activity that produce dorsal and ventral morphologies in appropriate regions.

A number of truly enigmatic fossils have been referred to the deuterostomes, mostly to the hemichordates and chordates (chap. 6; see also Aldridge et al. 2007; Donoghue and Purnell 2005; N. D. Holland and Chen 2001; Shu et al. 2001, 2003; Swalla and Smith 2008). These fossils illuminate the novelty and breadth of the radiation of early deuterostome diversification.

Because the protostome clade ancestor is a sister species to the deuterostome clade ancestor, it would closely resemble that form and the PDA as well.

CLADE, STEM, AND CROWN ANCESTORS OF THE ECDYSOZOA

Ecdysozoans—the arthropods and their allies—are the paradigmatic example of the diversity of complex morphologies. We do not know how the last common ecdysozoan/lophotrochozoan ancestor differed from the PDA. All ghosts are dim and obscured by the passage of so much time in this section of the bilaterian tree. All the descendant clades have a cuticle that was periodically molted, however, suggesting that the stem ecdysozoan ancestor differed from the stem lophotrochozoans in this character. Although we do not know for sure why such a molted cuticle originated, a plausible scenario would be that it was used to fortify the body wall for muscle attachments and against internal fluid pressures. Such strengthened body wall cuticles could also act as auxiliary locomotory devices to secure purchase in creeping or shallow burrowing. If cuticle was secreted over more or less the entire surface of the body and could not be enlarged, it effectively limited body size. To accommodate growth and also to permit body size increases, early ecdysozoans developed the ability to shed the cuticle and secrete a new, larger one. Thus, the onset of molting can be used to signal the presence of a stem ecdysozoan.

Once a cuticle was present, it could be sculpted to further enhance locomotion and other functions. Early priapulid cuticles, for example, are ringed by circular ridges and furrows, and some are spiny (fig. 6.23). These worms burrow by pushing ahead into the sediment, elongating and anchoring themselves anteriorly, and then shortening their body by pulling their more-posterior section forward, which is anchored in turn to support the next forward thrust. The cuticular extensions provide anchorage. Cuticles also provide support for muscle insertions. Further, a toughened cuticle can provide protection against predators, which may have been an important early function. Finally and perhaps most importantly, if cuticular coverings are stiff enough, they can act as exoskeletons to help support muscular activities for appendages, ranging from the simple legs of Cambrian lobopods to, eventually, the swimming and walking appendages of the arthropods. A stiff body covering inhibits body flexibility, but the articulation of exoskeletons into series of longitudinal segments permitted the segments to flex somewhat independently. Such a modular, segmented architecture also provides the opportunity to differentiate morphology along the body, providing for specialized functions in different segments. This specialization is particularly clear, with appendages that are usually paired on each segment and functionally differentiated into mouth parts, walking legs, wings, and so on. The molted cuticle allowed ecdysozoans to generate a large number of body plans that in arthropods have been exploited to produce a dazzling array of variations based partly on differences in segment number and on appendage types. Indeed, arthropods have been the most diverse of all phyla during most of if not all the Phanerozoic Era, and they remain so today. A current hypothesis of the branching pattern among ecdysozoans is shown in figure 9.6.

CLADE, STEM, AND CROWN ANCESTORS OF THE LOPHOTROCHOZOA

The lophotrochozoan clade ancestor, like its sister the ecdysozoan clade ancestor, cannot easily be differentiated from the hypothetical crowd of small mid-Neoproterozoic ghosts. For that matter, even the crown ancestor is difficult to reconstruct because of the diversity of body patterns shown by living lophotrochozoans. The Neoproterozoic fossils of *Kimberella* (fig. 5.9A,B), about 555 Ma, do not much help with understanding key ancestors because *Kimberella* appears more slug-like than wormlike and may have branched between the crown molluscan clade and its stem ancestors. Current hypotheses of branching among lophotrochozoan phyla are shown in figure 9.7.

For decades, many invertebrate zoologists viewed flatworms as the LCA of bilaterians and believed that they had given rise to mollusks and other clades as well. Evidence in favor of this model includes similarities in early development, such as spiral cleavage by quartets and the cell fates of blastomeres, between the nonacoelomorph flatworm group Rhabditophora and several other spiralian groups (chap. 4). Some but not all molecular phylogenies have placed rhabiditophorans outside the main lophotrochozoan phyla; that molecular position is not stable, however, and the developmental similarity with the other spiral-cleaving phyla suggests close affinity.

Given that the crown eumetazoan ancestor may have been a small, flatworm-like organism and that the PDA was probably a small-bodied worm (see above), one possibility is that a flatworm-like form was the stem ancestor for the spiralian phyla, with eventually the crown molluscan ancestor descending from within its stem branches. The canonical spiral cleavage system thus becomes a spiralian synapomorphy. On the other hand, the lophophorate brachiopods and phoronids do not share this cleavage system: their cleavage pattern is radial, in common with radiates and deuterostomes and also with chaetognaths, possible crown ancestors of lophophorates, and possibly

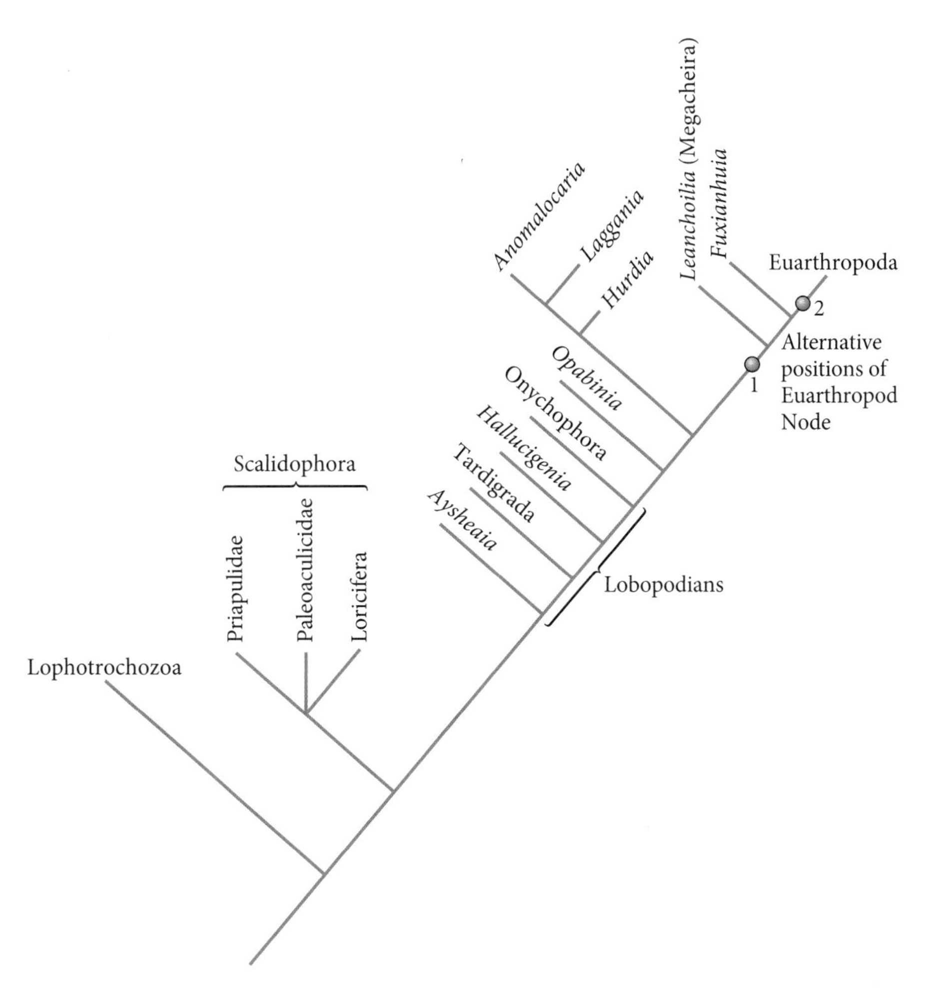

Figure 9.6 A hypothesis of divergences among the major clades of Ecdysozoa that have important explosion occurrences; notice the large number of distinctive but extinct body types that are present between the scalidophora (Priapulida, etc.) and the Euarthropoda. It is quite possible that trilobites were stem (rather than sisters to) euarthropods. From figures and discussions in Edgecombe (2010).

of lophotrochozoans in general. Perhaps the stem lophotrochozoan ancestor was a radially cleaving worm, and after chaetognaths branched off, further divergences produced the radially cleaving Lophophorata on one branch and Spiralia on another branch, with a derived cleavage pattern (fig. 9.5A). Alternatively, those lophophorates may have spiralian ancestors but may have modified their early developmental programs, and if so, radial cleavage evolved twice (fig. 9.5B). Lophotrochozoan trees have not stabilized, however, and there are still other lophotrochozoan branching patterns that remain possible.

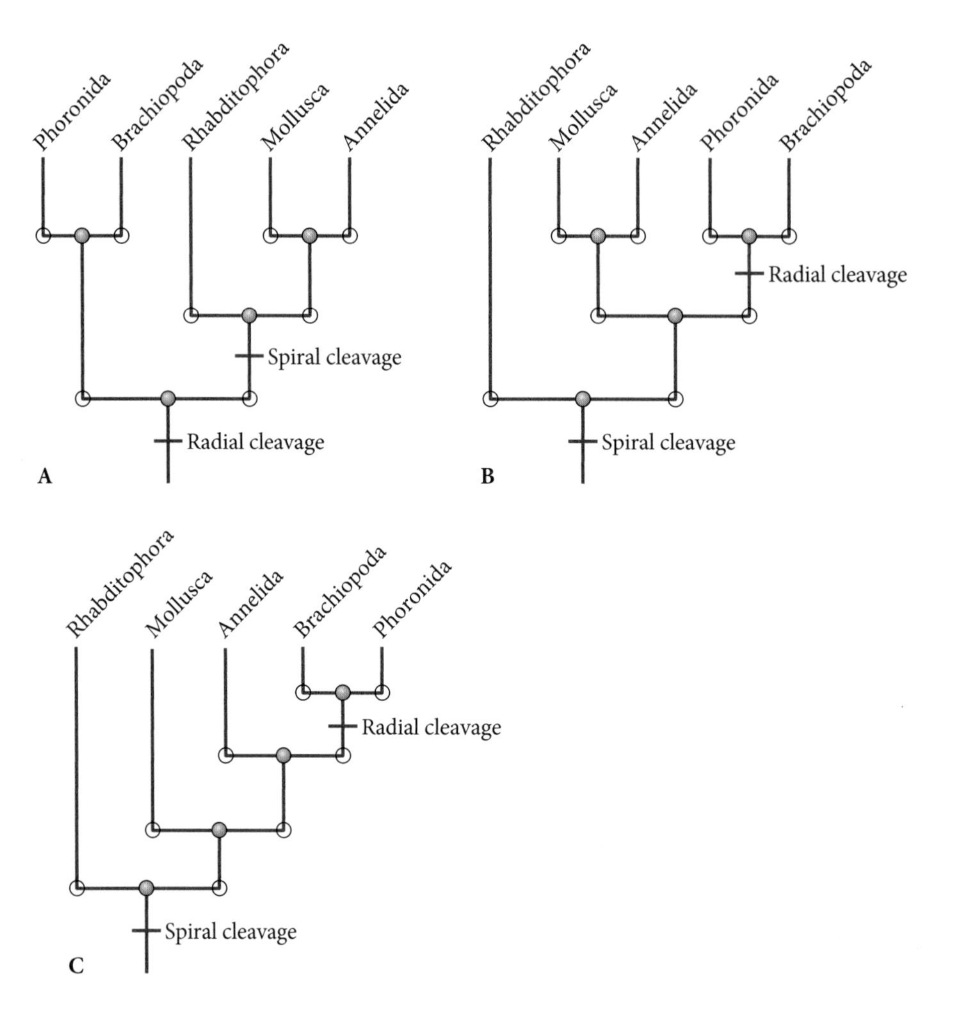

Figure 9.7 Hypotheses of divergences among the major clades of Lophotrochozoa that have important occurrences during the Cambrian explosion, with flatworms, although not found as fossils, shown to indicate alternative trees. (A) Divergences based on cleavage patterns, radial in Phoronida and Brachiopoda, with spiral derived in flatworms, Mollusca, and Annelida. (B) and (C) Divergences based on molecular phylogenies. (B) Flatworms with ancestral spiral cleavage, Phoronida + Brachiopoda with derived radial cleavage. (C) Flatworms with ancestral spiral cleavage, Phoronida + Brachiopoda with derived radial cleavage, but sister to Annelida.

SEEING GHOSTS

Generally, people who claim to have seen ghosts are not taken very seriously. In a similar vein, scientists who conjure hypothetical organisms have often had them discounted. From the foregoing speculations about ancestors, it is easy to see how alternate ideas can be entertained for many evolutionary pathways and many ghostly body plans. These Neoproterozoic and Cambrian ghosts are embedded in a phylogenetic tree, however, and the logic of cladistics enables us to form a general idea of their characteristics. Even the spotty available fossil record provides some help. Neoproterozoic traces were made not by ghosts but by living animals, and the remarkable animal fossils from the early Cambrian are anything but ghosts. Furthermore, the constraints and advantages that are well understood to accompany body-size changes are helpful in obtaining some sort of conceptual outline within which to frame the morphological changes that arose in early metazoan history. The general conclusions from our exercise in evolutionary séances support a growing body of evidence from the genomes of living groups as to how complexity evolves at the genetic level. Those data help clarify how the explosive appearance of many independently evolving lineages—occurring within a narrow geological time span—lay well within the capability of these incredibly flexible genomic systems. Even humans owe their interesting body plan and their sophisticated faculties to the character of the genomes that were evolved before and during the Cambrian explosion.

In the next chapter, we take up the task of integrating the perspective on developmental innovations discussed in this chapter and in chapter 8, within the context of the environmental and ecological changes during the Ediacaran and Cambrian, and suggest what it tells us about the evolutionary process.

CHAPTER TEN

Constructing the Cambrian

The early evolutionary history of metazoans was characterized by a range of innovations unmatched by subsequent Phanerozoic evolution, but the incomplete and spotty nature of the fossil and rock records of the Cambrian explosion have permitted the formulation of many hypotheses for the rates and causes of the explosion. Dozens of different explanations have been mooted from as many different scientific perspectives, and most were not, at least at the time they were proposed, inconsistent with the available data, although they tended to focus on some single cause. The intensive new studies we have discussed have increasingly illuminated the environmental and faunal context of the explosion and have narrowed the range of possible explanations. Indeed, the one conclusion we can draw with confidence is that it is futile to attempt to explain the explosion by invoking any single factor.

Explaining why the explosion took place after 550 Ma rather than earlier is a good example of the hypotheses that focus on only one aspect of the explosion. Simultaneous changes in so many different metazoan clades during the latest Ediacaran and early Cambrian, as well as associated changes in the plankton (chap. 6), point to either a pronounced environmental shift or ecological feedback. Thus, it is hardly surprising that increased concentrations of oxygen in shallow marine waters to levels that would sustain larger and more active animals have figured in several hypotheses, with evolutionary processes responding to this environmental opportunity. Even if historically true, as seems plausible, such an event hardly "explains" the biological diversification. Consequently, we have emphasized the interactions between the elements of the macroevolutionary triad throughout this book.

In previous chapters, we have traced the important changes in the physical marine environment as indicated by the geological record and by geochemical proxies; followed the modifications in the biological environment during the development of ecological interactions and the establishment of marine communities found in the fossil record; and recounted the increases in regulatory sophistication within metazoan genomes that were associated with the surprisingly rapid origins of novel morphologies, as indicated by molecular studies of living groups. As the explanations for the Cambrian explosion have become more complicated, and certainly more historically accurate, the feedbacks among the elements of this triad become more critical. Throughout this book, we have emphasized the construction of networks of interaction, both among developmental networks and feedbacks between activities of organisms that alter the physical environment and the subsequent response of selection to adapt organisms to the new environmental conditions. Perhaps the most outstanding single example in the history of Earth is the production of oxygen by photosynthetic algae to fill the oxygen sinks and eventually to create new metabolic systems within a relatively oxygen-rich ocean and atmosphere around 2.4 Ga.

Animals have created pervasive feedback effects as well, perhaps beginning with the pumping activity of sponges and its likely effects in spreading waters with higher oxygen concentrations into environments below the depths where photosynthesis is possible. Mat- and, probably, biofilm-forming bacteria and algae created a trophic basis for early benthic ecosystems that included a fauna of primitive, generally small bilaterians adapted to exploit the mats. The Mistaken Point biota reveals the invasion of deep-sea environments below the euphotic zone by a variety of rangeomorphs and other frond-like Ediacaran forms. The effects of those Ediacaran organisms on seawater composition and biological content remain to be understood. Crown diploblastic forms invaded the water column at some undetermined time, possibly in the late Neoproterozoic, although their early histories are still obscure. As oxygen concentrations in the oceans continued to increase, the burrowing activities of those bilaterians became so extensive that the mat communities were significantly reduced in importance, and new environmental conditions developed upon and beneath the seafloor.

Here we integrate the evidence from the preceding chapters on the history of metazoan clades from their origins through the Cambrian explosion. The available data suggest three phases: (1) the origin of Metazoa and the divergence into sponge and the stem eumetazoan lineage during the Cryogenian, prior to the Sturtian glaciation; (2) the origin of cnidarians and the divergence between protostomes and deuterostomes between the Sturtian and Marinoan glaciations and early divergence of major bilaterian clades, beginning probably in the late Cryogenian but largely in the early Ediacaran; and (3) the appearance of many crown bilaterian groups coincident with the Cambrian explosion documented in the fossil record. Landmarks in time along the unfolding of major metazoan morphological architectures are shown in figure 10.1.

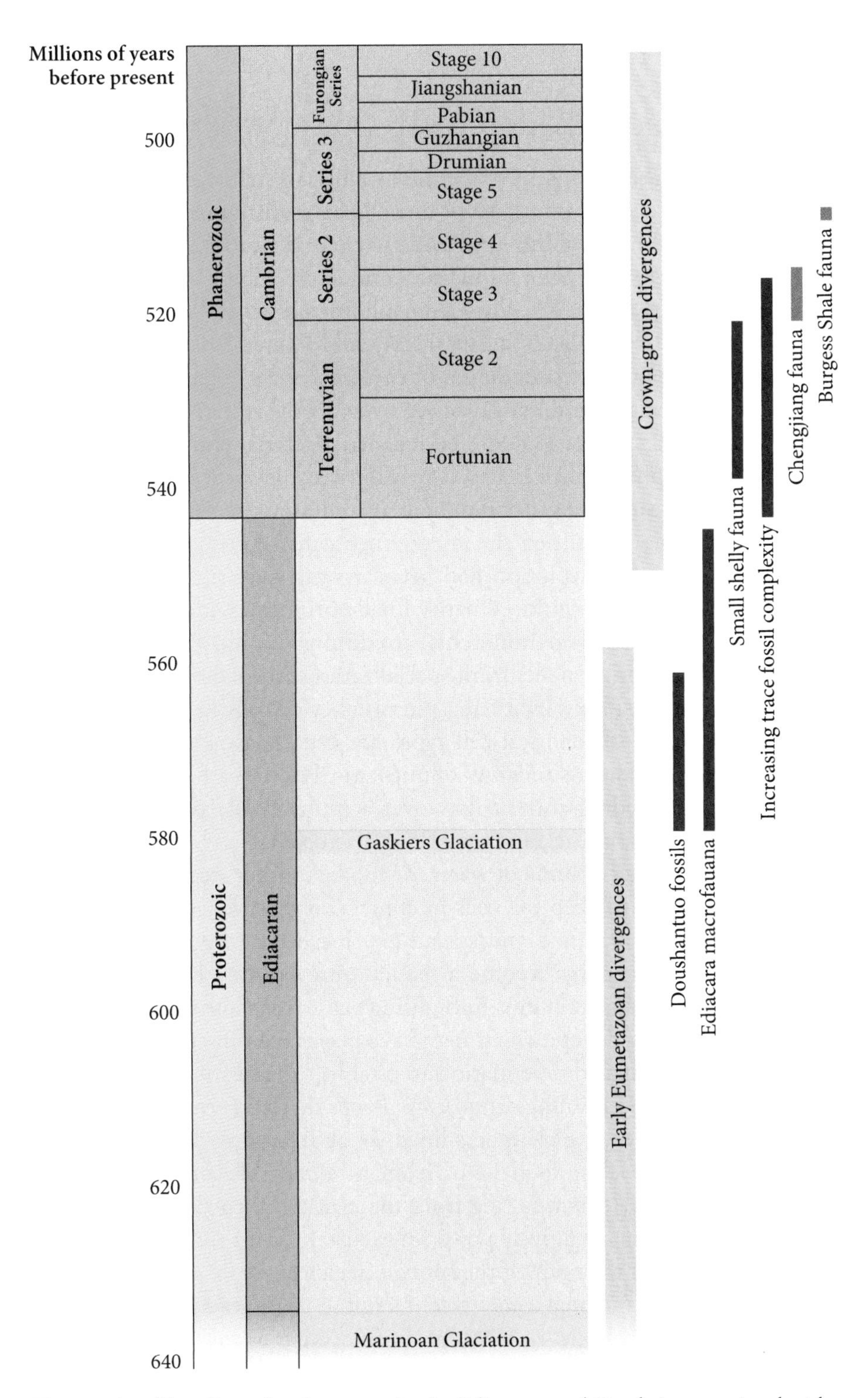

Figure 10.1 Time line of major events in the Ediacaran and Cambrian associated with the unfolding of major metazoan morphological architectures.

A SPONGE AND CNIDARIAN WORLD

Sponges, followed soon after by the earliest cnidarian stems, evolved within a biosphere composed of many clades of unicellular protists and eukaryotes and a variety of algae, but the sulfur- and iron-rich oceans of the Cryogenian have generally thought to have been inimical to animal life. Cryogenian ecosystems were based chiefly on energy flows from phototrophic prokaryotes such as cyanobacteria and from algae (Javaux, Knoll, and Walter 2001; Knoll 2011), with some contributions from sources of chemical energy (methane, reduced sulfur compounds, and hydrogen). At least regionally, the well-lit portion of the water column must have harbored archaea, bacteria, and algal cells that furnished the trophic resources for suspension feeders such as choanoflagellates and, on the seafloor, algae and microbial communities, including decomposers, as microbiota returned the energy in organic detritus back to living ecosystems. Why, after evolution had been at work over such vast spans of time, did metazoan-style multicellularity finally originate? And how did it do so under seemingly harsh environmental conditions?

Just as an economy benefits from specialization and division of labor, the main advantages of simply increasing the number of cells in a colony or an organism composed of identical cell types are usually ascribed to increased efficiency of feeding and assimilation of nutrients. For the earliest, suspension-feeding Metazoa, feeding-current flow over a multicellular colony could be enhanced by increasing the number of flagellae so that each individual cell could examine a larger volume of water for food. Another step toward multicellularity would be to keep the cells in contact so that they could exchange molecules derived from food and spread any localized energetic advantages across what had essentially become a tissue, a process that has been shown experimentally to enhance fitness in budding yeast (Koschwanez, Foster, and Murray 2011). A third step, which may have been taken in concert with the previous one, is cellular differentiation to establish the geometry of feeding-cell assemblages by nonfeeding supporting tissues that, in many sponges, help define the walls of feeding chambers lined by choanocytes. The suspension-feeding sponges can clear food items from a remarkable amount of water because the water currents resulting from the beating of choanocyte flagellae are channeled to form a relatively powerful excurrent stream, thereby drawing feeding currents into the body of the sponge (see chap. 4).

Presumably, many sponge genes were recruited from unicells in which the genes functioned in colonies or in simple multicellular forms (chap. 9). There are twenty-three cadherin genes (cell adhesion molecules) in the genome of the choanoflagellate *Monosiga brevicollis*, well within the number found in metazoans (Abedin and King 2008). Cadherin molecules have multiple domains, and choanoflagellate cadherins include domains that interact with bacteria, with cytoskeletal actins, and in metazoans interact within signaling networks (see Abedin and King 2008). Abedin and King suggest that cadherins may support intracellular signaling in choanoflagellates (to produce physiological responses) and may function in the adherence of food items such as bacteria to feeding surfaces such as the microvilli (Abedin and King 2008). These data suggest that the adherent properties of these transmembrane proteins were then co-opted along the metazoan stem for cell-cell adhesion and to mediate some intercellular signaling, among other functions.

The signaling systems employed for intracellular regulation were co-opted for intercellular developmental pathways along the metazoan stem, or before about 780 Ma. Judging from the demosponge genome (Srivastava et al. 2010), nearly 75% of the unique metazoan gene families evolved along the metazoan stem. These gene families included elements of metazoan signaling pathways and many transcription factors that are now widespread within animals. These changes led to a significant increase in the size of the genome, disproportionately so among protein-coding developmental regulatory genes, and included the appearance of regulatory miRNA genes as well. Many new transcription factors appeared, and new structural genes evolved as new functions (e.g., spicule formation) appeared. The genome growth surely led to more complex GRNs, although judging from the genomes of crown sponges, much of the increasing developmental complexity introduced as the number of cell types grew was encoded by new genes rather than by an increasing complexity of transcription patterns.

The production of collagen in metazoans requires oxygen; so, despite the geochemical proxies indicating that the oceans were largely sulfur- and iron-rich, sufficient oxygen must have existed in shallow waters, at least regionally, to allow these groups to persist. As discussed in chapter 3, the evidence from various geochemical proxies about when the oceans became better ventilated remains subject to conflicting interpretation. Butterfield (2009) argues that oxygen levels may have been sufficient, at least in shallow waters, throughout much of the Cryogenian and certainly the Ediacaran. In contrast, some interpretations of carbon isotopes suggest a rapid ventilation of the oceans associated with the end of the Shuram carbon isotope excursion around 555 Ma. Because geochemical proxies generally provide relatively poor information about shallow waters, these two different views may actually not be as far apart as they appear.

One scenario that accords with our understanding of Cryogenian environments is that the ancestral choanoflagellate populations may have invaded regions from which they had formerly been excluded, perhaps by the amelioration of anoxic or other adverse conditions, but where trophic supplies were low. Such a situation could arise when oxygenated waters expanded, at least regionally. Invading populations would face strong selective pressures to enhance feeding efficiency, an evolutionary pathway eventually leading to the multicellular bodies of stem sponges. We favor the view that as sponges fed, they transferred carbon from seawater (as bacteria and dissolved carbon) to their bodies (as collagen and other carbon-rich compounds). Upon burial, some of this carbon was transferred to sediments, and the cumulative result was a progressive increase in oxygen levels in seawater. The evolving sponge morphology thus permitted exploitation of suspended and perhaps dissolved material on a larger scale, probably enabling sponges to thrive in localities in which the density of food items was lower than optimal for prokaryote suspension feeders and eventually to form feeding tiers above the seafloor. Thus, sponges constructed new niches and expanded the scope of potential biodiversity.

The pathway from sponges to eumetazoans is the most enigmatic of any evolutionary transition in metazoans. This transition occurred during the Cryogenian, almost contemporaneously with the diversification of sponges. Many biologists concerned with metazoan phylogeny have been convinced that "ontogeny recapitulates phylogeny" and have therefore assumed that the planktonic larval stages of invertebrate phyla represented their ancestral forms. The benthic nature of sponges and the paraphyly of the major clades demonstrate that planktonic stages could not have been ancestral to eumetazoans (chap. 4). Furthermore, there are no living intermediates between sponges and eumetazoans, with the possible exception of placozoans, and no obvious hints from the fossil record.

The greatest conundrum about the origin of eumetazoans—more specifically, cnidarians—is, however, the evolution of the cnidocyst, the defining structure of the clade. As novel tissues evolved along the eumetazoan stem, the origin of endoderm and the gut during the transition to Eumetazoa required a major shift in structure and function that separated feeding from digestion and was associated with novel tissues and the origin of many cell type-specific genes (e.g., Hwang et al. 2007). Cnidarians are carnivorous, deploying their cnidae largely to catch zooplankton, a far cry from the microbial diet of sponges. Zooplankton did not appear until much later, however, and the energetically expensive, predatory functions of cnidae are wasted on the more passive food typical of the Cryogenian. The largely iron- and sulfur-rich environments of the Cryogenian, with some oxygen in shallow waters, evidently persisted after the Marinoan glaciation and perhaps as late as 550 Ma, exacerbating the question of the role of cnidae. It is possible that cnidae first evolved for other purposes, such as fighting off carnivores (which would have been other cnidarians or perhaps ctenophores), fending off competitors, or helping maintain purchase on the seafloor or within burrows, functions that they serve today in some lineages, but these functions are equally likely (perhaps even more likely) to be derived.

EARLY EUMETAZOAN DIVERGENCES

The protein-coding regulatory gene families are, as noted previously, largely similar among crown Metazoan phyla, and most of those gene families were introduced during evolution along the metazoan and eumetazoan stems (fig. 8.9) (Domazet-Loso, Brajkovic, and Tautz 2007; Marshall and Valentine 2010). For example, of the regulatory and structural genes found today in the *Drosphila melanogaster* genome, many originated among protists, and many others are first found along the metazoan or eumetazoan stems. Relatively few were introduced during the origin of the PDA or along of the stem of the phylum Arthropoda. Comparative developmental evidence suggests that the basic suite of eumetazoan proteins and the basic developmental toolkit date to the origin of Eumetazoa, perhaps 700 Ma. The latest molecular clock divergence estimates place the PDA at about 660 Ma, before the Marinoan glaciation, and the origins of deuterostomes, lophotrochozoans, and ecdysozoans seem to largely date to the early Ediacaran, between the Marinoan and the Gaskiers glaciations.

We do not, however, have fossils to tell us how disparate the body plans of those preexplosion bilaterian clades were; indeed, it is not clear that paleontologists have even appropriate search images for fossils of this interval. Although it appears that stems of nearly or just about all phyla had appeared well before the explosion, many may have lacked the diagnostic morphology of crown phyla. The overall pattern is consistent with the inference that differences in genomes among the early stems of bilaterian phyla were relatively small before the explosion; their divergences had not required much genomic (or perhaps morphological) innovation beyond what had occurred previously. In this case, it would have been the distinctive morphologies—those that came to characterize the later stems and crown body plans and that we first see during the explosion—that were associated with the evolution of diagnostic features of phylum-level genomes, arising long after the origination of the PDA.

The reconstructed PDA genome tells us clearly that the developmental genomic architecture that had been evolved along the eucoelomate stem became increasingly elaborate well before the Cambrian explosion, with gene families enlarged through domain swaps and gene duplications. Several brave attempts to infer the morphology of the PDA from its developmental toolkit have produced a range of reconstructions, from very simple worms only slightly more complex than acoels to moderately complex coelomates. As we have seen, however, many of the genes that function to mediate the development of complexly patterned animals with three (or four) germ layers and advanced organ systems are also found in quite simple organisms that lack such features, even animals as simple as sponges, within which many such genes are functioning today. Before we bestow the PDA with a complex morphology, we should learn the functions of those genes in simple animals as well as in complex ones.

In addition to the rather rich supply of genes inherited from stem eumetazoans, new families of regulatory genes, and in all probability novel GRN architectures, made their appearances in stem bilaterians. The relative importance of changes in protein-coding genes to the evolution of *cis*-regulatory elements began to change at this time because morphological changes were increasingly underpinned by the evolution of *cis*-regulatory networks. The morphological differences between different early bilaterians increased faster than differences in gene composition. This tendency was evidently reinforced by a growth in the richness of miRNA regulatory elements, particularly in lineages that were increasing greatly in morphological complexity, which helped regulate the development of that complexity without increasing the richness of protein-coding genes very much. Some of the evolutionary steps involved in the expansion of gene regulatory systems have been identified in yeasts, which are useful experimental models for the study of such features. For example, the "shuffling" of domains among regulatory genes (C. Vogel et al. 2004) has been suggested as a likely source of some of the gene diversification along the metazoan stem (e.g., Srivastava et al. 2010). Domain shuffling has been identified experimentally in yeast as a process of rapid evolution of cell signaling genes by Peisajovich et al. (2010). Intercalary evolution—in which a connection between the regulatory and target genes may be broken and a new, indirect connection established that expands the regulatory network while preserving the logic of the old circuitry and giving rise to a new regulatory outcome—has been demonstrated in yeast (Booth, Tuch, and Johnson 2010) and proposed as a major factor in regulatory evolution (Erwin and Davidson 2009).

The earliest macroscopic metazoans are the various clades of Ediacara macro-fossils. Because the frond-like lineages, along with their possible Ediacaran allies, are extinct, we can place them phylogenetically only by the timing of their appearances in the fossil record and on morphologic evidence, which as we have seen is quite equivocal. Most ediacarans exhibit both axial and regional patterning in a manner comparable to undoubted metazoans, suggesting that they possessed metazoan-style genomes. Many of the Ediacaran fossils are found in communities that suggest metazoan-like environmental partitioning, with a variety of benthic forms among microbial mats and intermixed with various levels of fronds. Although the rangeomorph and ernettiomorph-like forms fit nicely as branches along the eumetazoan stem below cnidarians, it is possible that some solitary types such as the sprigginids and dickinsoniids branched after cnidarians but before bilaterians. The Ediacara macrofossils enter the record at 579 Ma, and some persist up to (and possibly a bit beyond) the basal Cambrian boundary at 542 Ma.

THE WORLD OF THE CAMBRIAN EXPLOSION

Relatively large horizontal burrows (1 cm or more in diameter) that churned sediments on the seafloor appeared just before the earliest Cambrian, about the time that vertical burrowing first appeared as well. These signs of relatively vigorous activity imply a wormlike organism furnished with a fluid skeleton, presumably either a hemocoel or coelom. We have as yet no confirmed body fossils of those animals, but an animal constructed like an early hemichordate, priapulan, or annelid could have produced such trace fossils and is consistent with what we know of metazoan divergences (e.g., Dzik 2005; Vannier et al. 2010). The tubes, shells, and sclerites of the small shelly fauna, followed by trilobites and other arthropod groups, are the earliest fossils of the Cambrian explosion. When the taphonomic window finally opened to the preservation of soft-bodied faunal elements, such as in the Chengjiang and Burgess Shale deposits, the rich fauna contained larger-bodied and complex morphologies that spanned the structural range from simple worms to elaborate arthropods no less complicated than those of today. Furthermore, the benthic fauna was accompanied by an equally impressive pelagic fauna, which included vertebrates. Molecular clock evidence suggests that the appearance of so many bilaterian clades in the fossil record during the Cambrian explosion coincides with the appearance of most of the bilaterian crown groups.

The loci of change in metazoan developmental systems shifted during the rise of complex multicellular organisms. As we have seen, the lineage from choanoflagellates to sponges encompassed an expansion of the regulatory system through the evolution of new genes, whereas early metazoan development included the establishment of the major elements of the metazoan developmental toolkit. Following the origin of Eumetazoa around 700 Ma, the diversification of bilaterian clades, and particularly the origin of the bilaterian crown groups, appears to have primarily involved the rise of complicated transcriptional regulatory networks through the addition of more complex *cis*-regulatory features (Davidson and Erwin 2010b) and through the diversification of miRNAs to stabilize the multitude of new cell types in different lineages. More complex metazoans experienced a disproportionate rise of rich RNA regulatory components.

The interactions among the environment, development, and ecology began with the first metazoans and created conditions that not only affected the events leading up to the explosion but also expanded during the explosion itself. That is why we have argued that ecosystem engineering, the active modification of their environment by organisms, has played a significant role in driving the Cambrian explosion. Interactions among the burgeoning clades of the explosion fauna are more easily linked to studies of modern ecological interactions, particularly because principles of trophodynamics can be applied to both Cambrian and modern faunas (e.g., Dunne et al. 2008). It seems likely that the diversity and complexity of body types generated during the explosion were partly products of the increasingly complex selective regimes created by the increase of ecological interactions among the diversifying clades within Cambrian communities (Erwin 2008; Marshall 2006; Marshall and Valentine 2010; Niklas 1994). It is possible that the speeding up of selective responses implied by the shift in genome evolution from novel genes to novel *cis*-regulatory changes in gene networks was partly a reflection of an increasingly complex selective regime as well. As the effects of the evolution of development on the environment were transmitted through morphological and behavioral changes in phenotypes, however, feedbacks from genomic changes were indirect; the exception was that the response time tended to be reduced, leading to potentially more rapid evolutionary change.

It is to the explosion, and not to the origin of phylum-level clades, that we are indebted for the body plans of living phyla, those familiar morphological themes that are still with us and that are almost taken for granted as being appropriately representative of the living biosphere. Those body plans have—so far, at least—proven robust to very severe environmental perturbations over geological time (see below). An important caveat to our reading of the explosion is that we have learned about it by peering through newly opened taphonomic windows that have surely made the explosion appear to be more abrupt than was actually the case. In peering at Burgess Shale–type faunas, we are looking through only one window and seeing the inhabitants of only a relatively restricted range of ecological and preservational conditions. So, we are bound to be missing many of the products of the explosion and can hardly be exaggerating its significance in the history of animal life.

FORWARD INTO THE PHANEROZOIC

Most elements of the Cambrian explosion faunas do not readily fossilize and consequently play little role in the later fossil record of the Cambrian or, indeed, of later intervals of the Phanerozoic. The Cambrian fauna is dominated by the stem arthropod group Trilobita along with members of two other clades: the Inarticulata (an abandoned brachiopod grouping) and the Monoplacophora (a minor molluscan group, now known as the Class Tergomya). The extinct Trilobita certainly dominates the Cambrian fossil record in terms of species, genus, and family diversities. Both the Inarticulata and the Tergomya are extant (although their Cenozoic diversity is so low that it is hard to distinguish it from zero) and were relatively minor elements even in the explosion faunas.

The major biodiversification event of the Paleozoic occurred chiefly during the Ordovician; by the Middle Ordovician, marine fossil communities had become dominated by a host of articulate brachiopods, crinoids, and other stalked echinoderms and colonial bryozoans. These taxonomically disparate clades share some ecological similarities: the adults of all three groups chiefly lived attached to the sea bottom and filtered their food from the water column (chap. 6). They also share calcareous skeletons and therefore a somewhat similar taphonomic window. We see few of the softer-bodied taxa of the time. Planktonic larvae probably evolved during the Cambrian explosion, and planktotrophic larval development appeared in several clades of marine invertebrates at least by the late Cambrian or Early Ordovician (Peterson 2005; Peterson and Butterfield 2005). The indirect-developing larval morphologies clearly evolved so as to be appropriate to the adult body plans that preceded them. Important Paleozoic-style reef ecosystems appeared, commonly rich in species (as are reefs today); they included rugose and tabulate corals, now extinct, and many highly specialized lineages within the benthic fauna. Pelagic communities included hemichordates, cephalopod mollusks, arthropods, and vertebrates. Clearly, a rich planktonic biota was required to support the diversification of these communities. There are relatively few fossil localities with unskeletonized remains that represent the post-Cambrian faunas. Thus, groups such as arthropods, annelids, and other worms are surely underrepresented by fossils, and they may have made up the bulk of bottom feeders, both algal and detritus feeders, and of carnivores. The recent discovery of a Burgess Shale–like assemblage in Ordovician rocks from Morocco demonstrates that some stem groups that characterized the explosion lived on, at least within the lower Paleozoic (Van Roy et al. 2010). The bulk of the Ordovician fauna, however, comprised species belonging to crown classes and increasing numbers of crown orders, although the dominant groups were largely different from those that are dominant today.

The great extinctions near the Permian-Triassic boundary (Erwin 2006b) forever changed the balance of diversity among marine invertebrate clades, with the rise of mollusks to rival arthropods for dominance in marine communities and with modern types of such groups as bryozoans and corals replacing their Paleozoic relatives. It can be argued that the extinctions accelerated a trend that was already in place, and there was certainly an increasing dominance of clades that had become more prominent in the late Paleozoic. The legacy of the explosion lives on today in the body plans of animal phyla.

A TRACTABLE BUT UNRESOLVED PROBLEM

The patterns of disparity observed during the Cambrian pose two unresolved questions. First, what evolutionary processes produced the gaps between the morphologies of major clades? Second, why have the morphologic boundaries of these body plans remained relatively stable over the past half a billion years? After all, there is no a priori reason clades could not display a pattern of rapid exploration of morphologic space, coupled by subsequent expansion of that space during the Phanerozoic, but it is not a pattern commonly observed among the bilaterian metazoans.

Evidence from comparative developmental studies indicates that the answer to the question of morphologic gaps must involve discontinuities in the patterns of the gene regulatory networks. The morphological evolution of body plans during the latest Neoproterozoic and early Cambrian chiefly involved massive and phylogenetically widespread *cis*-regulatory evolution of the networks, the first and only time that genomic network disparity was achieved at such a scale among metazoans. Thus, there is a question as to whether the morphological disparities arising from the discontinuities between major clades are due to evolutionary processes acting on interactions between morphology and the environment, which would be chiefly mediated by selection, or, alternatively, are due to the evolution of properties within the gene regulatory networks that could pose morphological limits because of restrictions imposed on changes in the architecture of the networks. We have presented evidence to show that the regulatory networks within bilaterian genomes include a number of characteristic elements, many of which are highly conserved among phyla. These elements include kernels that have recursive control circuits used for very basic features such as heart development, which would be difficult to alter while maintaining circulatory functions. As we learn more about developmental gene regulatory networks, it appears possible that much of the constraint of metazoan body plans may reflect the structure of these networks, in particular the hardening against evolutionary change of core elements of pattern formation. Thus, one hypothesis about the conservation of morphological boundaries postulates that the morphological body plans of major clades at the levels of phyla and classes may have become permanently constrained because critical parts of GRNs became increasingly inflexible. The core elements of pattern formation are conserved because they are resistant to evolutionary modification. This hypothesis shunts evolutionary change to other parts of the network. In other words, the patterns and trends of morphology reviewed above entailed similar patterns and trends among the gene regulatory networks. Thus, the developmental factors that generated the discrete body architectures during the Cambrian explosion may have served to limit subsequent morphological evolution.

The critical point here is a distinction between internal selection and external selection, first elucidated by Schwenk (Schwenk 1995; Schwenk and Wagner 2001). Schwenk noted that evolutionary processes must work together to produce a viable embryo, various developmental stages, and an adult. Variation that disrupts these processes will be rapidly eliminated, regardless of the nature of the external environment. External selection, in contrast, encompasses all those interactions between organisms and the external environment. Experimental evidence suggests that the recursive nature of regulatory interactions within developmental kernels means that after formation, the kernel as a whole is subject to external selection, but the regulatory interactions within the kernel are subject entirely to internal selection; there simply may not be any viable phenotypic variation within development kernels for selection to act upon. The evolutionary consequence is to stabilize the formation of regional patterning that makes important contributions to the body plan: evolution may eliminate parts of the body plan, as with hearts and their GRN kernels in some acoelomate- and pseudocoelomate-grade organisms, or may modify GRNs upstream and downstream of the kernel, but the kernels, once evolved, necessarily change the range of variation upon which selection can easily act.

Selection has doubtless also been important in crafting the distal ends of GRNs and the exquisite adaptations seen in many of the Cambrian animals. The entrancing Cambrian body plans are clearly adaptive to distinctive modes of life, and some seem quite specialized. The conservation of a kernel of regulatory genes in heart development is not necessarily an indication that body plans were limited in their capacity to evolve novelty. Indeed, it could be argued that such developmental kernels actually enhance the evolutionary potential of bilaterians by ensuring that basic aspects of regional patterning are maintained and can be deployed in diverse morphological contexts as evolution proceeds, just as is true of regulatory gene cassettes in general.

Another plausible hypothesis may explain the conservative and clumpy nature of body plans besides that of developmental constraint, one that relies on the adaptive zone aspect of ecospace. In chapter 7, the relations of niches and adaptive zones to theoretical ecospace models were discussed. Ecospaces are multidimensional regions that are defined by the physical and biological requirements and resources of populations, or species, or higher taxa. Empty ecospace represents multidimensional combinations of the defining factors that are not inhabited by organisms but that are available to occupation through evolutionary change (combinations of physical and biological parameters that are not habitable are not considered to be ecospace), which raises the question of what is a full ecospace. What happens when the occupying organisms approach and then reach the limit of one or more of the resources in the region of ecospace available to them? In other words, what happens when the carrying capacity of their ecospace has been reached? Clearly, as their ecospace fills, both the origin and the immigration of lineages would be slowed and eventually stopped unless the limiting resources could be more finely partitioned among lineages. Such limitations on diversity appear to be operating in the modern biosphere, where some incumbent species evidently preclude the introduction of additional ones into their ecospaces (Philimore and Price 2008, 2009; Rabosky 2009; Ricklefs 2009; Valentine et al. 2008).

The processes vital to evolution during the latest Neoproterozoic and earliest Cambrian, however, involved the expansion of ecospace: oxygen levels had increased, the activity of sponges and then burrowing worms ventilated larger areas, and pelagic production was increasingly linked to benthic assimilation by metazoans. Lineages from among the relatively small-bodied but diversified benthic metazoan fauna responded to new ecological opportunities by increasing in body size and in morphological complexity, each diversifying clade evolving features appropriate to the accessible modes of life and producing adaptive zones. Perhaps the diversification eventually filled the more accessible ecospace regions in a metazoan biosphere. The increases in body-plan complexity and disparity that were achieved during the explosion may have slowed significantly in the later Cambrian (Bambach, Knoll, and Wang 2004). In subsequent diversifications, as during the great Ordovican biodiversification event, or following severe extinction events during the Phanerozoic, however, new phylum-level body plans did not evolve, perhaps (in this hypothesis) because the newly available ecological opportunities were accessible to branches evolving from one of the established body plans. In other words, there were enough adaptive types present in the biosphere to respond to the subsequent opportunities and challenges. The possible ecospace regions in the sea are hardly limitless, and once the biosphere contained a large variety of body plans (chiefly originating during the explosion interval), open ecospace would be filled by branches from established phyla long before it would have been possible to evolve an entirely new body plan. In this formulation, constraints on the evolution of new body plans reflect the great potentials of existing body plans and therefore their genomes to produce appropriate modifications to exploit new evolutionary opportunities.

These hypotheses—genomic and ecologic—are not mutually exclusive, and each has an explanatory role in metazoan history. In fact, the authors of this book, who are usually in agreement, happen to be divided on the question of the relative importance of each. Considering the great advances in the understanding of macroevolution and macroecology that are in full swing, we look forward to a resolution of these particular problems sooner rather than later.

OTHER PROBLEMS RAISED BY THE CAMBRIAN EXPLOSION

Although the constraint problem seems likely to yield to research in the near future, it is only one of many questions associated with the Cambrian explosion that remain open. Indeed, in some of the preceding chapters, it must seem as if every few paragraphs pose a different, unsolved question of some importance for which important new results are announced every few months. Of course, when dealing with biological events that occurred many hundreds of millions of years ago and that involved processes whose operations are incompletely understood even among the living biota, this situation is not unexpected. Many of the unanswered questions involve fields in which we have little or no expertise and are therefore difficult for us to evaluate. A few of those problems, however, lie within our own fields or significantly overlap with them and seem open to attack using currently available tools and theory. Here, then, quite briefly and exercising significant restraint, we ask ourselves what questions we would most like to address if we were just starting out in our fields in the hope that some zealous young workers will answer them for us.

As paleontologists, the most obvious course of action is to go back to the field in search of more new fossils. Each field season finds our colleagues working in South Australia, Newfoundland, northern Russia, China, and many other localities in search of fossils and more information from previously studied areas. There are always more fossils to discover, new approaches to try, and students with a different perspective. (The one constant among good students seems to be the conviction that their advisors are at best misguided and at worst wrong, and it is the students' obligation to set things right. Some things never change.) There are always new regions to explore as well. For example, deposits in Oman have only recently received much attention, and some of the best data on geochemical patterns in the late Neoproterozoic are emerging from drilling through sections there and in northern Canada. Rich opportunities are known to be present in the Ediacaran rocks of southern Namibia and the lower Cambrian rocks of Morocco. It is also worth remembering that the Chengjiang fauna in Yunnan Province, China, was only discovered in 1985. The only safe conclusion that we can reach is that we are far from discovering all the riches of the Ediacaran-Cambrian fossil record. Not all new fossil discoveries come from new fieldwork, however. The first reconstruction of an important explosion fossil, the Cambrian anomolacaridid *Hurdia*, was based on Burgess Shale specimens found in two collections, those of the National Museum of Natural History in Washington, DC, and Canada's Royal Ontario Museum (Daley et al. 2009). *Hurdia* had been known from the carapace alone, and the new reconstruction brings together pieces previously thought to represent other organisms. Although some of these specimens were collected years ago, in the early twentieth century, this work shows that museum collections still have many secrets left to reveal to observant students.

The development of very-high-resolution U-Pb radiometric dating of single crystals of zircon has already revolutionized our understanding of the distribution of time throughout the Ediacaran-Cambrian interval (chap. 2)—gone is the idea that there was a long interval barren of fossils between the Ediacaran fauna and the base of the Cambrian, for example—but many opportunities still remain. The interval between the early Doushantuo Formation, about 580 Ma, and the 555 Ma ash bed at the White Sea in Russia has very few reliable published radiometric dates. There is certainly material during that interval that can be dated, but the only way to find it is to get out to the field and walk the rocks. Increasing the accuracy and precision of radiometric dating will, for example, constrain the duration of the Shurum carbon isotope anomaly and provide new information on the timing and causes of the oxidation of the deep oceans during the late Neoproterozoic. There is no reason that within a few years we should not be able to construct a temporal framework for all the biological and geochemical events of the Ediacaran-Cambrian to time bins smaller than 5 million years and perhaps in favorable circumstances smaller than 1 million years. Achieving such refinement requires combining geochronology with new statistical correlation techniques, an area of active progress.

The application of carbon isotopic studies of Ediacaran and Cambrian rocks has a significant history, but it has now been joined by an almost dizzying array of new geochemical techniques, with iron speciation (Canfield et al.

2008) being one of the more recent. As was evident from chapter 3, we still have much to learn about the pattern and timing of the oxygenation of the oceans during the Ediacaran and early Cambrian. Current evidence from carbon isotopes suggests that this process was discontinuous or episodic over tens of millions of years, with some hints of geographic variation. Sulfur isotopes provide a confusing picture, currently difficult to interpret, although in principle they should yield significant clues. As geochemical methods continue to develop and be applied to new sections in different areas, they will certainly continue to provide new insights into changing chemical cycles and particularly into the timing and causes of the oxygenation of the oceans.

An area with possibilities for new insights into the diversification of animals involves the identification of clade-specific lipid biomarkers, which may provide a new source of information on the times of origin of clades. Biomarker studies also depend on the identification of new localities where organic sediments have been preserved and where the rocks have not been subject to much heating or diagenesis during burial. The sedimentary conditions involved in the accumulation and preservation of biomarkers are not yet well understood and provide important opportunities to advance this young field. Geologists are increasingly turning to cores drilled through interesting sections to retrieve relatively pristine samples.

We have argued for a significant role for ecological factors in producing the breadth and rapidity of the diversification, particularly during its most generative phase in the early Cambrian, yet there is much we still do not know on both the empirical and theoretical levels. How many ecological roles were present among the Ediacaran organisms? How did Ediacaran and early Cambrian ecosystems actually function? What was the role of predation, ecosystem engineering, and niche construction in increasing the complexity of Cambrian ecosystems? Are there biogeographic signals to be found in the record, and what roles did biogeographic deployments have on explosion communities? On a theoretical level, models of ecological innovation beyond adaptive radiations remain in their infancy. Indeed, for many evolutionary biologists, adaptive radiations are the most important problem of diversification. As we have explained, the Cambrian explosion was not a simple adaptive radiation but a diversification that encompassed many different clades in a relatively brief span of time and to far greater morphologic, developmental, and ecological effect than any subsequent period of innovation in Earth history. It is hardly surprising that models of adaptive radiations at smaller scales may not be sufficient to explain these events. More sophisticated models of the expansion of ecological networks and of the growth of biodiversity that involve interactions between critical variables are sorely needed.

Comparative developmental biology is advancing so rapidly that some of the data (although we hope not the interpretations) in chapters 8 and 9 will soon be outdated. Whole-genome sequences are being published rapidly, so that we have a growing body of sequences of evolutionarily interesting clades rather than simply those taxa of interest to the biomedical community. Within the next few years, there is every reason to expect that sequences will be available for most phyla and many of their major clades (sequences of lophotrochozoan groups have been relatively slow to appear, but are greatly needed).

Sequences alone do not solve the problems, as we discussed in chapter 9, but in the past few years, developmental biologists have moved from simply identifying the occurrence of similar regulatory sequences in different groups and of being able to "light up" the activity of particular genes during development to experimentally exploring the regulatory networks that underlie development. We can anticipate rapid advances in our understanding of the development of different clades of living organisms, and through application of the comparative method this information will reveal much of what happened some 550 million years ago during the early evolution of animals. Although surprises doubtless await, we are comfortable with our conclusions that most of the developmental innovations that allowed the generation of animal diversity long predate the explosion of animal body plans. Studies of miRNAs have revealed a new level of regulatory control, and much of the zoo of small regulatory RNAs remains to be explored. A lesson from the miRNA work is that new forms of control over development evolved in various clades to ensure cell type specification and allow other forms of regulation, but that the basic developmental toolkit evolved early in the history of animals, and this lesson, of course, raises a new question. Why was such an apparently sophisticated regulatory system needed for what appear to be fairly simple animals? Sponges and cnidarians appear to be "overdesigned." Does that mean, as has been suggested by several workers, that existing clades are the remnants of now-extinct clades with greater morphologic complexity or, alternatively (and more plausibly), that those molecular tools were initially evolved for simpler developmental tasks and were later co-opted within early bilaterians as developmental hierarchies became increasingly elaborate? Quantitative models of regulatory interactions are become increasingly sophisticated, and they hold great promise for revealing how these networks have evolved over time.

Although we would like to be able to predict that views of metazoan phylogenetic relationships will finally stabilize in the next few years, we must confess a certain degree of pessimism. Although we both feel quite strongly that the metazoan phylogeny presented in chapter 4 and used throughout the book is about right, the polytomies on the tree show that a number of issues remain unresolved. There has been ongoing debate about the topology of relationships at the base of the tree, including the paraphyly of sponges and the position of the ctenophores and acoels. The relationship of many of the lophotrochozoan clades remains very uncertain, and some of them may not even be monophyletic. There are also some groups of deuterostomes and ecdysozoans for which the phylogenetic placement continues to be debated. These phylogenetic issues seem unlikely to greatly affect our main conclusions here, but we hope that they will flesh out our understanding of the morphology of ancient animals at early nodes on the tree.

We have emphasized the importance of integrating our understanding of the environmental, developmental, and ecologic aspects of the origin and early diversification of animals. No one element of this triad is isolated from the others. We have argued that the extent of the evolutionary innovation among bilaterians that forms the Cambrian radiation, sensu stricto, was driven by the construction of ecological interactions and the introduction of new methods of developmental regulation, and that environmental changes—for example, the ventilation of the ocean—were essential elements of the diversification. What we find particularly compelling about the pattern of innovation is the extent to which it was biologically mediated: critical environmental changes were the result of biological activity that probably included the action of sponges in facilitating the oxygenation of the oceans by sequestering carbon during the Ediacaran, the substrate revolution of the early Cambrian, and, as Butterfield (2009) has persuasively argued, the diversification of pelagic mesozooplankton in linking pelagic and benthic ecosystems and further shifting the nature of the carbon cycle.

THE CAMBRIAN DIVERSIFICATION

The "standard model" of biological diversification involves adaptive radiation in which a single species exploits an ecological opportunity to diversify as different populations adapt to specialize on parts of the available resources (Schluter 2000). Competition between the diverging populations drives speciation to form a suite of new species, and the adaptive radiation eventually stabilizes as the opportunities for further resource specialization decline. Losos (2009) emphasizes speciation within an initially resource-rich environment, followed by resource depletion, partitioning of resources among the different species to reduce competition, and subsequent adaptation to the new, more specialized niches. In either case, current discussions of adaptive radiations involve multiplication of species within a clade, that is, adaptation driven by natural selection and generating extraordinary diversification (Glor 2010). Classic examples of such adaptive radiations include the Galapagos finches, cichlid fish from the East African Rift lakes, the silverswords of Hawaii, and Miocene horses. The Cambrian explosion is frequently described as an adaptive radiation. So, to try to establish whether the model truly applies to this episode—and, if not, what implications it may have for our understanding of evolutionary radiations in particular and for evolutionary processes more generally—it is worth exploring this concept further.

Ecological opportunity plays a key role in the concept of adaptive radiations (see Yoder et al. 2010) and continues to play a critical role in discussions on the causes of the Cambrian explosion. As described by George Gaylord Simpson (1944, 1953), ecological opportunities can involve the appearance of new resources, the extinction of species previously exploiting resources, the colonization of an area where resources were previously underused, or evolution of a new trait, often described as a "key innovation" that permits the use of resources in ways that were not previously possible. Whatever the sources of the opportunity, they open new avenues for evolutionary diversification. In the context of Ediacaran-Cambrian metazoans, apparent opportunities have been suggested in all these classes: new resources in the form of increased oxygen levels and of phytoplankton diversification, the apparent extinction of the Ediacaran fauna to provide opportunities in the early Cambrian, colonization of the infauna by burrowing organisms, and a variety of key innovations, from coeloms to the jointed appendages of arthropods. The relative importance of these and other opportunities remains unclear, but there is seemingly no lack of them. As Losos (2010) noted, however, there is a more general problem with the common invocation of opportunity in adaptive radiation: a large number of clades fail to radiate despite the presence of ecological opportunity. He suggested that the concept has little utility in the absence of methods for identifying and quantifying the opportunities. Opportunities must have abounded during the Ediacaran-Cambrian interval, but it is difficult to evaluate the relative importance of the various possibilities that underlay the evolutionary pathways actually taken during the Cambrian diversifications.

The concept of a key innovation, one of Simpson's four classes of opportunity, has received considerable attention, both theoretically and through the description of candidate key innovations. In A. H. Miller's (1949) original definition, key innovations were closely associated with the invasion of new adaptive zones and the establishment of new higher taxa (as previously discussed by G. G. Simpson 1944; see also Mayr 1960). Adaptive zones proved difficult to identify rigorously, and testing claims of key innovations has often been difficult (if not impossible). With the spread of phylogenetic methods, greater emphasis was placed on the identification of characters associated with extensive taxic diversification rather than with adaptive zones (Galis 2001). There has also been dispute about whether key innovations are necessarily associated with increased taxic diversity or could represent an increase in the *potential* for diversification (e.g., Rosenzweig and McCord 1991). Many authors have subsequently identified the potential for increased diversification as relying on the number of independent parameters available to evolution, and most recently this notion has involved a focus on the importance of modularity (Galis 2001; McShea 2000; Vermeij 1974; G. P. Wagner, Pavlicev, and Cheverud 2007) with some applications to the Cambrian explosion (Dewel, Dewel, and McKinney 2001; Jacobs et al. 2005).

The sorts of radiations involved in diversification of Hawaiian silverswords or African cichlids are not in evidence in the Cambrian explosion; there are no species swarms. Indeed, at the higher levels, as of phyla, the pattern is one of rather cryptic origination of the body plans and body types—*radiations*, if the term is applicable—that achieved significant morphological and therefore developmental disparity but for which there is no evidence of speciation as a driver or even an engaged participant, aside from the evolving branches being unsurprisingly composed of species. The clearest explosion trends involve diverging branches that lead to higher taxa in a Linnean sense. Although morphospace within the early disparity ranges of higher taxa is commonly filled by later branchings, the separation and distinctiveness of higher taxa seem on balance no less today than in the Cambrian, although cases of convergence do occur.

One criticism of the application of theories of adaptive radiation, or of adaptation more broadly, to episodes such as the Ediacaran-Cambrian diversification is that such major organizational transitions are fundamentally a different class of phenotypic change than most adaptations because they involve evolutionary novelties. G. B. Muller and Wagner, for example, defined a morphologic novelty as "a new constructional element in a body plan that neither has a homologous counterpart in the ancestral species nor in the same organism (serial homolog)" (G. B. Muller and Wagner 1991, 243). By this standard, novelty is rampant in the Ediacaran and Cambrian, but because so few intermediate species have been preserved, we are not able to assess whether these novelties are more apparent than real. The critical issue is the claim that evolutionary novelties may arise from different mechanisms than adaptive change. The literature on major evolutionary transitions, for example, argues that these innovations involve the incorporation of smaller entities into larger evolutionary units, as with the origin of the eukaryotic cell or the appearance of multicellularity (Maynard Smith and Szathmary 1995; Michod 2007). One way to assess the differences between the Cambrian explosion and adaptive radiations is to examine the patterns of morphologic diversification.

Morphologic evolution is commonly depicted with lineages more or less gradually diverging from their common ancestor. New features arise along the evolving lineages, and diversification turns those features into synapomorphies of new clades while new apomorphies appear among the morphologically diverging branches. Gould (1989, 38) characterized this pattern as the "cone of increasing diversity," but neither the fauna of the Cambrian nor the living marine fauna display this pattern. In fact, metazoan morphologies are quite clumped—underdispersed is the technical term—into clades with unique body plans and with significant gaps in architectural style between them, and this pattern continues among classes within phyla and to some extent even among orders within classes. Indeed, Lewontin (2003) has described the clumpy pattern of morphology as one of the outstanding unsolved problems in evolution. This situation was as true of the explosion fauna as it is of the modern one. To be sure, all pairs of crown phyla had common ancestors; as far as we know, however, none of those bilaterian LCAs had features that would cause them to be diagnosed as members of living phyla, although that could be the case in a few instances.

In other words, the morphological distances—gaps—between body plans of crown phyla were present when body fossils first appeared during the explosion and have been with us ever since. The morphological disparity is so great between most phyla that the homologous reference points or landmarks required for quantitative studies of comparative morphology are absent. To be sure, there was a radiation of bilaterian lineages into the stems of crown phyla, which is likely to have had a spreading architecture in the phylogenetic tree, and radiations at lower taxonomic categories within phyla and classes most commonly display a spreading architecture as well. At the level of crown phyla, however, the pattern is not one of divergence but rather of simply ringing changes on their distinctive body plans. Crown bilaterian phyla, and perhaps Phanerozoic clades of the radiate phyla as well, are essentially a heritage of the Cambrian explosion (Budd 2008; Budd and Jensen 2000).

Moreover, this pattern of early exploration of morphological space is commonly repeated at lower levels of the taxonomic hierarchy, where common landmarks do exist; thus, we can quantitatively assess patterns of disparity and compare them with taxonomic diversity. Two robust results emerge from such studies (reviewed in Erwin 2007 and Foote 1997). First, the morphological disparity among Cambrian clades is not noticeably lower than among their living counterparts. Marine arthropods, for example, are the most diverse group in Cambrian faunas and are rivaled in marine diversity today only by mollusks, yet studies indicate that their disparity was probably as great or greater in the Cambrian than it is today (Briggs, Fortey, and Wills 1992; Erwin 2007; Foote 1997; Fortey, Briggs, and Wills 1996). Second, in the majority of Cambrian and early Paleozoic clades studied, generally at the class level, morphologic disparity initially greatly exceeds taxonomic diversity (fig. 6.40). (There are some exceptions, largely among unstudied clades such as some of the lower Paleozoic echinoderm classes that have neither high disparity nor high diversity.) In other words, most clades of organisms initially explore the morphologic limits of their body plans at low taxonomic diversity and subsequently fill in the morphologic space as new species evolve. Thus, the morphologic distance between successful evolutionary steps must decline through time. It is just the opposite of the expectation of gradually expanding morphologic diversity through time.

Because the Cambrian explosion involved a significant number of separate lineages, achieving remarkable morphological breadth over millions of years, the Cambrian explosion can be considered an adaptive radiation only by stretching the term beyond all recognition. Rather, the Cambrian explosion seems to represent a very different sort of evolutionary diversification, one involving ecological feedback between multiple, evolutionarily independent lineages, with both morphologic and taxic diversification within each of these different lineages (Erwin 1992). Most importantly, the scale of morphologic divergence during the Cambrian is wholly incommensurate with that seen in other adaptive radiations. Recall that classic adaptive radiations represent specialization upon a newly available resource (for whatever reason) and that the radiation ends as the resource is fully used. In other words, adaptive radiations are expected to display negative feedback, eventually limiting the diversification. During the Cambrian explosion, the opposite appears to have happened, at least into the mid-Cambrian. The diversifications continued far beyond specialization on an existing resource and can only be explained, as we discussed in chapter 7, by the successive acquisition of new resources. Losos (2009, 625) has called it a "self-perpetuating" evolutionary radiation (based on Erwin 2008; in fairness, Losos seemed somewhat dubious that such radiations actually occur). Self-propagating radiations provide new resources to feed the diversification and may involve the alteration of ecosystem properties via ecosystem engineering and niche construction. As we explored in chapter 7, the advent of bioturbation may have entrained a series of changes in sediment redox, enhanced primary productivity, and expanded resources for other clades. This element of positive feedback is wholly absent from adaptive radiations and distinguishes the events of the Cambrian from adaptive radiations. We by no means wish to claim that the Cambrian is unique in this regard. Indeed, a number of other evolutionary radiations recorded by fossils likely reflect a similar process, including the Ordovician radiation, the diversification of a variety of vertebrate lineages in the Triassic following the end-Permian mass extinction, and the expansion of terrestrial clades in the Devonian and Carboniferous.

OUTLOOK

Historical events are difficult to explain; complex histories do not have single, easily understandable (or easily testable) causes. Indeed, if we think about most turning points in human history, it is usually difficult for historians to identify a single, straightforward cause. There is no simple explanation for the Renaissance, the Industrial Revolution, or World War I; each has a complex series of causes, many contingent on other events. The history of life's evolution is no different; it is just vastly broader in scope. Complex patterns of causality, the importance of contingency, and the interaction of many different processes are the norm. Clearly, the biosphere has promoted its own evolutionary trajectory, and the Cambrian explosion was a once-in-an-era happening; it could hardly have been more complicated and could hardly be more tantalizing. In addition, there can hardly be more of a challenge to paleobiologists, evolutionary biologists, and many other scientists than to describe and interpret the confluence of history and process responsible for events during that remote and critical time in life's history.

APPENDIX

First Appearances of Major Metazoan Clades in the Fossil Record

Prepared by Sarah Tweedt

First occurrences for all phyla and classes: first-appearing genus (or representative genus if there are multiple coeval appearances) for each phylum and class, the period + Stage of appearance, the primary references, and Paleobiology Database (PBDB) reference identification number (if applicable) (see www.paleodb.org).

Phylum	Class	Genus	First Occurrence	Source(s); PBDB Ref. No
PORIFERA				
Demospongiae			Cam 2 [Meis/Tom]	
	Class stem	*Choia*	Cam 2 [Meis/Tom]	Xiao, Hu et al. 2005; 32708
(Subclass)	Tetractinomorpha	*Geodia*	Cam 3 [Atd]	Reitner and Worheide 2002
	unassigned	*Hamptonia*	Cam 3 [Atd]	Steiner et al. 2005; 29233
Homoscleromorpha			Carb [early]	
	Homoscleromorpha[1]		Carb [early]	Reitner and Worheide 2002
Hexactinellida			Cam 2 [Meis/Tom]	
	Class stem	*Hunanospongia*	Cam 2/3 [Meis-Atd]	Steiner et al. 1993; 32714
(Subclass)	Amphidiscophora	*Protospongia*	Cam 2	Xiao, Hu et al. 2005
(Subclass)	Hexasterophora	*Calcihexactina*	Cam 2/3 [Meis-Atd]	G. X. Li et al. 2007
Calcarea			Cam 2 [Meis/Tom]	
	Class stem	*Gravestockia*	Cam 3 [Atd]	Reitner 1992
	Heteractinida	*Eiffelia*	Cam 2 [Meis/Tom]	Bengtson et al. 1990; 13290
(Subclass)	Calcinea		Recent	
(Subclass)	Calcaronea	*Protoleucon*	Carb [Vise]	Sepkoski 2002
Archaeocyatha			Cam 2 [Tom]	
	Regulares	*Coscinocyathus*	Cam 2 [Tom]	Gandin, Debrenne, and Debrenne 2007; 25805
	Irregulares	*Okulitchicyathus*	Cam 2 [Tom]	Rozanov et al. 1969; 13330
	Cribricyathea	*Leibaella*	Cam 3 [Atd]	R. Wood, Zhuravlev, and Anaaz 1993; 18191
CNIDARIA				
(Stem class)	"Anabaritids"[2]	*Anabarites*	Cam 1 [N-D]	Kouchinsky et al. 2009
(Stem class)	Hydroconozoa	*Hydroconus*	Cam 3 [Atd-Bot]	Kruse et al. 1996; 6600

Phylum	Class	Genus	First Occurrence	Source(s); PBDB Ref. No
Subphylum Anthazoa			Cam 3 [Atd]	
	Class stem	*Arrowipora*	Cam 3	Fuller and Jenkins 2007; 26866
(Subclass)	Zoantharia	*Xianguangia*	Cam 3 [Atd]	Sepkoski 2002
(Subclass)	Alcyonarea	*Petilavenula*	Ord [Aren-Mori]	Cope 2005; 29486
Subphylum Medusozoa			Cam 1 [N-D]	
	Class stem	*Cordubia*	Cam 1 [N-D]	Mayoral et al. 2004; 25892
	Scyphozoa		Recent	
	Conulata	*Carinachites*	Cam 1 [N-D]	Steiner et al. 2004; 29166
	Staurozoa		Recent	
	Cubozoa	*Unnamed*	Cam 6/7 [Marj]	Cartwright et al. 2008
	Hydrozoa	*Cambrohydra*	Cam 3 [Atd]	Hu 2005; 30233
CTENOPHORA				
Ctenophora			Cam 3 [Atd]	
	Stem class	*Maotianoascus*	Cam 3 [Atd]	Hou et al. 2004
LOPHOTROCHOZOA				
Chaetognatha			Cam 1 [N-D]	
	"Protoconodonts"	*Protohertzina*	Cam 1 [N-D]	G. X. Li et al. 2007; Landing et al. 1989; 29312
	Class stem	*Protosagitta*	Cam 3 [Atd]	G. X. Li et al. 2007
	Sagittoidea		Recent	
Rotifera			Paleo [Eocene]	
	Monogononta	*Notholca*	Early Holocene	Swadling et al. 2001
	Digononta		Recent	
	Bdelloidea	*Habrotrocha*	Paleo [Eocene]	Waggoner and Poinar 1993
	Seisonidea		Recent	

Phylum	Class	Genus	First Occurrence	Source(s); PBDB Ref. No
Platyhelminthes			Paleo [Eocene]	
	Turbellaria	*Micropulaeosoma*	Paleo [Eocene]	Poinar 2003; 33073
	Monogenea		Recent	
	Trematoda		Recent	
	Cestoda		Recent	
Entoprocta			Jur [Kim]	
(Family)	Barentsiidae	*Barentsia*	Jur [Kim]	Todd and Taylor 1992; Sepkoski 2002
Phoronida			Cam 1	
	Class stem	*Eccentrotheca*	Cam 1	Landing et al. 1989; Skovsted et al. 2008
	Extant phoronids		Recent	
Brachiopoda			Cam 1/2	
	Class stem	*Camenella*	Cam 1/2	Kouchinsky et al. 2012; Skovsted, Balthasar et al. 2009
	Stem Linguliformea[3]	*Mickwitzia*	Cam 3 [Atd]	Holmer and Popov 2007
	Lingulata	*Obolus*	Cam 2	Landing 1991; 430
	Paterinata	*Aldanotreta*	Cam 2	Kruse, Zharavlev, and James 1995; 6607
	Craniata	*Fengzuella*	Cam 1	Steiner et al. 2007; 29183
	Stem Rhyncholiformea[4]	*Salanygolina*	Cam 3 [Bot]	Holmer et al. 2009
	Chileata	*Kotujella*	Cam 3 [Atd]	Sepkoski 2002
	Obolellata	*Nochoriella*	Cam 2 [Tom]	Gregoryeva, Melnikova, and Pelman 1983; 900
	Kutorginata	*Khasagtina*	Cam 2 [Tom]	Ushatinskaya 1987; 876
	Strophomenata	*Billingsella*	Cam 3 [Bot]	Sepkoski 2002; 751
	Rhynchonellata	*Wangyuia*	Cam 3 [Atd]	Hu 2005; 30233

Phylum	Class	Genus	First Occurrence	Source(s); PBDB Ref. No
Bryozoa			Cam 9	
	Stenolaemata	*Prophyllodictya*	Ord [Trem]	Xia, Zhang, and Wang 2007
	Gymnolaemata	*Callopora*	Ord [Aren]	Allen and Lester 1957; 8741
	Phylactolaemata		Recent	
Hyolitha			Cam 1 [N-D]	
	Hyolithamorpha	*Ovalitheca*	Cam 1 [N-D]	Khomentovsky and Karlova 1993; 13517
	Orthothecimorpha	*Loculitheca*	Cam 1 [N-D]	Khomentovsky and Karlova 1993; 13517
Mollusca			Cam 1	
	Class stem	*Mobergella*	Cam 2	G. X. Li et al. 2007; Rozanov et al. 1969; 13330
		Odontogriphus	Cam 5	Caron et al. 2006
	Halwaxiids	*Halkieria*	Cam 1	Landing et al. 1989; 29312
	Polyplacophora	*Ocruranus/Eohalobia*	Cam 1	Vendrasco et al. 2009; 32126
		Lopochites	Cam 1	Steiner et al. 2004; 29166
	Aplacophora	*Matthevia*	Cam 8	Sigwart and Sutton 2007; Sepkoski 2002
		Acaenoplax	Sil [Wenlock]	Sutton et al. 2004
	Caudofoveata		Recent	
	Solenogastres		Recent	
	Rostroconcha	*Watsonella*	Cam 1	Landing et al. 1989; 29312
	Helcionelloida	*Helcionella*	Cam 1	Landing et al. 1989; 29312
	Tergomya	*Canopoconus*	Cam 1	Feng, Sun, and Qian 2001; 15906
	Scaphapoda	*Rhytiodentalium*	Ord [Cara-mid]	Sepkoski 2002
	Bivalvia	*? Fordilla*	Cam 2 [N. Scotia]	Landing 1991; 430
		Pojetaia	Cam 3	Parkhaev 2004; 13185
	Gastropoda	*Chippewaella*	Cam 7	Gunderson 1993; 566

Phylum	Class	Genus	First Occurrence	Source(s); PBDB Ref. No
		? Aldanella	Cam 1	Landing et al. 1989; 29312
	Paragastropoda	*Yuwenia*	Cam 3	Elicki 1994; 13333
	Cephalopoda	*? Nectocaris pteryx*	Cam 5	M. R. Smith and Caron 2010
		Plectronoceras	Cam 9	Mutvei, Zhang, and Dunca 2007
	Stenothecoida	*Manikai*	Cam 1	Missarzhevsky 1989
		Stenothecoides	Cam 2	Brasier et al. 1996
	Tentaculita	*Tentaculites*	Ord [Trem]	Fisher and Young 1955; 26506
Coeloscleritophora[5]			Cam 1	
	Chancelloriids	*Chancelloria*	Cam 1 [Fortunian]	Kouchinsky et al. 2012
	Siphonogonuchitids	*Spingonuchites*	Cam 1 [Fortunian]	Kouchinsky et al. 2012
		Drepanochites	Cam 2 [China]	G. X. Li et al. 2007
	Engimatic sclerites	*Zhiiginites*	Cam 1	Conway Morris and Menge 1991; 506
	Enigmatic tubular fossils		Cam 2	G. X. Li et al. 2007
Sipuncula			Cam 3 [Atd]	
	Class stem	*Archaeogolfingia*	Cam 3 [Atd]	Huang et al. 2004
	Phascolosomida		Recent	
	Sipunculida		Recent	
Annelida			Cam 3 [Atd]	
	Class stem	*Maotianchaeta*	Cam 3 [Atd]	G. X. Li et al. 2007
	Machaeridia	*Plumulites*	Ord [Trem]	Vinther, Van Roy, and Briggs 2008
	Polychaeta	*Phragmochaeta*	Cam 3 [Atd]	Conway Morris and Peel 2010
	Echiura	*Coprinoscolex*	Carb [Mazon]	D. Jones and Thompson 1977
	Myzostomiida	*Myzostomites*	Ord [late]	Warn 1974
	Oligochaeta	*Pronaidites*	Carb [Kas/Gze]	Wills 1993
	Hirudinea	*Burejospermum*	Jur [Toar-Plie]	Jansson, McLoughlin, and Vajda 2008

Phylum	Class	Genus	First Occurrence	Source(s); PBDB Ref. No
Nemertea			Carb [Serp-l]	
	Class stem	*Archisymplectes*	Carb [Serp-l]	Schram 1973
	Anolpla		Recent	
	Enopla		Recent	
ECDYSOZOA				
Priapulida			Cam 2 [Tom]	
	Palaeoscolecida	*Maotianshania*	Cam 2 [Tom]	Sun and Hou 1987; 419
	Priapulimorpha	*Ancalagon*	Cam 5 [Burgess]	Caron and Jackson 2008; 28283
	Halicryptomorpha		Recent	
	Seticoronaria		Recent	
Nematomorpha			Cam 3	
	Class stem	*Cricocosmia*	Cam 3 [Chengjiang]	Hou and Sun, 1988; 847
	Nectonematoidea		Recent	
	Gordioidea		Recent	
Loricifera			Cam 3	
	Class stem	*Sirilorica*	Cam 3 [Sirius Passet]	Peel 2010
Nematoda			Cret [Barr]	
	Class stem	*Heleidomermis*	Cret [Barr]	Poinar, Acra, and Acra 1994
	Adenophorea		Recent	
	Secernentea		Recent	
Panarthropoda				
Phylum indet.	Stem arthropod traces	*Rusophycus*	Cam 2 [Tom]	Edgecombe 2010
Unranked stem	Cambrian lobopods	*Luolishania*	Cam 3	E.g., J. Y. Chen and Zhou 1997

Phylum	Class	Genus	First Occurrence	Source(s); PBDB Ref. No
Tardigrada			Cam 5? [Siberia]	
	Class stem	*Unnamed tardigrade*	Cam 5? [Siberia]	K. J. Muller, Walossek, and Zakharov 1995
	Heterotardigrada		Recent	
	Mesotardigrada		Recent	
	Eutardigrada	*Milnesium*	Cret (Turo)	Bertolani and Grimaldi 2000
Lobopodia			Cam 3	
	Class stem	*Microdictyon*	Cam 3	Hinz 1987; 15995; Kouchinsky et al. 2012
		Hadranax	Cam 3	Budd and Peel 1998; 546
	Gilled lobopods	*Kerygmachela*	Cam 3	Budd 1993; 30407
Euarthropoda			Cam 3	
	Class stem	*Fuxianhuia*	Cam 3	Hou and Bergström 1997
		Perspicaris	Cam 3	Steiner et al. 1993; 32714
		Tamisiocaris	Cam 3	Daley and Peel 2010
	Lamellipedia	*Naraoia*	Cam 3 [Atd]	Steiner et al. 2005; 29233
		Retifacies	Cam 3	Hou and Bergström 1997
		Kuamaia	Cam 3	Hou and Bergström 1997
		Xandarella	Cam 3	Hou and Bergström 1997
		Fallotaspis	Cam 3 [Atd-l]	Hollingsworth 1999; 3887
		Kwanyinaspis	Cam 3	X. L. Zhang and Shu 2005
Subphylum Chelicerata				
	Class stem	*Sanctacaris*	Cam 5 [Burgess]	Dunlop and Selden 1998
	Pycnogonida	*Cambropycnogon*	Cam 8 [Maent]	Waloszek and Dunlop 2002
	Megacheira	*Haikoucaris*	Cam 3 [Atd]	J. Y. Chen, Waloszek, and Maas 2004
	Merostomata	*Paleomerus*	Cam 3 [Atd]	Jensen 1990; 858
	Xiphosura	*Unnamed specimen*	Ord [Trem]	Van Roy et al. 2010
	Arachnida	*Land scorpions*	Sil [early]	Sepkoski 2002

Phylum	Class	Genus	First Occurrence	Source(s); PBDB Ref. No
Subphylum indet.				
	Thylacocephala	*Isoxys*	Cam 3	Hu et al. 2007; Vannier et al. 2006
Subphylum Crustacea				
	Class stem	*Kunmingella*	Cam 3	Hou et al. 2010; Steiner et al. 2005; 29233
		Marrella	Cam 5 [Burgess]	
		Cambropachycope	Cam 8	Waloszek and Muller 1990
	Crown stem	*Pectocaris*	Cam 3 [Atd]	Hou, Bergström et al. 2004
		Yicaris	Cam 3 [Atd]	X. G. Zhang et al. 2007
	Remipedia	*Tesnusocaris*[6]	Carb [Pen-lower]	Brooks 1955
	Cephalocarida	*Dala*	Cam 8 [Maent]	K. J. Muller 1983; 860
	Branchiopoda	*Rehbachiella*	Cam 8 [Maent]	Walossek 1995
	Maxillopoda	*Priscansermarinus*	Cam 5 [Burgess]	
		Heymonsicambria	Cam 8/9 [Maent]	Walossek, Repetski, and Muller 1994
	Ostracoda	*Kimsella*	Ord [Trem]	M. Williams et al. 2008
	Malacostraca	*Proboscicaris*	Cam 5 [Solvan]	Chlupac and Kordule 2002; 57206
Subphylum Hexapoda				
	Insecta	*Leverhulmia*	Dev [Emsi]	Fayers and Trewin 2005
	Collembola	*Rhyniella*	Dev [Emsi]	Greenslade and Whalley 1986
DEUTEROSTOMIA				
Vetulicolia			Cam 3 [Atd]	
	Vetulicolata	*Pomatrum*	Cam 3 [Atd]	Aldridge et al. 2007
	Heteromorphida	*Heteromorphus*	Cam 3 [Atd]	Aldridge et al. 2007
		Banffia	Cam 5 [Burgess]	Caron 2006

Phylum	Class	Genus	First Occurrence	Source(s); PBDB Ref. No
Unranked stem	"Cambroernids"	*Eldonia*	Cam 3 [Atd]	Zhu, Zhao, and Chen 2002
		Herpetogaster	Cam 5 [Burgess]	Caron, Conway Morris, and Shu 2010
Hemichordata			Cam 3 [Atd-Bot]	
	Graptolithina	*Chaunograptus*	Cam 5 [Burgess]	Caron and Jackson 2008; 28283
	Pterobranchia	*Galeaplumosus*	Cam 3 [Atd-Bot]	Hou et al. 2011
	Enteropneusta	*Ottoia tenuis*	Cam 5 [Burgess]	
		Megaderaion	Jur [Sine]	Arduini, Pinna, and Teruzzi 1981
Echinodermata			Cam 3	
	Class stem	*Ventulocystis*	Cam 3 [Chengjiang]	G. X. Li et al. 2007
		Echinoderm plates	Cam 3	Kouchinsky et al. 2012
Subphylum Homalozoa	Stylophora	*Ceratocystis*	Cam 5	Zamora 2010; Fatka and Kordule 2001; 19350
	Homoiostelea	*Castericystis*	Cam 6 [Marj]	Ubaghs and Robison 1985; 32653
	Homostelea	*Asturicystis*	Cam 5?	Fatka and Kordule 2001; 19350
		Undescribed form	Cam 5	Zamora 2010
	Ctenocystoidea	*Ctenosystis*	Cam 5	Sepkoski 2002
		Undescribed form	Cam 5	Zamora 2010
	Cincta	*Protocinctus*	Cam 5	Rahman and Zamora 2009
Subphylum Blastazoa	Eocrinoidea	*Alanisicystis*	Cam 4 [Bot]	Ubaghs and Vizcaino 1990; 544
		Gogia	Cam 4/5	Zamora 2010; Durham 1978; 32669
	Rhombifera	*Cuniculocystis*	Ord [Aren-l]	Sepkoski 2002
	Diploporita	*? Lichenoides*	Mid Cam	Chlupac 1993; 25868
		Sinocystis	Ord [Trem]	Bruton, Wright, and Hamedi 2004; 19028
	Parablastoidea	*Blastoidocrinus*	Ord [Aren]	Sepkoski 2002
	Blastoidea	*Decaschisma*	Sil [Wenl-l]	Frest, Brett, and Witzke 1999; 4379
	Coronoidea	*Cupulocorona*	Ord [Ashgill-l]	Sepkoski 2002

Phylum	Class	Genus	First Occurrence	Source(s); PBDB Ref. No
Subphylum indet.	Edrioasteroidea	*Cambraster*	Cam 5 [lMid]	Zamora et al. 2007; 30458
		Stromatocystites	Cam 3/4 [Bot]	Chulpac 1993; 25868
	Helioplacoidea	*Helioplacus*	Cam 3/4 [Atd-Bot]	Wilbur 2006; 30495
	Ophiocistoidea	*Volchovia*	Ord [Aren-u]	Sepkoski 2002
	Cyclocystoidea	*Cyclosystoidea*	Ord Blackriveran	Kolata, Brower, and Frest 1987; 6707
	Camptostromoidea	*Camptostroma*	Cam 3/4	Sprinkle 1973; 22459; Sepkoski 2002
Subphylum Crinozoa	Class stem	*Echmatocrinus*	Mid Cam	Sepkoski 2002
	Crinoidea	*Hybocrinus*	Ord [Aren]	Sepkoski 2002
	Paracrinoidea	*Malocysites*	Ord [Llde]	Kobluk 1981; 26898
Subphylum Asterozoa	Somasteroida	*Apullaster*	Ord [Trem-u]	Sepkoski 2002
	Asteroidea	*Eriaster*	Ord [Trem]	Blake and Guensberg 2005; 28538
	Ophiuroidea	*Pradesura*	Ord [Aren-l]	Sepkoski 2002
Subphylum Echinozoa	Echinoidea	*Neobothriocidaris*	Ord [Llvi]	Kolata, Brower, and Frest 1987; 6707; Sepkoski 2002
	Holothuroidea	*Thuroholia*	Ord [Cara]	Gutschick 1954; 31864
Urochordata			Cam 3	
	Class stem	*Shankouclava*	Cam 3 [Chengjiang]	J. Y. Chen et al. 2003
	Ascidiacea	*Permosoma*	Perm [Leon]	Sepkoski 2002
	Thaliacea		Recent	
	Appendicularia		Recent	
	Sorberacea		Recent	
Cephalochordata			Cam 3	
	Class stem	*Cathaymyrus*	Cam 3 [Chengjiang]	Shu, Conway Morris, and Zhang 1996

Phylum	Class	Genus	First Occurrence	Source(s); PBDB Ref. No
Craniata			Cam 3 [Atd]	
	Cephalaspidomorphi	*Unnamed*	Ord [Cara]	Sepkoski 2002
		Tremataspis	Sil [Wenl]	Mark-Kurik 1969; 6155
	Pteraspidomorphi	*Arandaspis*	Ord [Aren]	Ritchie and Gilbert-Tomlinson 1977; 30365
	Agnatha	*Haikouichthys*	Cam 3 [Atd]	X. G. Zhang and Hou 2004
		Various agnathans	Cam 3 [Chengjiang]	G. X. Li et al. 2007
		Anatolepis	Cam 9 [upper]	M. P. Smith, Sansom, and Chochrane 2001
	Chondrichthyes	*Areyonga*	Ord [Llvi]	G. C. Young 1997; 30377

NOTES

1. Homoscleromorpha has recently been elevated to class rank; see Gazave et al. (2012).

2. The affinities of the "Anabaratids" are uncertain; we treat this grouping as a cnidarian grade.

3. *Mickwitzia* is a stem brachiopod but may have Linguliformea affinities.

4. *Salanygolina*, also a stem brachiopod, shows affinities with the Rhyncholiformea.

5. Coeloscleritophora: halwaxiids form an accepted clade and have thus been removed from Coeloscleritophora. The phylogenetic affinities of remaining taxa remain uncertain; thus, they have been treated as a paraphyletic (but morphologically disparate) group.

6. *Tesnusocaris* is very unlikely to be stem Remipedia; see Neiber et al. 2011. As yet, it is the only fossil taxon known, we have left it in our compilation pending further analyses.

References

Abedin, M., and N. King. 2008. The premetazoan ancestry of cadherins. *Science* 319:946–48.

Adams, M. D., S. E. Celniker, R. A. Holt, C. A. Evans, J. D. Gocayne et al. 2000. The genome sequence of *Drosophila melanogaster*. *Science* 287:2185–95.

Adamska, M., B. M. Degnan, K. Green, and C. Zwafink. 2011. What sponges can tell us about the evolution of developmental processes. *Zoology* 114:1–10.

Adamska, M., S. M. Degnan, K. M. Green, M. Adamski, A. Craigie et al. 2007. Wnt and TGF-b expression in the sponge *Amphimedon queenslandica* and the origin of metazoan embryonic patterning. *PLoS ONE* 2(10):e1031.

Ader, M., M. Macouin, R. I. F. Trindade, M. H. Hadrien, Z. Yang et al. 2009. A multilayered water column in the Ediacaran Yangtze platform? Insights from carbonate and organic matter paired $\delta^{13}C$. *Earth and Planetary Science Letters* 288:213–27.

Aguinaldo, A. M. A., J. M. Turbeville, L. S. Linford, M. C. Rivera, J. R. Garey et al. 1997. Evidence for a clade of nematodes, arthropods and other moulting animals. *Nature* 387:489–93.

Aldridge, R. J., D. E. Briggs, E. N. K. Clarkson, and M. P. Smith. 1986. The affinities of conodonts: New evidence from the Carboniferous of Edinburgh, Scotland. *Lethaia* 19:279–91.

Aldridge, R. J., X. G. Hou, D. J. Siveter, D. J. Siveter, and S. E. Gabbott. 2007. The systematics and phylogenetic relationships of vetulicolians. *Palaeontology* 50:131–68.

Allen, A. T., and J. G. Lester. 1957. Zonation of the Middle and Upper Ordovician strata in northwestern Georgia. *Georgia State Division of Conservation, Geological Survey Bulletin* 66:1–104.

Amthor, J. E., J. P. Grotzinger, S. Schröder, S. A. Bowring, J. Ramezani et al. 2003. Extinction of *Cloudina* and *Namacalathus* at the Precambrian-Cambrian boundary in Oman. *Geology* 31:431–34.

Amundson, R. 2005. *The Changing Role of the Embryo in Evolutionary Thought*. Cambridge, UK: Cambridge University Press.

Anbar, A. D., and A. H. Knoll. 2002. Proterozoic ocean chemistry and evolution: A bioinorganic bridge? *Science* 297:1137–42.

Anderson, M. M., S. Conway Morris, and T. P. Crimes. 1982. A review, with descriptions of four unusual forms, of the soft-bodied fauna of the conception and St. John's Group (late Precambrian), Avalon Peninsula, Newfoundland. *Third North American Paleontological Convention Proceedings* 1:1–8.

Antcliffe, J. B., and M. D. Brasier. 2008. *Charnia* at 50: Developmental models for Ediacaran fronds. *Palaeontology* 51:11–26.

Arduini, P., G. Pinna, and G. Teruzzi. 1981. *Megaderaion sinemuriense* n.g. n.sp., a new fossil enteropneust of the Sinemurian of Osteno in Lombardy. *Atti Societa Italiana di Scienze Naturale e Museo Civico di Storia Naturale Milan* 122:462–68.

Arendt, D. 2008. The evolution of cell types in animals: Emerging principles from molecular studies. *Nature Reviews Genetics* 8:868–82.

Arendt, D., A. S. Denes, G. Jekely, and K. Tessmar-Raible. 2008. The evolution of nervous system centralization. *Philosophical Transactions of the Royal Society B* 363:1523–28.

Arendt, D., H. Hausen, and G. Purschke. 2009. The "division of labour" model of eye evolution. *Philosophical Transactions of the Royal Society B* 364:2809–17.

Arendt, D., and K. Nubler-Jung. 1994. Inversion of dorso-ventral axis. *Nature* 371:26.

———. 1999. Comparison of early nerve cord development in insects and vertebrates. *Development* 126:2309–25.

Arendt, D., U. Technau, and J. Wittbrodt. 2001. Evolution of the bilateria larval foregut. *Nature* 409:81–85.

Ashworth, J. H. 1912. *Catalogue of the Chaetopoda in the British Museum (Natural History). A. Polychaeta: Part I. Arinicolidae.* London: Trustees of the British Museum.

Awramik, S. M., and J. W. Valentine. 1985. Adaptive aspects of the origin of autotrophic eukaryotes. In *Geological Factors and the Evolution of Plants*, edited by B. H. Tiffney, 11–21. New Haven, CT: Yale University Press.

Ax, P. 1996. *Multicellular Animals: A New Approach to the Phylogenetic Order in Nature.* Berlin: Springer-Verlag.

Babcock, L. E. 2003. Trilobites in Paleozoic predator-prey systems, and their role in reorganization of early Paleozoic ecosystems. In *Predator-Prey Interactions in the Fossil Record*, edited by P. H. Kelley, M. Kowalewski, and A. J. Hansen, 55–92. New York: Kluwer Academic Press / Plenum.

Babcock, L. E., and S. C. Peng. 2007. Cambrian chronostratigraphy: Current state and future plans. *Palaeogeography, Palaeoclimatology, Palaeoecology* 254:62–66.

Badano, E. I., P. A. Marquet, and L. A. Cavieres. 2010. Predicting effects of ecosystem engineering on species richness along primary productivity gradients. *Acta Oecologica* 36:46–54.

Baguña, J., P. Martinez, J. Paps, and M. Riutort. 2008. Back in time: A new systematic proposal for the Bilateria. *Philosophical Transactions of the Royal Society B* 363:1481–91.

Baguña, J., and M. Riutort. 2004. The dawn of bilaterian animals: The case of acoelomorph flatworms. *BioEssays* 26:1046–57.

Baguña, J., I. Ruiz-Trillo, J. Paps, M. Loukota, C. Ribera et al. 2001. The first bilaterian organisms: Simple or complex? New molecular evidence. *International Journal of Developmental Biology* 45(SI):S133–34.

Bailey, J. V., S. B. Joye, K. M. Kalanetra, B. E. Flood, and F. A. Corsetti. 2007. Evidence of giant sulphur bacteria in Neoproterozoic phosphorites. *Nature* 445:198–201.

Balavoine, G., and A. Adoutte. 2003. The segmented *Urbilateria*: A testable scenario. *Integrative and Comparative Biology* 43:137–47.

Balavoine, G., R. de Rosa, and A. Adoutte. 2002. *Hox* clusters and bilaterian phylogeny. *Molecular Phylogenetics and Evolution* 24:366–73.

Balthasar, U., C. B. Skovsted, L. E. Holmer, and G. A. Brock. 2009. Homologous skeletal secretion in tommotiids and brachiopods. *Geology* 37:1143–46.

Bambach, R. K., A. M. Bush, and D. H. Erwin. 2007. Autecology and the filling of ecospace: Key metazoan radiations. *Palaeontology* 50:1–22.

Bambach, R. K., A. H. Knoll, and S. C. Wang. 2004. Origination, extinction, and mass depletions of marine diversity. *Paleobiology* 30:522–42.

Bao, H. M., J. R. Lyons, and C. M. Zhou. 2008. Triple oxygen isotope evidence for elevated CO_2 levels after a Neoproterozoic glaciation. *Nature* 453:504–6.

Barfod, G. H., F. Albarede, A. H. Knoll, S. H. Xiao, P. Telouk et al. 2002. New Lu-Hf and Pb-Pb age constraints on the earliest animal fossils. *Earth and Planetary Science Letters* 201:203–12.

Barnes, R. S. K., P. Calow, and P. J. W. Olive. 1993. *The Invertebrates: A New Synthesis*. Oxford, UK: Blackwell.

Bayer, F. M., and H. B. Owre. 1968. *The Free-Living Lower Invertebrates*. New York: Macmillan.

Bengtson, S. 1983. The early history of the Conodonta. *Fossils and Strata* 15:5–19.

———. 2002. Origins and early evolution of predation. *Paleontological Society Papers* 8:289–318.

———. 2005. Mineralized skeletons and early animal evolution. In *Evolving Form and Function: Fossils and Development*, edited by D. E. G. Briggs, 101–24. New Haven, CT: Peabody Museum of Natural History, Yale University.

Bengtson, S., and S. Conway Morris. 1992. Early radiation of biomineralizing phyla. In *Origin and Early Evolution of the Metazoa*, edited by J. H. Lipps and P. W. Signor, 447–81. New York: Plenum.

Bengtson, S., S. Conway Morris, B. Cooper, P. A. Jell, and B. N. Runnegar. 1990. Early Cambrian fossils from South Australia. *Association of Australasian Palaeontologists Memoir* 9.

Bengtson, S., and Y. Zhou. 1992. Predatorial borings in late precambrian mineralized exoskeletons. *Science* 257:367–69.

Bengtson, S., and Y. Zhao. 1997. Fossilized metazoan embryos from the earliest Cambrian. *Science* 277:1645–48.

Berke, S. K. 2010. Functional groups of ecosystem engineers: A proposed classification with comments on current issues. *Integrative and Comparative Biology* 50:147–57.

Berkner, L. V., and L. C. Marshall. 1964. The history of oxygenic concentration in the Earth's atmosphere. *Discussions of the Faraday Society* 37:122–41.

———. 1965. On the origin and rise of oxygen concentration in the Earth's atmosphere. *Journal of the Atmospheric Sciences* 22:225–61.

Bertolani, R., and D. Grimaldi. 2000. A new eutardigrade (Tardigrada: Milnesiidae) in amber from the Upper Cretaceous (Turonian) of New Jersey. In *Studies on Fossils in Amber, with Particular Reference to the Cretaceous of New Jersey*, edited by D. Grimaldi, 103–10. Leiden, Netherlands: Backhuys.

Blake, B. D., and T. E. Guensburg. 2005. Implications of a new Early Ordovician asteroid (Echinodermata) for the phylogeny of asterozoans. *Journal of Paleontology* 79:395–99.

Boardman, R. S., A. H. Cheetham, and A. J. Rowell. 1987. *Fossil Invertebrates*. Palo Alto, CA: Blackwell Scientific.

Bodmer, R., and T. V. Venkatesh. 1998. Heart development in *Drosophila* and vertebrates: Conservation of molecular mechanisms. *Developmental Genetics* 22:181–86.

Boero, F., B. Schierwater, and S. Piraino. 2007. Cnidarian milestones in metazoan evolution. *Integrative and Comparative Biology* 47:693–700.

Bone, Q., H. Karp, and A. C. Pierrot-Bults, 2011. *The Biology of Chaetognaths*. Oxford, UK: Oxford University Press (reprint of 1991 ed.).

Booth, L. N., B. B. Tuch, and A. D. Johnson. 2010. Intercalation of a new tier of transcription regulation into an ancient circuit. *Nature* 468:959–63.

Bosak, T., D. J. G. Lahr, S. B. Pruss, F. A. Macdonald, A. J. Gooday et al. 2012. Possible early foraminiferans in post-Sturtian (716–635 Ma) cap carbonates. *Geology* 40:67–70.

Botting, J. P., and N. J. Butterfield. 2005. Reconstructing early sponge relationships by using the Burgess Shale fossil *Eiffelia globosa*. *Proceedings of the National Academy of Sciences, USA* 102:1554–59.

Bottjer, D. J., J. W. Hagadorn, and S. Q. Dornbos. 2000. The Cambrian substrate revolution. *GSA Today* 10(9):1–7.

Bourlat, S. J., T. Juliusdottir, C. J. Lowe, R. Freeman, J. Aronowicz et al. 2006. Deuterostome phylogeny reveals monophyletic chordates and the new phylum Xenoturbellida. *Nature* 444:85–88.

Bourlat, S. J., C. Nielsen, A. E. Lockyer, D. T. Littlewood, and M. J. Telford. 2003. Xenoturbella is a deuterostome that eats molluscs. *Nature* 424:925–28.

Boury-Esnault, N., S. Efremova, C. Bezac, and J. Vacelet. 1999. Reproduction of a hexactinellid sponge: First description of gastrulation by cellular delamination in the Porifera. *Invertebrate Reproduction and Development* 35:187–201.

Boury-Esnault, N., A. Ereskovsky, C. Bezac, and D. Tokina. 2003. Larval development in the Homoscleromorpha (Porifera, Demospongiae). *Invertebrate Biology* 122:187–202.

Boute, N., J. Y. Exposito, N. Boury-Esnault, J. Vacelet, N. Noro et al. 1996. Type IV collagen in sponges, the missing link in basement membrane ubiquity. *Biology of the Cell* 88:37–44.

Bowring, S. A., J. P. Grotzinger, D. Condon, J. Ramezani, M. Newall et al. 2007. Geochronologic constraints of the chronostratigraphic framework of the Neoproterozoic Huqf Supergroup, Sultanate of Oman. *American Journal of Science* 307:1097–115.

Bowring, S. A., J. P. Grotzinger, C. E. Isachsen, A. H. Knoll, S. M. Pelechaty et al. 1993. Calibrating rates of Early Cambrian evolution. *Science* 261:1293–98.

Bowring, S. A., P. M. Myrow, E. Landing, and J. Ramezani. 2003. Geochronological constraints on terminal Neoproterozoic events and the rise of metazoans. *Geophysical Research Abstracts* 5:219.

Brasier, M. D. 1990. Nutrients in the Early Cambrian. *Nature* 347:521–22.

———. 1991. Nutrient flux and the evolutionary explosion across the Precambrian-Cambrian boundary interval. *Historical Biology* 5:85–93.

———. 1992. Nutrient-enriched waters and the early skeletal fossil record. *Journal of the Geological Society of London* 149:621–29.

Brasier, M. D., J. C. Cowie, and M. Taylor. 1994. Decision on the Precambrian-Cambrian boundary. *Episodes* 17:95–100.

Brasier, M. D., O. Green, and G. Shields. 1997. Ediacarian sponge spicule clusters from southwestern Mongolia and the origins of the Cambrian fauna. *Geology* 25:303–6.

Brasier, M. D., and J. F. Lindsay. 2001. Did supercontinental amalgamation trigger the "Cambrian explosion"? In *The Ecology of the Cambrian Radiation*, edited by A. Y. Zhuralev and R. Riding, 69–89. New York: Columbia University Press.

Brasier, M. D., G. McCarron, T. Tucker, J. Leather, P. Allen et al. 2000. New U-Pb zircon dates for the Neoproterozoic Ghubrah glaciation and for the top of the Hugf supergroup, Oman. *Geology* 28:175–78.

Brasier, M. D., A. Y. Rozanov, A. Y. Zhuravlev, R. M. Corfield, and L. A. Derry. 1994. A carbon isotope reference scale for the Lower Cambrian succession in Siberia: Report of IGCP Project 303. *Geological Magazine* 131:767–83.

Brasier, M. D., G. Shields, V. N. Kuleshov, and E. A. Zhegallo. 1996. Integrated chemo- and biostratigraphic calibration of early animal evolution: Neoproterozoic–Early Cambrian of southwest Mongolia. *Geological Magazine* 133:445–85.

Brasier, M. D., and S. S. Sukhov. 1998. The falling amplitude of carbon isotopic oscillations through the Lower to Middle Cambrian: Northern Siberia data. *Canadian Journal of Earth Sciences* 35:353–73.

Brennan, S. T., T. K. Lowenstein, and J. Horita. 2004. Seawater chemistry and the advent of biocalcification. *Geology* 32:473–76.

Briggs, D. E. G., D. H. Erwin, and F. J. Collier. 1994. *The Fossils of the Burgess Shale*. Washington, DC: Smithsonian Institution Press.

Briggs, D. E. G., R. A. Fortey, and M. A. Wills. 1992. Morphological disparity in the Cambrian. *Science* 256:1670–73.

Bristow, T. F., and M. J. Kennedy. 2008. Carbon isotope excursions and the oxidant budget of the Ediacaran atmosphere and ocean. *Geology* 25:863–66.

Brocks, J. J., and N. J. Butterfield. 2009. Early animals out in the cold. *Nature* 457:672–73.

Brooke, N. M., J. Garcia-Fernandez, and P. W. H. Holland. 1998. The *ParaHox* gene cluster is an evolutionary sister of the *Hox* gene cluster. *Nature* 392:920–22.

Brooks, H. K. 1955. A crustacean from the Tesnus Formation (Pennsylvanian) of Texas. *Journal of Paleontology* 29:852–56.

Brown, F. A., ed. 1950. *Selected Invertebrate Types*. New York: Wiley.

Brown, J. H., J. F. Gillooly, A. P. Allen, V. M. Savage, and G. B. West. 2004. Toward a metabolic theory of ecology. *Ecology* 85:1771–89.

Brusca, R. C., and G. J. Brusca. 1990. *Invertebrates*. Sunderland, MA: Sinauer.

Bruton, D. L., A. J. Wright, and M. A. Hamedi. 2004. Ordovician trilobites from Iran. *Palaeontographica Abteilung A* 271:111–49.

Buitois, L. A., and M. G. Mángano. 2012. An Early Cambrian shallow-marine ichnofauna from the Puncoviscana Formation of northwest Argentina: The interplay between sophisticated feeding behaviors, matgrounds and sea-level changes. *Journal of Paleontology* 86:7–18.

Budd, G. 2001. Why are arthropods segmented? *Evolution and Development* 3:332–42.

Budd, G., and M. J. Telford. 2009. The origin and evolution of the arthropods. *Nature* 457:812–16.

Budd, G. E. 1993. A Cambrian gilled lobopod from Greenland. *Nature* 364:709–11.

———. 1996. The morphology of *Opabinia regalis* and the reconstruction of the arthropod stem-group. *Lethaia* 29:1–14.

———. 1999. The morphology and phylogenetic significance of *Kergamachela kierkegaardi* Budd (Buen Formation, Lower Cambrian, N Greenland). *Transactions Royal Society of Edinburgh: Earth Sciences* 89:249–90.

———. 2002. A palaeontological solution to the arthropod head problem. *Nature* 417:271–74.

———. 2008. The earliest fossil record of the animals and its significance. *Philosphical Transactions Royal Society of London B* 363:1425–34.

Budd, G. E., and A. C. Daley. 2012. The lobes and lobopods of *Opabinia regalis* from the Middle Cambrian Burgess Shale. *Lethaia* 45:83–95.

Budd, G. E., and S. Jensen. 2000. A critical reappraisal of the fossil record of the bilaterian phyla. *Biological Reviews* 75:253–95.

Budd, G. E., and J. S. Peel. 1998. A new xenusiid lobopod from the Early Cambrian Sirius Passet fauna of North Greenland. *Palaeontology* 41:1201–13.

Bulman, O. M. B. 1970. Graptolithina. In *Treatise on Invertebrate Paleontology Part V*, 2nd ed., edited by C. Teichert, V1–149. Lawrence: Geological Society of America / University Press of Kansas.

Bush, A. M., R. K. Bambach, and G. M. Daley. 2007. Changes in theoretical ecospace utilization in marine fossil assemblages between the mid-Paleozoic and Late Cenozoic. *Paleobiology* 33:76–97.

Bush, A. M., R. K. Bambach, and D. H. Erwin. 2011. Ecospace utilization during the Ediacaran radiation and the Cambrian eco-explosion. In *Quantifying the Evolution of Early Life*, edited by M. Laflamme, J. D. Schiffbauer, and S. Q. Dornbos, 111–33. Dordrecht, Netherlands: Springer.

Buss, L. W. 1987. *The Evolution of Individuality*. Princeton, NJ: Princeton University Press.

Buss, L. W., and A. Seilacher. 1994. The phylum Vendobionta: A sister group of Eumetazoa. *Paleobiology* 20:1–4.

Butterfield, N. J. 1990. Organic preservation of non-mineralizing organisms and the taphonomy of the Burgess Shale. *Paleobiology* 16:272–86.

———. Secular distribution of Burgess Shale–type preservation. *Lethaia* 28:1–13.

———. 1997. Plankton ecology and the Proterozoic-Phanerozoic transition. *Paleobiology* 23:247–62.

———. 2001. Ecology and evolution of Cambrian plankton. In *The Ecology of the Cambrian Radiation*, edited by A. Y. Zhuravlev and R. Riding, 200–216. New York: Columbia University Press.

———. 2002. *Leanchoilia* guts and the interpretation of three-dimensional structures in Burgess Shale–type deposits. *Paleobiology* 28:155–71.

———. 2003. Exceptional fossil preservation and the Cambrian explosion. *Integrative and Comparative Biology* 43:166–77.

———. 2004. A vaucherian alga from the middle Neoproterozoic of Spitsbergen: Implications for the evolution of Proterozoic eukaryotes and the Cambrian explosion. *Paleobiology* 30:231–52.

———. 2006. Hooking some stem-group "worms": Fossil lophotrochozoans in the Burgess Shale. *BioEssays* 28:1161–66.

———. 2009. Oxygen, animals and oceanic ventilation: An alternative view. *Geobiology* 7:1–7.

Butterfield, N. J., and C. J. Nicholas. 1996. Burgess Shale–type preservation of both non-mineralizing and "shelly" Cambrian organisms from the Mackenzie Mountains, northwestern Canada. *Journal of Paleontology* 70:893–99.

Callow, R. H. T., and M. D. Brasier. 2009. Remarkable preservation of microbial mats in Neoproterozoic siliciclastic settings: Implications for Ediacaran taphonomic models. *Earth-Science Reviews* 96:207–19.

Calloway, C. B. 1988. Priapulida. In *Introduction to the Study of Meiofauna*, edited by R. P. Higgins and H. Thiel, 322–27. Washington, DC: Smithsonian Institution Press.

Campbell, L. I., O. Rota-Stabelli, G. D. Edgecombe, T. Marchioro, S. J. Longhorn et al. 2011. MicroRNAs and phylogenomics resolve the relationships of Tardigrada and suggest that velvet worms are the sister group of Arthropoda. *Proceedings of the National Academy of Sciences, USA* 108:15920–24.

Canfield, D. E., and J. Farquhar. 2009. Animal evolution, bioturbation, and the sulfate concentration of the oceans. *Proceedings of the National Academy of Sciences, USA* 106:8123–27.

Canfield, D. E., S. W. Poulton, A. H. Knoll, G. M. Narbonne, G. Ross et al. 2008. Ferruginous conditions dominated later Neoproterozoic deep-water chemistry. *Science* 321:949–52.

Canfield, D. E., and A. Teske. 1996. A new model for Proterozoic ocean chemistry. *Nature* 396:450–53.

Cardenas, A. L., and P. J. Harries. 2010. Effect of nutrient availability on marine origination rates throughout the Phanerozoic eon. *Nature Geoscience* 3:430–34.

Cardinale, B. J., D. S. Srivastava, J. E. Duffy, J. P. Wright, A. L. Downing et al. 2006. Effects of biodiversity on the functioning of trophic groups and ecosystems. *Nature* 443:989–92.

Caron, J. B. 2006. *Banffia constricta*, a putative vetulicolid from the Middle Cambrian Burgess Shale. *Transactions of the Royal Society of Edinburgh: Earth Sciences* 96:95–111.

Caron, J. B., S. Conway Morris, and D. Shu. 2010. Tentaculate fossils from the Cambrian of Canada (British Columbia) and China (Yunnan) interpreted as primitive deuterostomes. *PLoS ONE* 5(3):e9586.

Caron, J. B., and D. A. Jackson. 2008. Paleoecology of the Greater Phyllopod Bed community, Burgess Shale. *Palaeogeography, Palaeoclimatology, Palaeoecology* 258:222–56.

Caron, J. B., A. Scheltema, C. Schander, and D. Rudkin. 2006. A soft-bodied mollusc with radula from the Middle Cambrian Burgess Shale. *Nature* 442:159–63.

Carr, M., B. S. Leadbeater, R. Hassan, M. Nelson, and S. L. Baldauf. 2008. Molecular phylogeny of choanoflagellates, the sister group to Metazoa. *Proceedings of the National Academy of Sciences, USA* 105:16641–46.

Carrera, M. G. 1998. First Ordovician sponge from the Puna region, northwestern Argentina. *Ameghiniana* 35:205–10.

Carrera, M. G., and J. P. Botting. 2008. Evolutionary history of Cambrian spiculate sponges: Implications for the Cambrian evolutionary fauna. *Palaios* 23:124–38.

Carroll, S. B. 2008. Evo-devo and an expanding evolutionary synthesis: A genetic theory of morphological evolution. *Cell* 134:25–36.

Carroll, S. B., J. K. Grenier, and S. D. Weatherbee. 2001. *From DNA to Diversity*. Malden, MA: Blackwell.

Cartwright, P., S. L. Halgedahl, J. R. Hendricks, R. D. Jarrard, A. C. Marques et al. 2008. Exceptionally preserved jellyfishes from the Middle Cambrian. *PLoS ONE* (10):1–7.

Casenove, D., T. Goto, and J. Vannier. 2011. Relation between anatomy and lifestyle in Early Cambrian chaetognaths. *Paleobiology* 37:563–76.

Catling, D. C., C. R. Glein, K. J. Zahnle, and C. P. McKay. 2005. Why O_2 is required by complex life on habitable planets and the concept of planetary "oxygenation time." *Astrobiology* 5:415–37.

Cavalier-Smith, T. 1998. A revised six-kingdom system of life. *Biological Reviews* 73:203–66.

Chapman, J. A., E. F. Kirkness, O. Simakov, S. E. Hampson, T. Mitros et al. 2010. The dynamic genome of *Hydra*. *Nature* 464:592–96.

Chen, J. Y. 2004. *Dawn of the Animal World*. Nanjing, China: Publishing House of Jiangsu Science and Technology.

Chen, J. Y., D. J. Bottjer, E. H. Davidson, G. Li, F. Gao et al. 2009. Phase contrast synchrotron X-ray microtomography of Ediacaran (Doushantuo) metazoan microfossils: Phylogenetic diversity and evolutionary implications. *Precambrian Research* 173:191–200.

Chen, J. Y., J. Dzik, G. D. Edgecombe, L. Ramskold, and G. Q. Zhou. 1995. A possible Early Cambrian chordate. *Nature* 377:720–22.

Chen, J. Y., and D. Y. Huang. 2002. A possible Lower Cambrian chaetognath (arrow worm). *Science* 298:187.

Chen, J. Y., D. Y. Huang, Q. Q. Peng, H. M. Chi, X. Q. Want et al. 2003. The first tunicate from the Early Cambrian of south China. *Proceedings of the National Academy of Sciences, USA*. 100:8314–18.

Chen, J. Y., P. Oliveri, F. Gao, S. Q. Dornbos, C. W. Li et al. 2002. Precambrian animal life: Probable developmental and adult cnidarian forms from southwest China. *Developmental Biology* 248:182–96.

Chen, J. Y., P. Oliveri, C.-W. Li, G.-Q. Zhou, F. Gao et al. 2000. Precambrian animal diversity: Putative phosphatized embryos from the Doushantuo Formation of China. *Proceedings of the National Academy of Sciences, USA* 97:4457–62.

Chen, J. Y., J. W. Schopf, D. J. Bottjer, C. Y. Zhang, A. B. Kudryavtsev et al. 2007. Raman spectra of a Lower Cambrian ctenophore embryo from southwestern Shaanxi, China. *Proceedings of the National Academy of Sciences, USA*. 104:6289–92.

Chen, J. Y., D. Waloszek, and A. Maas. 2004. A new "great appendage" arthropod from the Lower Cambrian of China and homology of chelicerate chelicerae and raptorial antero-ventral appendages. *Lethaia* 37:3–20.

Chen, J. Y., D. Waloszek, A. Maas, A. Braun, D. Y. Huang et al. 2007. Early Cambrian Yangtze plate Maotianshan shale macrofauna biodiversity and the evolution of predation. *Palaeogeography, Palaeoclimatology, Palaeoecology* 254:250–72.

Chen, J. Y., and G. Q. Zhou. 1997. Biology of the Chengjiang fauna. *Bulletin of the National Museum of Natural Science* 10:11–105.

Chen, Y. Q., S. Y. Jiang, H. F. Ling, and J. H. Yang. 2009. Pb-Pb dating of black shales from the Lower Cambrian and Neoproterozoic strata, south China. *Chemie der Erde* 69:183–89.

Chiori, R., M. Jager, E. Denker, P. Wincker, C. Da Silva et al. 2009. Are *Hox* genes ancestrally involved in axial patterning? Evidence from the hydrozoan *Clytia hemisphaerica* (Cnidaria). *PLoS ONE* 4(1):e4231.

Chipman, A. D. 2010. Parallel evolution of segmentation by co-option of ancestral gene regulatory networks. *BioEssays* 32:60–70.

Chlupac, I. 1993. *Geology of the Barrandian: A Field Trip Guide.* Frankfurt: Senckenberg-Buchen.

Chlupac, I., and V. Kordule. 2002. Arthropods of Burgess Shale type from the Middle Cambrian of Bohemia (Czech Republic). *Bulletin of the Czech Geological Survey* 77:167–82.

Christodoulou, F., F. Raible, R. Tomer, O. Simakov, K. Trachana et al. 2010. Ancient animal microRNAs and the evolution of tissue identity. *Nature* 463:1084–88.

Clapham, M. E., and G. M. Narbonne. 2002. Ediacaran epifaunal tiering. *Geology* 30:627–30.

Clapham, M. E., G. M. Narbonne, and J. G. Gehling. 2003. Paleoecology of the oldest known animal communities: Ediacaran assemblages at Mistaken Point, Newfoundland. *Paleobiology* 29:527–44.

Clapham, M. E., G. M. Narbonne, J. G. Gehling, C. Greentree, and M. M. Anderson. 2004. *Thectardis avalonensis*: A new Ediacaran fossil from the Mistaken Point biota, Newfoundland. *Journal of Paleontology* 78:1031–36.

Clausen, S., X. G. Hou, J. Bergstrom, and C. Franzen. 2010. The absence of echinoderms from the Lower Cambrian Chengjiang fauna of China: Palaeoecological and palaeogeographical implications. *Palaeogeography, Palaeoclimatology, Palaeoecology* 294:133–41.

Cohen, P. A., A. Bradley, A. H. Knoll, J. P. Grotzinger, S. Jensen et al. 2009. Tubular compression fossils from the Ediacaran Nama Group, Namibia. *Journal of Paleontology* 83:110–20.

Cohen, P. A., A. H. Knoll, and R. Kodner. 2009. Large spinose microfossils in Ediacaran rocks as resting stages of early animals. *Proceedings of the National Academy of Sciences, USA* 106:6519–24.

Colbourne, J. K., M. E. Pfrender, D. Gilbert, W. K. Thomas, A. Tucker et al. 2011. The ecoresponsive genome of *Daphnia pulex*. *Science* 331:555–61.

Collins, A. G., J. H. Lipps, and J. W. Valentine. 2000. Modern mucociliary creeping trails and the bodyplans of Neoproterozoic trace-makers. *Paleobiology* 26:47–55.

Collins, D. 1996. The "evolution" of *Anomalocaris* and its classification in the arthropod Class Dinocardia (nov.) and Order Radiodonta (nov.). *Journal of Paleontology* 70:280–93.

Condon, D., M. Zhu, S. Bowring, W. Wang, A. Yang et al. 2005. U-Pb ages from the Neoproterozoic Doushantuo Formation, China. *Science* 308:95–98.

Condon, D. J., and S. A. Bowring. 2011. A user's guide to Neoproterozoic geochronology. In *The Geological Record of Neoproterozoic Glaciations*, edited by E. Arnaud, G. P. Halverson, and G. Shields-Zhou, 135–49. London: Geological Society.

Conway Morris, S. 1979a. The Burgess Shale (Middle Cambrian) fauna. *Annual Review Ecology and Systematics* 10:397–49.

———. 1979b. Middle Cambrian polychaetes from the Burgess Shale of British Columbia. *Philosophical Transactions of the Royal Society B* 285:227–74.

———. 1986. The community structure of the Middle Cambrian Phyllopod Bed (Burgess Shale). *Palaeontology* 29:423–67.

———. 1993. Ediacaran-like fossils in Cambrian Burgess Shale–type faunas of North America. *Palaeontology* 36:593–36.

Conway Morris, S., ed. 1982. *Atlas of the Burgess Shale*. London: Palaeontological Association.

Conway Morris, S., and J. B. Caron. 2007. Halwaxiids and the early evolution of the lophotrochozoans. *Science* 315:1255–58.

Conway Morris, S., and D. H. Collins. 1996. Middle Cambrian ctenophores from the Stephen Formation, British Columbia. *Philosophical Transactions of the Royal Society B* 351:279–308.

Conway Morris, S., and C. Menge. 1991. Cambroclaves and paracarinachitids, early skeletal problematica from the Lower Cambrian of south China. *Palaeontology* 34:357–97.

Conway Morris, S., and J. S. Peel. 1990. Articulate halkieriids from the Lower Cambrian of north Greenland. *Nature* 345:802–5.

———. 1995. Articulated halkeriids from the Lower Cambrian of north Greenland and their role in early protostome evolution. *Philosophical Transactions of the Royal Society B* 347:305–58.

———. 2010. New palaeoscolecidan worms from the lower Cambrian: Sirius Passet, Latham Shale and Kinzers Shale. *Acta Palaeontologica Polonica* 55:141–56.

Cook, C. E., E. Jimenez, M. Akam, and E. Salo. 2004. The *Hox* gene complement of acoel flatworms, a basal bilaterian clade. *Evolution and Development* 6:154–63.

Cook, P. J., and J. E. Shergold. 1984. Phosphorus, phosphorites and skeletal evolution at the Precambrian-Cambrian boundary. *Nature* 308:231–23.

Cope, J. C. W. 2005. Octocorallian and hydroid fossils from the lower Ordovician of Wales. *Palaeontology* 48:433–445.

Copf, T., R. Schroder, and M. Averof. 2004. Ancestral role of caudal genes in axis elongation and segmentation. *Proceedings of the National Academy of Sciences, USA* 101:17711–15.

Cornell, H. V. 1999. Unsaturation and ecological influences on species richness in ecological communities: A review of the evidence. *Ecoscience* 6:303–15.

Corsetti, F. A., S. M. Awramik, and D. Pierce. 2003. A complex microbiota from snowball Earth times: Microfossils from the Neoproterozoic Kingston Peak Formation, Death Valley, USA. *Proceedings of the National Academy of Sciences, USA* 100:4399–404.

Corsetti, F. A., and J. P. Grotzinger. 2005. Origin and significance of tube structures in Neoproterozoic post-glacial cap carbonates: Examples from Noonday Dolomite, Death Valley, United States. *Palaios* 20:348–62.

Crimes, T. P. 1987. Trace fossils and correlation of late Precambrian and early Cambrian strata. *Geological Magazine* 124:97–119.

———. 1994. The period of early evolutionary failure and the dawn of evolutionary success: The record of biotic changes across the Precambrian-Cambrian boundary. In *The Palaeobiology of Trace Fossils*, edited by S. K. Donovan, 105–33. London: Wiley.

Crimes, T. P., and M. A. Fedonkin. 1996. Biotic changes in platform communities across the Precambrian Phanerozoic boundary. *Rivista Italiana de Paleontologia e Stratigrafia* 102:317–32.

Cuddington, K., J. E. Byers, W. G. Wilson, and A. Hastings, eds. 2007. *Ecosystem Engineers: Plants to Protists*. London: Academic Press.

Cunningham, J. A., C.-W. Thomas, S. Bengtson, F. Marone, M. Stampanoni et al. 2012. Experimental taphonomy of giant sulfur bacteria: Implications for the interpretation of the embryo-like Ediacaran Doushantuo fossils. *Proceedings of the Royal Society of London, Series B* [published online 7 December 2011].

Dahl, T. W., D. E. Canfield, M. T. Rosing, R. E. Frei, G. W. Gordon et al. 2011. Molybdenum evidence for expansive sulfidic water masses in ~ 750 Ma oceans. *Earth and Planetary Science Letters* 311:264–74.

Dahl, T. W., E. U. Hammarlund, A. D. Anbar, D. P. Bond, B. C. Gill et al. 2010. Devonian rise in atmospheric oxygen correlated to the radiations of terrestrial plants and large predatory fish. *Proceedings of the National Academy of Sciences, USA* 107:17911–15.

Daley, A. C., G. E. Budd, J. B. Caron, G. D. Edgecombe, and D. Collins. 2009. The Burgess Shale anomalocaridid *Hurdia* and its significance for early euarthropod evolution. *Science* 323:1597–1600.

Daley, A. C., and J. S. Peel. 2010. A possible anomalocaridid from the Cambrian Sirius Passet lagerstätte, north Greenland. *Journal of Paleontology* 84:352–55.

Darroch, S. A., M. Laflamme, J. D. Schiffbauer, and D. E. G. Briggs. 2012. Experimental formation of a microbial death mask. *Palaios* 27:293–303.

Davidson, E. H. 2001. *Genomic Regulatory Systems*. San Diego: Academic Press.

———. 2006. *The Regulatory Genome*. San Diego: Academic Press.

———. 2009. Network design principles from the sea urchin embryo. *Current Opinion in Genetics and Development* 19:535–40.

Davidson, E. H., and D. H. Erwin. 2006. Gene regulatory networks and the evolution of animal body plans. *Science* 311:796–800.

———. 2010a. Evolutionary innovation and stability in animal gene networks. *Journal of Experimental Zoology Part B: Molecular and Developmental Evolution* 312B:1–5.

———. 2010b. An integrated view of precambrian eumetazoan evolution. *Cold Spring Harbor Symposia on Quantitative Biology* 79:65–80.

Davidson, E. H., and M. Levine. 2008. Properties of developmental gene regulatory networks. *Proceedings of the National Academy of Sciences, USA* 105:20063–66.

Degnan, B. M., S. P. Leys, and C. Larroux. 2005. Sponge development and antiquity of animal pattern formation. *Integrative and Comparative Biology* 45:335–41.

Degnan, B. M., M. Vervoort, C. Larroux, and G. S. Richards. 2009. Early evolution of metazoan transcription factors. *Current Opinion in Genetics and Development* 19:591–99.

Delsuc, F., H. Brinkmann, D. Chourrout, and H. Philippe. 2006. Tunicates and not cephalochordates are the closest living relatives of vertebrates. *Nature* 439:965–68.

De Robertis, E. M. 2008. Evo-devo: Variations on ancestral themes. *Cell* 132:185–95.

De Robertis, E. M., and Y. Sasai. 1996. A common plan for dorsoventral patterning in Bilateria. *Nature* 380:37–40.

de Rosa, R., J. K. Grenier, T. Andreeva, C. E. Cook, A. Adoutte et al. 1999. *Hox* genes in brachiopods and priapulids and protostome evolution. *Nature* 399:772–76.

de Rosa, R., B. Prud'homme, and G. Balavoine. 2005. Caudal and even-skipped in the annelid *Platynereis dumerilii* and the ancestry of posterior growth. *Evolution and Development* 7:574–87.

Derry, L. A. 2006. Fungi, weathering and the emergence of animals. *Science* 311:1386–87.

———. 2010a. A burial diagenesis origin for the Ediacaran Shuram-Wonoka carbon isotope anomaly. *Earth and Planetary Science Letters* 294:152–62.

———. 2010b. On the significance of $\delta^{13}C$ correlations in ancient sediments. *Earth and Planetary Science Letters* 296:497–501.

Derry, L. A., A. J. Kaufman, and S. B. Jacobsen. 1992. Sedimentary cycling and environmental change in the Late Proterozoic: Evidence from stable and radiogenic isotopes. *Geochemica et Cosmochemica Acta* 56:1317–29.

Derstler, K. L. 1981. Morphological diversity of Early Cambrian echinoderms. Papers for the Second International Symposium on the Cambrian System. USGS Open File Report 81-743:71–75.

Des Marais, D. J., H. Strauss, R. E. Summons, and J. M. Hayes. 1992. Carbon isotope evidence for the stepwise oxidation of the Proterozoic environment. *Nature* 359:605–9.

deVos, L., K. Rutzler, N. Boury-Esnalt, C. Donadey, and J. Vacelet. 1991. *Atlas of Sponge Morphology*. Washington, DC: Smithsonian Institution Press.

Dewel, R. A., W. C. Dewel, and F. K. McKinney. 2001. Diversification of the Metazoa: Ediacarans, colonies and the origin of eumetazoan complexity by nested modularity. *Historical Biology* 15:193–218.

Domazet-Loso, T., J. Brajkovic, and D. Tautz. 2007. A phylostratigraphy approach to uncover the genomic history of major adaptations in metazoan lineages. *Trends in Genetics* 23:533–39.

Dong, X. P., S. Bengtson, N. J. Gostling, J. A. Cummingham, and P. C. Donoghue. 2011. The anatomy, taphonomy, taxonomy and systematic affinity of *Markuelia*: Early Cambrian to Early Ordovician scalidophorans. *Palaeontology* 53:1291–313.

Donoghue, P. C. J. 2007. Embryonic identity crisis. *Nature* 445:155–56.

Donoghue, P. C. J., S. Bengtson, X. P. Dong, N. J. Gostling, T. Huldtgren et al. 2006. Synchrotron X-ray tomographic microscopy of fossil embryos. *Nature* 442:680–83.

Donoghue, P. C. J., and X. P. Dong. 2005. Embryos and ancestors. In *Evolving Form and Function: Fossils and Development*, edited by D. E. G. Briggs, 81–99. New Haven, CT: Peabody Museum of Natural History, Yale University.

Donoghue, P. C. J., and M. A. Purnell. 2005. Genome duplication, extinction and vertebrate evolution. *Trends in Ecology and Evolution* 20:312–19.

Dornbos, S. Q. 2006. Evolutionary paleoecology of epifaunal echinoderms: Response to increasing bioturbation levels during the Cambrian radiation. *Palaeogeography, Palaeoclimatology, Palaeoecology* 237:225–39.

Dornbos, S. Q., and D. J. Bottjer. 2000. Evolutionary paleoecology of the earliest echinoderms: Helioplacoids and the Cambrian substrate revolution. *Geology* 28:839–42.

Dornbos, S. Q., D. J. Bottjer, I. A. Chen, P. Oliveri, F. Gao et al. 2005. Precambrian animal life: Taphonomy of phosphatized metazoan embryos from southwest China. *Lethaia* 38:101–9.

Dornbos, S. Q., D. J. Bottjer, and J. Y. Chen. 2004. Evidence for seafloor microbial mats and associated metazoan lifestyles in Lower Cambrian phosphorites of southwest China. *Lethaia* 37:127–39.

Dornbos, S. Q., D. J. Bottjer, and J. Y. Chen. 2005. Paleoecology of benthic metazoans in the Early Cambrian Maotianshan Shale biota and the Middle Cambrian Burgess Shale biota: Evidence for the Cambrian substrate revolution. *Palaeogeography, Palaeoclimatology, Palaeoecology* 220:47–67.

Dornbos, S. Q., D. J. Bottjer, J. Y. Chen, F. Gao, P. Oliveri et al. 2006. Environmental controls on the taphonomy of phosphatized animals and animal embryos from the Nanjing Doushantuo Formation, southwest China. *Palaios* 21:3–14.

Droser, M. L., and D. J. Bottjer. 1986. A semiquantitative field classification of ichnofabric. *Journal of Sedimentary Petrology* 56:558–59.

———. 1988. Trends in depth and extent of bioturbation in Cambrian carbonate marine environments. *Geology* 16:233–36.

Droser, M. L., and J. G. Gehling. 2008. Synchronous aggregate growth in an abundant new Ediacaran tubular organism. *Science* 319:1660–62.

Droser, M. L., J. G. Gehling, and S. Jensen. 1999. When the worm turned: Concordance of Early Cambrian ichnofabric and trace fossil record in siliciclastic rocks of South Australia. *Geology* 27:625–28.

Droser, M. L., J. G. Gehling, and S. R. Jensen. 2005. Ediacaran trace fossils: True and false. In *Evolving Form and Function: Fossils and Development*, edited by D. E. G. Briggs, 125–38. New Haven, CT: Peabody Museum of Natural History, Yale University.

———. Jensen. 2006. Assemblage palaeoecology of the Ediacara biota? The unabridged edition. *Palaeogeography, Palaeoclimatology, Palaeoecology* 232:131–47.

Droser, M. L., S. Jensen, and J. G. Gehling. 2002. Trace fossils and substrates of the terminal Proterozoic-Cambrian transition: Implications for the record of early bilaterians and sediment mixing. *Proceedings of the National Academy of Sciences, USA* 99:12572–76.

Droser, M. L., S. Jensen, J. G. Gehling, P. M. Myrow, and G. M. Narbonne. 2002. Lowermost Cambrian ichnofabrics from the Chapel Island Formation, Newfoundland: Implications for Cambrian substrates. *Palaios* 17:3–15.

Duboule, D. 2007. The rise and fall of *Hox* gene clusters. *Development* 134:2549–60.

Dunlop, J. A., and P. A. Selden. 1998. The early history and phylogeny of the chelicerates. In *Arthropod Relationships*, edited by R. A. Fortey and R. Thomas, 221–35. London: Chapman and Hall.

Dunn, C. W., A. Hejnol, D. Q. Matus, K. Pang, W. E. Browne et al. 2008. Broad phylogenomic sampling improves resolution of the animal tree of life. *Nature* 452:745–49.

Dunne, J. A., R. J. Williams, N. D. Martinez, R. A. Wood, and D. H. Erwin. 2008. Compilation and network analysis of Cambrian food webs. *PLoS Biology* 6:e102.

Durham, J. W. 1978. A Lower Cambrian eocrinoid. *Journal of Paleontology* 52:195–99.

———. Observations on the Early Cambrian helioplacoid echinoderms. *Journal of Paleontology* 67:590–604.

Durham, J. W., and K. E. Caster. 1965. Helioplacoids. In *Treatise on Invertebrate Paleontology, Part U*, edited by C. H. Moore, 131–36. Lawrence: Geological Society of America / University Press of Kansas.

Dzik, J. 1999. Organic membranous skeleton of the Precambrian metazoans from Namibia. *Geology* 27:519–22.

———. 2002. Possible ctenophoran affinities of the Precambrian "sea-pen" *Rangea*. *Journal of Morphology* 252:315–34.

———. 2003. Anatomical information content in the Ediacaran fossils and their possible zoological affinities. *Integrative and Comparative Biology* 43:114–26.

———. 2005. Behavioral and anatomical unity of the earliest burrowing animals and the cause of the "Cambrian explosion." *Paleobiology* 31:503–21.

Dzik, J., and G. Krumbigel. 1989. The oldest "onycophoran" *Xenusion*: A link connecting phyla? *Lethaia* 22:169–82.

Edgecombe, G. D. 2010. Arthropod phylogeny: An overview from the perspectives of morphology, molecular data and the fossil record. *Arthropod Structure and Development* 39:74–87.

Ehlers, U., and B. Sopott-Ehlers. 1997. Ultrastructure of the subepidermal musculature of *Xenoturbella bocki*, the adelphotaxon of the Bilateria. *Zoomorphology* 117:71–79.

Elicki, O. 1994. Lower Cambrian carbonates from eastern Germany: Palaeontology, stratigraphy and palaeogeography. *Neues Jahrbuch für Geologie und Paläontologie, Abhandlungen* 191:69–93.

Elser, J. J., J. Watts, J. H. Schampel, and J. Farmer. 2006. Early Cambrian food webs on a trophic knife-edge: A hypothesis and preliminary data from a stromatolite-based ecosystem. *Ecology Letters* 9:295–303.

Embley, T. M., and W. Martin. 2006. Eukaryotic evolution, changes and challenges. *Nature* 440:623–30.

Emerson, B. C., and N. Kolm. 2005. Species diversity can drive speciation. *Nature* 434:1015–17.

Emig, C. C. 1979. *British and Other Phoronids*. London: Academic Press.

Endler, J. A. 1986. *Natural Selection in the Wild*. Princeton, NJ: Princeton University Press.

Erwin, D. H. 1992. A preliminary classification of radiations. *Historical Biology* 6:133–147.

———. 2000. Macroevolution is more than repeated rounds of microevolution. *Evolution and Development* 2:78–84.

———. 2006a. The developmental origins of animal body plans. In *Neoproterozoic Geobiology and Paleobiology*, edited by S. H. Xiao and A. J. Kaufman, 157–97. Dordrecht, Netherlands: Klewer Press.

———. 2006b. *Extinction: How Life Nearly Died 250 Million Years Ago*. Princeton, NJ: Princeton University Press.

———. 2007. Disparity: Morphological pattern and developmental context. *Palaeontology* 50:57–73.

———. 2008. Macroevolution of ecosystem engineering, niche construction and diversity. *Trends in Ecology and Evolution* 23:304–10.

———. 2009. Early genomic origins of the bilaterian developmental toolkit. *Philosophical Transactions of the Royal Society B* 264:2253–61.

———. 2011. Evolutionary uniformitarianism. *Developmental Biology* 357:27–34.

Erwin, D. H., and E. H. Davidson. 2002. The last common bilaterian ancestor. *Development* 129:3021–32.

———. 2009. The evolution of hierarchical gene regulatory networks. *Nature Reviews Genetics* 10:141–48.

Erwin, D. H., M. Laflamme, S. M. Tweedt, E. A. Sperling, D. Pisani et al. 2011. The Cambrian conundrum: Early divergence and later ecological success in the early history of animals. *Science* 334:1091–97.

Erwin, D. H., and S. M. Tweedt. 2012. Ecosystem engineering and the Ediacaran-Ordovician diversification of Metazoa. *Evolutionary Ecology* 26:417–33.

Erwin, D. H., J. W. Valentine, and J. J. Sepkoski, Jr. 1987. A comparative study of diversification events: The early Paleozoic versus the Mesozoic. *Evolution* 41:1177–86.

Evans, D. A. 2000. Stratigraphic, geochronological and paleomagnetic constraints upon the Neoproterozoic climatic paradox. *American Journal of Science* 300:347–433.

Fahey, B., and B. M. Degnan. 2010. Origin of animal epithelia: Insights from the sponge genome. *Evolution and Development* 12:601–17.

Fahey, B., C. Larroux, B. J. Woodcroft, and B. M. Begnan. 2008. Does the high gene density in the sponge NK homeobox gene cluster reflect limited regulatory capacity? *Biological Bulletin* 214:205–17.

Fatka, O., and V. Kordule. 2001. *Asturicystis havliceki* sp. nov. (Echinodermata, Homostelea) from the Middle Cambrian of Bohemia (Barrandian area, Czech Republic). *Journal of the Czech Geological Society* 46:189–94.

Fayers, S. R., and N. H. Trewin. 2005. A hexapod from the Early Devonian Windyfield Chert, Rhynie, Scotland. *Palaeontology* 48:1117–30.

Fedonkin, M. A. 2003. The origin of the Metazoa in light of the Proterozoic fossil record. *Paleontological Research* 7:9–41.

Fedonkin, M. A., J. G. Gehling, K. Grey, G. M. Narbonne, and P. Vickers-Rich. 2007. *The Rise of Animals*. Baltimore: Johns Hopkins University Press.

Fedonkin, M. A., A. Simonetta, and A. Y. Ivantsov. 2007. New data on *Kimberella*, the Vendian mollusc-like organism (White Sea region, Russia): Paleontological and evolutionary implications. In *The Rise and Fall of the Ediacaran Biota*, edited by P. Vickers-Rich and P. Komarower, 157–79. London: Geological Society.

Fedonkin, M. A., and B. M. Waggoner. 1997. The Late Precambrian fossil *Kimberella* is a mollusc-like bilaterian organism. *Nature* 388:868.

Felsenstein, J. 1978. Cases in which parsimony or compatibility methods will be positively misleading. *Systematic Zoology* 27:401–10.

Feng, W. M., W. G. Sun, and Y. Qian. 2001. Skeletalization characters, classification and evolutionary significance of Early Cambrian Monoplacophoran maikhanellids. *Acta Palaeontologica Sinica* 40:195–213.

Ferrier, D. E., and C. Minguillon. 2003. Evolution of the *Hox/ParaHox* gene clusters. *International Journal of Developmental Biology* 47:605–11.

Field, K. G., G. J. Olsen, D. J. Lane, S. J. Giovannoni, M. T. Ghiselin et al. 1988. Molecular phylogeny of the animal kingdom. *Science* 239:748–53.

Fike, D. A., J. P. Grotzinger, L. M. Pratt, and R. E. Summons. 2006. Oxidation of the Ediacaran ocean. *Nature* 444:744–47.

Finnerty, J. R., and M. Q. Martindale. 1999. The evolution of the *Hox* cluster: Insights from outgroups. *Current Opinion in Genetics and Development* 8:681–87.

Fisher, D. W., and R. S. Young. 1955. The oldest known tentaculitid: From the Chepultepec Limestone (Canadian) of Virginia. *Journal of Paleontology* 29:871–75.

Foote, M. 1993. Discordance and concordance between morphological and taxonomic diversity. *Paleobiology* 19:185–204.

———. 1996. Models of morphological diversification. In *Evolutionary Paleobiology*, edited by D. Jablonski, D. H. Erwin, and J. H. Lipps, 62–86. Chicago: University of Chicago Press.

———. 1997. Evolution of morphological diversity. *Annual Review of Ecology and Systematics* 28:129–52.

Force, A., W. A. Cresko, F. B. Pickett, S. R. Proulx, C. Amemiya et al. 2005. The origin of subfunctions and modular gene regulation. *Genetics* 170:433–46.

Forey, P. L., R. A. Fortey, P. R. Kenrick, and A. B. Smith. 2004. Taxonomy and fossils: A critical appraisal. *Proceedings of the Royal Society London B* 359:639–53.

Fortey, R. A., D. E. G. Briggs, and M. A. Wills. 1996. The Cambrian evolutionary "explosion": Decoupling cladogenesis from morphological disparity. *Biological Journal of the Linnean Society* 57:13–33.

Fraiser, M. L., and F. A. Corsetti. 2003. Neoproterozoic carbonate shrubs: Interplay of microbial activity and unusual environmental conditions in post-snowball Earth oceans. *Palaios* 18:378–87.

Frest, T. J., C. E. Brett, and B. J. Witzke. 1999. Caradocian-Gedinnian echinoderm associations of central and eastern North America. In *Paleocommunities—A Case Study from the Silurian and Lower Devonian*, edited by A. J. Boucot and J. D. Lawson, 638–783. Cambridge, UK: Cambridge University Press.

Fuller, M., and R. Jenkins. 2007. Reef corals from the Lower Cambrian of the Flinders Ranges, South Australia. *Palaeontology* 50:961–80.

Gaines, R. R., D. E. G. Briggs, and Y. L. Zhao. 2008. Cambrian Burgess Shale–type deposits share a common mode of fossilization. *Geology* 36:755–58.

Gains, R. R., E. U. Mammarlund, X. G. Hou, S. E. Gabbott, Y. L. Zhao et al. 2012. Mechanism for Burgess Shale–type preservation. *Proceedings of the National Academy of Sciences* 109:5180–84.

Galis, F. 2001. Key innovations and radiations. In *The Character Concept in Evolutionary Biology*, edited by G. P. Wagner, 581–605. San Diego: Academic Press.

Galliot, B., M. Quiquand, L. Ghila, R. de Rosa, M. Miljkovic-Lucina et al. 2009. Origins of neurogenesis, a cnidarian view. *Developmental Biology* 332:2–24.

Gandin, A., F. Debrenne, and M. Debrenne. 2007. Anatomy of the Early Cambrian La Sentinella reef complex, Serra Scoris, SW Sardinia, Italy. In *Paleozoic Reefs and Bioaccumulations: Climatic and Evolutionary Controls*, edited by J. J. Alvaro, M. Aretz, F. Boulvain, A. Munnecke, D. Vachard, and E. Vennin, Geological Society Special Publication No. 275:29–50.

Gazave, E., P. Lapébie, A. V. Ereskovsky, J. Vacelet, E. Renard et al. 2012. No longer Demospongiae: Homoscleromorpha formal nomination as a fourth class of Porifera. *Hydrobiologia* 687:3–10.

Gehling, J. G. 1988. A cnidarian of actinian-grade from the Ediacaran Pound Subgroup, South Australia. *Alcheringa* 12:299–314.

———. 1991. The case for Ediacaran fossil roots to the metazoan tree. *Memoirs of the Geological Society of India* 20:181–223.

———. 1999. Microbial mats in terminal Proterozoic siliciclastics: Ediacaran death masks. *Palaios* 14:40–57.

Gehling, J. G., and M. L. Droser. 2009. Textured organic surfaces associated with the Ediacara biota in South Australia. *Earth-Science Reviews* 96:196–206.

Gehling, J. G., M. L. Droser, S. R. Jensen, and B. N. Runnegar. 2005. Ediacara organisms: Relating form to function. In *Form and Function: Fossils and Development*, edited by D. E. G. Briggs, 43–66. New Haven, CT: Peabody Museum of Natural History, Yale University.

Gehling, J. G., G. M. Narbonne, and M. M. Anderson. 2000. The first named Ediacaran body fossil, *Aspidella terranovica*. *Palaeontology* 43:427–56.

Gehling, J. G., and J. K. Rigby. 1996. Long expected sponges from the Neoproterozoic Ediacara fauna of South Australia. *Journal of Paleontology* 70:185–95.

Gehring, W. J. 2004. Historical perspective on the development and evolution of eyes and photoreceptors. *International Journal of Developmental Biology* 48:707–17.

Gehring, W. J., and K. Ikeo. 1999. *Pax 6*: Mastering eye morphogenesis and eye evolution. *Trends in Genetics* 15:371–75.

Gehring, W. J., U. Kloter, and H. Suga. 2009. Evolution of the *Hox* gene complex from an evolutionary ground state. *Current Topics in Developmental Biology* 88:35–61.

Geological Society of America. 1999. *1999 Geologic Time Scale*. Boulder, CO: Geological Society of America.

Gerhart, J., C. Lowe, and M. Kirschner. 2005. Hemichordates and the origin of chordates. *Current Opinion in Genetics and Development* 15:461.

Geyer, G., and J. Shergold. 2000. The quest for internationally recognized divisions of Cambrian time. *Episodes* 23:188–95.

Gingeras, T. R. 2009. Implications of chimaeric non-co-linear transcripts. *Nature* 461:206–11.

Giribet, G., D. L. Distel, M. Polz, W. Sterrer, and W. C. Wheeler. 2000. Triploblastic relationships with emphasis on the acoelomates and the position of Gnathostomulida, Cycliophora, Plathelminthes, and Chaetognatha: A combined approach of 18S rDNA sequences and morphology. *Systematic Biology* 49:539–62.

Glaessner, M. F. 1958. New fossils from the base of the Cambrian in South Australia. *Transactions of the Royal Society of South Australia* 81:185–89.

———. 1984. *The Dawn of Animal Life: A Biohistorical Study*. Cambridge, UK: Cambridge University Press.

Glaessner, M. F., and M. Wade. 1966. The Late Precambrian fossils from Ediacara, South Australia. *Palaeontology* 9:599–628.

Glor, R. E. 2010. Phylogenetic insights on adaptive radiation. *Annual Review of Ecology, Evolution and Systematics* 41:251–70.

Gostling, N. J., C. W. Thomas, J. M. Greenwood, X. Dong, S. Bengtson et al. 2008. Deciphering the fossil record of early bilaterian embryonic development in light of experimental taphonomy. *Evolution and Development* 10:339–49.

Gould, S. J. 1965. Is uniformitarianism necessary? *American Journal of Science* 263:223–28.

———. 1989. *Wonderful Life*. New York: Norton.

Gradstein, F. M., J. G. Ogg, and A. G. Smith, eds. 2005. *A Geologic Time Scale 2004*. Cambridge, UK: Cambridge University Press.

Grant, S. W. F. 1990. Shell structure and distribution of *Cloudina*, a potential index fossil for the terminal Proterozoic. *American Journal of Science* 290-A:261–94.

Grazhdankin, D. 2004. Patterns of distribution in the Ediacaran biotas: Facies versus biogeography and evolution. *Paleobiology* 30:203–21.

Grazhdankin, D. V., U. Balthasar, K. E. Nagovitsin, and B. B. Kochnev. 2008. Caarbonate-hosted Avalon-type fossils in arctic Siberia. *Geology* 36:803–6.

Grazhdankin, D. V., and A. Seilacher. 2002. Underground vendobionta from Namibia. *Palaeontology* 45:57–78.

Greenslade, P., and P. E. S. Whalley. 1986. The systematic position of *Rhyniella praecursor* Hirst & Maulik (Collembola), the earliest known hexapod. In *Second International Seminar on Apterygota*, edited by R. Dallai, 319–23. Siena, Italy: University of Siena.

Gregoryeva, N. V., L. M. Melnikova, and L. Yu Pelman. 1983. Brachiopods, ostracodes (bradoriids) and problematical fossils from the stratotype region of the Lower Cambrian stages. *Paleontological Journal* 17:51–56.

Grey, K., M. R. Walter, and C. R. Calver. 2003. Neoproterozoic biotic diversification: Snowball Earth or aftermath of the Acraman impact? *Geology* 31:459–62.

Grice, K., C. Cao, G. D. Love, M. E. Bottcher, R. J. Twitchett et al. 2005. Photic zone euxinia during the Permian-Triassic superanoxic event. *Science* 307:706–9.

Grimson, A., M. Srivastava, B. Fahey, B. J. Woodcroft, H. R. Chiang et al. 2008. Early origins and evolution of microRNAs and Piwi-interacting RNAs in animals. *Nature* 455:1193–97.

Grotzinger, J. P., S. A. Bowring, B. Z. Saylor, and A. J. Kaufman. 1995. Biostratigraphic and geochronologic constraints on early animal evolution. *Science* 270:598–604.

Grotzinger, J. P., D. A. Fike, and W. W. Fischer. 2011. Enigmatic origin of the largest-known carbon isotope excursion in Earth's history. *Nature Geoscience* 4:285–92.

Grotzinger, J. P., and A. H. Knoll. 1995. Anomalous carbonate precipitates: Is the Precambrian the key to the past? *Palaios* 10:578–96.

Grotzinger, J. P., W. A. Watters, and A. H. Knoll. 2000. Calcified metazoans in thrombolite-stromatolite reefs of the terminal Proterozoic Nama Group, Namibia. *Paleobiology* 26:334–59.

Guerroue, E. L., P. A. Allen, A. Cozzi, J. Ettienne, and C. M. Fanning. 2006. 50 Myr recovery from the larggest negative $\delta^{13}C$ excursion in the Ediacaran ocean. *Terra Nova* 18:147–53.

Gunderson, G. O. 1993. New genus of Late Cambrian gastropod. *Journal of Paleontology* 67:1083–84.

Gutierrez, J. L., and C. G. Jones. 2008. Ecosystem engineers. *Encyclopedia of Life Sciences*. Chichester: Wiley.

Gutschick, R. C. 1954. Holothurian sclerites from the Middle Ordovician of northern Illinois. *Journal of Paleontology* 28:827–29.

Hagadorn, J. W., and D. J. Bottjer. 1997. Wrinkle structures: Microbially mediated sedimentary structures common in subtidal siliciclastic settings at the Proterozoic-Phanerozoic transition. *Geology* 25:1047–50.

Hagadorn, J. W., S. Xiao, P. C. Donoghue, S. Bengtson, N. J. Gostling et al. 2006. Cellular and subcellular structure of Neoproterozoic animal embryos. *Science* 314:291–94.

Halanych, K. M., J. D. Bacheller, A. M. A. Aguinaldo, S. M. Liva, D. M. Hillis et al. 1995. Evidence of 18S ribosomal DNA that the lophophorates are protostome animals. *Science* 267:1641–43.

Halder, G., P. Callaerts, and W. J. Gehring. 1995. Induction of ectopic eyes by targeted expression of the eyeless gene in *Drosophila*. *Science* 267:1788–92.

Hall, B. G. 2007. *Phylogenetic Trees Made Easy*. Sunderland, MA: Sinauer.

Halverson, G. P., F. O. Dudas, A. C. Maloof, and S. A. Bowring. 2007. Evolution of the $^{87}Sr/^{86}Sr$ composition of Neoproterozoic seawater. *Palaeogeography, Palaeoclimatology, Palaeoecology* 256:103–29.

Halverson, G. P., P. F. Hoffman, D. P. Schrag, and A. J. Kaufman. 2002. A major perturbation of the carbon cycle before the Ghaub glaciation (Neoproterozoic) in Namibia: Prelude to snowball Earth? *Geochemistry Geophysics Geosystems* 3:1–24.

Halverson, G. P., P. F. Hoffman, D. P. Schrag, A. C. Maloof, and A. H. N. Rice. 2006. Toward a Neoproterozoic composite carbon isotope record. *Geological Society of America Bulletin* 117:1181–1207.

Halverson, G. P., and M. T. Hurtgen. 2007. Ediacaran growth of the marine sulfate reservoir. *Earth and Planetary Science Letters* 263:32–44.

Halverson, G. P., M. T. Hurtgen, S. M. Porter, and A. S. Collins. 2009. Neoproterozoic-Cambrian biogeochemical evolution. In *Neoproterozoic-Cambrian Tectonics, Global Change and Evolution: A Focus on Southwestern Gondwana*, edited by C. Gaucher, A. N. Sial, G. P. Halverson, and H. E. Frimmel, 351–65. Amsterdam: Elsevier.

Halverson, G. P., B. S. Wade, M. T. Hurtgen, and K. M. Barovich. 2010. Neoproterozoic chemostratigraphy. *Precambrian Research* 182:337–50.

Han, J., S. Kubota, H. O. Uchida, G. D. Stanley, Jr., X. Yao et al. 2010. Tiny sea anemone from the Lower Cambrian of China. *PLoS ONE* 5(10):e13276.

Hansen, T. F. 2006. The evolution of genetic architecture. *Annual Review of Ecology, Evolution and Systematics* 37:123–57.

Hardisty, M. W. 1979. *Biology of the Cyclostomes*. London: Chapman and Hall.

Harland, W. B. 1982. *A Geologic Time Scale*. Cambridge, UK: Cambridge University Press.

———. 2007. Origins and assessment of snowball Earth hypotheses. *Geological Magazine* 144:633–42.

Harland, W. B., R. L. Armstrong, A. V. Cox, L. E. Craig, A. G. Smith et al. 1989. *A Geologic Time Scale 1989*. Cambridge, UK: Cambridge University Press.

Hartman, O. 1961. Polychaetous annelids from California. *Alan Hancock Pacific Expeditions* 22:69–216.

Harvey, R. P. 1996. NK-2 homeobox genes and heart development. *Developmental Biology* 178:203–16.

Harvey, T. H., and N. J. Butterfield. 2008. Sophisticated particle-feeding in a large Early Cambrian crustacean. *Nature* 452:868–71.

Harvey, T. H. P. 2010. Carbonaceous preservation of Cambrian hexactinellid sponge spicules. *Biology Letters* 6:834–37.

Harvey, T. H. P., X. P. Dong, and P. C. J. Donoghue. 2010. Are palaeoscolecids ancestral ecdysozoans? *Evolution and Development* 12:177–200.

Haszprunar, G. 1996. The mollusca: Coelomate turbellarians or mesenchymate annelids? In *Origin and Evolutionary Radiation of the Mollusca*, edited by J. D. Taylor, 3–28. Oxford, UK: Oxford University Press.

Hayes, J. M., H. Strauss, and A. J. Kaufman. 1999. The abundance of ^{13}C in marine organic matter and isotopic fractionation in the global biogeochemical cycle of carbon during the past 800 Ma. *Chemical Geology* 161:103–25.

Hejnol, A., and M. Q. Martindale. 2008. Acoel development supports a simple planula-like urbilaterian. *Philosphical Transactions Royal Society of London B* 363:1493–96.

———. 2009. Coordinated spatial and temporal expression of *Hox* genes during embryogenesis in the acoel *Convolutriloba longifissura*. *BMC Biology* 7:65.

Helfenbein, K. G., H. M. Fourcade, R. G. Vanjani, and J. L. Boore. 2004. The mitochondrial genome of *Paraspadella gotoi* is highly reduced and reveals that chaetognaths are a sister group to protostomes. *Proceedings of the National Academy of Sciences, USA* 101:10639–43.

Hennig, W. 1950. *Grundzüge einer Theorie der phylogenetischen Systematik*. Berlin: Zentralverag.

———. 1966. *Phylogenetic Systematics*. Urbana: University of Illinois Press.

Henry, J. Q., M. Q. Martindale, and B. C. Boyer. 2000. The unique developmental program of the acoel flatworm, *Neochilda fusca*. *Developmental Biology* 220:285–95.

Hernandez-Nicaise, M. L., G. Nicaise, and L. Malaval. 1984. Giant smooth muscle fibers of the ctenophore *Mnemiopsis leydii*: Ultrastructural study of *in situ* and isolated cells. *Biological Bulletin* 167:210–28.

Higgins, J. A., and D. P. Schrag. 2003. Aftermath of a snowball Earth. *Geochemistry Geophysics Geosystems* 4(3).

Hill, D. 1972. Archaeocyatha. In *Treatise on Invertebrate Paleontology, Part E*, edited by C. Teichert, 1–158. Lawrence: Geological Society of America / University Press of Kansas.

Hinman, V. F., and E. H. Davidson. 2007. Evolutionary plasticity of developmental gene regulatory network architecture. *Proceedings of the National Academy of Sciences, USA* 104:19404–9.

Hinman, V. F., A. T. Nguyen, R. A. Cameron, and E. H. Davidson. 2003. Developmental gene regulatory network architecture across 500 million years of echinoderm evolution. *Proceedings of the National Academy of Sciences, USA* 100:13356–61.

Hinman, V. F., A. Nguyen, and E. H. Davidson. 2007. Caught in the evolutionary act: Precise *cis*-regulatory basis of difference in the organization of gene networks of sea stars and sea urchins. *Developmental Biology* 312:584–95.

Hinman, V. F., K. A. Yankura, and B. S. McCauley. 2009. Evolution of gene regulatory network architectures: Examples of subcircuit conservation and plasticity between classes of echinoderms. *Biochimica et Biophysica Acta* 1789:326–32.

Hinz, I. 1987. The Lower Cambrian microfauna of Comley and Rushton, Shropshire, England. *Palaeontographica Abteilung A* 198:41–100.

Hirth, F., L. Kammermeier, E. Frei, U. Walldorf, M. Noll et al. 2003. An urbilaterian origin of the tripartite brain: Developmental genetic insights from *Drosophila*. *Development* 130:2365–73.

Hoffman, P. F., A. J. Kaufman, G. P. Halverson, and D. P. Schrag. 1998. A Neoproterozoic snowball Earth. *Science* 281:1342–46.

Hoffman, P. F., and D. P. Schrag. 2002. The snowball Earth hypothesis: Testing the limits of global change. *Terra Nova* 14:129–55.

Hofmann, H. J., and E. W. Mountjoy. 2001. *Namacalathus-Cloudina* assemblage in Neoproterozoic Miette Group (Byng Formation), British Columbia: Canada's oldest shelly fossils. *Geology* 29:1091–94.

Hofmann, H. J., E. K. O'Brian, and A. F. King. 2008. Ediacaran biota on Bonavista Peninsula, Newfoundland, Canada. *Journal of Paleontology* 82:1–36.

Holland, L. Z., R. Albalat, K. Azumi, E. Benito-Gutierrez, M. J. Blow et al. 2008. The amphioxus genome illuminates vertebrate origins and cephalochordate biology. *Genome Research* 18:1100–1111.

Holland, N. D., and J. Y. Chen. 2001. Origin and early evolution of the vertebrates: New insights from advances in molecular biology, anatomy and palaeontology. *BioEssays* 23:142–51.

Holley, S. A., P. D. Jackson, Y. Sasai, B. Lu, E. M. De Robertis et al. 1995. A conserved system for dorsal-ventral patterning in insects and vertebrates involving sog and chordin. *Nature* 376:249–53.

Hollingsworth, J. S. 1999. Proposed stratotype section and point for the base of the Dyeran Stage. *Laurentai V Field Conference of the Cambrian Stage and Subdivision* 99:38–42.

Holmer, L. E., and L. E. Popov. 2007. Incertae sedis organophosphatic bivalved stem-group brachiopods. In *Treatise on Invertebrate Paleontology, Part H, Brachiopoda [Revised]*, edited by A. Williams, S. J. Carlson, and C. H. C. Brunton, 2580–90. Lawrence: Geological Society of America / University of Kansas.

Holmer, L. E., S. P. Stolk, C. B. Skovsted, U. Balthasar, and L. Popov. 2009. The enigmatic Early Cambrian *Salanygolina*—A stem group of rhynchonelliform chileate brachiopods? *Palaeontology* 52:1–10.

Holmes, A. C. 1960. A revised geological time scale. *Transactions of the Edinburgh Geological Society* 17:183–216.

Hou, X. G., R. J. Aldridge, J. Bergstrom, D. J. Siveter, D. J. Siveter et al. 2004. *The Cambrian Fossils of Chengjiang, China*. Oxford, UK: Blackwell.

Hou, X. G., R. J. Aldridge, D. J. Siveter, M. Williams, J. Zalasiewicz et al. 2011. An Early Cambrian hemichordate zooid. *Current Biology* 21:612–16.

Hou, X. G., and J. Bergström. 1995. Cambrian lobopodians—Ancestors of extant onycophorans? *Zoological Journal of the Linnean Society* 114:3–19.

———. 1997. Arthropods of the Lower Cambrian Chengjiang fauna, southwest China. *Fossils and Strata* 45:1–116.

Hou, X. G., J. Bergström, and G. H. Xu. 2004. The Lower Cambrian crustacean *Pectocaris* from the Chengjiang biota, Yunnan, China. *Journal of Paleontology* 78:700–708.

Hou, X. G., G. D. Stanley, Jr., J. Zhao, and X. Y. Ma. 2005. Cambrian anemones with preserved soft tissue from the Chengjiang biota, China. *Lethaia* 38:193–203.

Hou, X. G., M. Williams, D. J. Siveter, D. J. Siveter, R. J. Aldridge et al. 2010. Soft-part anatomy of the Early Cambrian bivalved arthropods *Kunyangella* and *Kunmingella*: Significance for the phylogenetic relationships of Bradoriida. *Proceedings of the Royal Society of London B* 277:1835–41.

Hou, X., and W. Sun. 1989. Discovery of Chengjiang fauna at Meishucun, Jinning, Yunnan. *Acta Palaeontologica Sinica* 27:1–12.

Hu, S., M. Steiner, M. Zhu, B. D. Erdtmann, H. Luo et al. 2007. Diverse pelagic predators from the Chengjiang lagerstätte and the establishment of modern-style pelagic ecosystems in the Early Cambrian. *Palaeogeography, Palaeoclimatology, Palaeoecology* 254:307–16.

Hu, S. X. 2005. Taphonomy and palaeoecology of the Early Cambrian Chengjiang biota from eastern Yunnan, China. *Berliner Paläobiologische Abhandlungen* 7:1–197.

Hua, H., Z. Chen, X. Yuan, L. Zhang, and S. Xiao. 2005. Skeletogenesis and asexual reproduction in the earliest biomineralizing animal *Cloudina*. *Geology* 33:277–80.

Huang, D. Y., J. Y. Chen, J. Vannier, and J. I. Saiz Salinas. 2004. Early Cambrian sipunculan worms from southwest China. *Proceedings of the Royal Society of London B* 271:1671–76.

Hui, J. H. L., P. W. H. Holland, and D. E. K. Ferrier. 2008. Do cnidarians have a *ParaHox* cluster? Analysis of synteny around a *Nematostella* homeobox cluster. *Evolution and Development* 10:725–30.

Huntley, J. W., S. H. Xiao, and M. Kowalewski. 2006. 1.3 billion years of acritarch history: An empirical morphospace approach. *Precambrian Research* 144:52–68.

Hurst, L. D. 2009. Genetics and the understanding of selection. *Nature Reviews Genetics* 10:83–93.

Hurtgen, M. T., M. A. Arthur, and G. P. Halverson. 2005. Neoproterozoic sulfur isotopes, the evolution of microbial sulful species and the burial efficiency of sulfide as sedimentary pyrite. *Geology* 33:41–44.

Hurtgen, M. T., G. P. Halverson, M. A. Arthur, and P. Hoffman. 2006. Sulfur cycling in the aftermath of a Neoproterozoic (Marinoan) snowball glaciation: Evidence for a syn-glacial sulfidic deep ocean. *Earth and Planetary Science Letters* 245:551–70.

Hutchinson, G. E. 1957. Concluding remarks. *Cold Spring Harbor Symposium on Quantitative Biology* 22:415–27.

———. 1967. *Treatise on Limnology. II. Introduction to Lake Biology and the Limnoplankton*. New York: Wiley.

Hwang, J. H., H. Ohyanagi, S. Hayakawa, N. Osato, C. Nichimiya-Fujisawa et al. 2007. The evolutionary emergence of cell type-specific genes inferred from the gene expression analysis of *Hydra*. *Proceedings of the National Academy of Sciences, USA* 104:14735–40.

Hyman, L. H. 1940. *The Invertebrates. I. Protozoa through Ctenophora*. New York: McGraw-Hill.

———. 1951. *The Invertebrates. II. Platyhelminthes and Rhynchocoela*. New York: McGraw-Hill.

Innan, H., and F. A. Kondrashov. 2010. The evolution of gene duplications: Classifying and distinguishing between models. *Nature Reviews Genetics* 11:97–108.

Isachsen, C. E., S. A. Bowring, E. Landing, and S. D. Samson. 1994. New constraint on the division of Cambrian time. *Geology* 22:496–98.

Ivantsov, A. Y. 2009. New reconstruction of *Kimberella*, problematic Vendian metazoan. *Paleontological Journal* 43:601–11.

Ivantsov, A. Y., and D. V. Grazhdankin. 1997. A new representative of the Petaloamae from the Upper Vendian of the Arkhangelsk region. *Paleontological Journal* 311:1–16.

Ivantsov, A. Y., and Y. E. Malakhovskaya. 2002. Giant traces of Vendian animals. *Doklady Earth Sciences* 385A:618–22.

Jablonski, D. 2007. Scale and hierarchy in macroevolution. *Palaeontology* 50:87–109.

Jackson, D. J., L. Macis, J. Reitner, and G. Worheide. 2011. A horizontal gene transfer supported the evolution of an early metazoan biomineralization strategy. *BMC Evolutionary Biology* 11:238.

Jackson, J. B. C., and D. H. Erwin. 2006. What can we learn about ecology and evolution from the fossil record? *Trends in Ecology and Evolution* 21:322–28.

Jacobs, D. K., N. C. Hughes, S. T. Fitz-Gibbon, and C. J. Winchell. 2005. Terminal addition, the Cambrian radiation and the Phanerozoic evolution of bilaterian form. *Evolution and Development* 7:498–514.

Jacquier, A. 2009. The complex eukaryotic transcriptome: Unexpected pervasive transcription and novel small RNAs. *Nature Reviews Genetics* 10:833–44.

James, N. P., G. M. Narbonne, and T. K. Kyser. 2001. Late Neoproterozoic cap carbonates: Mackenzie Mountains, northwestern Canada: Precipitation and global glacial meltdown. *Canadian Journal of Earth Sciences* 38:1229–62.

Janssen, R., B. J. Eriksson, G. E. Budd, M. Akam, and N. M. Prpic. 2010. Gene expression patterns in an onycophoran reveal that regionalization predates limb segmentation in pan-arthropods. *Evolution and Development* 12:363–72.

Jansson, I.-M., S. McLoughlin, and V. Vajda. 2008. Early Jurassic annelid cocoons from eastern Australia. *Alcheringa* 32:285–96.

Janussen, D., M. Steiner, and M. Y. Zhu. 2002. New well-preserved scleritomes of Chancelloridae from Early Cambrian Yuanshan Formation (Chenjiang, China) and the Middle Cambrian Wheeler Shale (Utah, USA) and paleobiological implications. *Journal of Paleontology* 76:596–606.

Javaux, E. J., A. H. Knoll, and M. R. Walter. 2001. Morphological and ecological complexity in early eukaryotic ecosystems. *Nature* 412:66–69.

Jefferies, R. P. S. 1979. The origin of chordates—A methodological essay. In *The Origin of Major Invertebrate Groups*, edited by M. R. House, 443–47. London: Academic Press.

———. 1986. *The Ancestry of the Vertebrates*. Cambridge, UK: Cambridge University Press.

———. 1990. The solute *Dendrocystoides scoticus* from the Upper Ordovician of Scotland and the ancestry of chordates and echinoderms. *Palaeontology* 33:631–79.

Jenkins, R. J. F. 1985. The enigmatic Ediacaran (late Precambrian) genus *Rangea* and related forms. *Paleobiology* 11:336–55.

———. 1992. Functional and ecological aspects of Ediacaran assemblages. In *Origin and Early Evolution of the Metazoa,* edited by J. H. Lipps and P. Signor, 131–76. New York: Plenum.

———. 2007. "Ediacaran" as a name for a newly designated terminal Proterozoic period. In *The Rise and Fall of the Ediacaran Biota,* edited by P. Vickers-Rich and P. Komarower, 137–42. London: Geological Society.

Jenkins, R. J. F., C. H. Ford, and J. G. Gehling. 1983. The Ediacaran Member of the Rawnsley Quartzite: The context of the Ediacara assemblage (late Precambrian, Flinders Range). *Journal of the Geological Society of Australia* 30:101–19.

Jensen, S. 1990. Predation by early Cambrian trilobites on infaunal worms—evidence from the Swedish Mickwitzia sandstone. *Lethaia* 23:29–42.

———. 2003. The Proterozoic and earliest Cambrian trace fossil record: Patterns, problems and perspectives. *Integrative and Comparative Biology* 43:219–28.

Jensen, S., M. L. Droser, and J. G. Gehling. 2005. Trace fossil preservation and the early evolution of animals. *Palaeogeography, Palaeoclimatology, Palaeoecology* 220:19–29.

———. 2006. A critical look at the Ediacaran trace fossil record. In *Neoproterozoic Geobiology and Paleobiology,* edited by S. Xiao and A. J. Kaufman, 115–57. Berlin: Springer.

Jensen, S., J. G. Gehling, M. L. Droser, L. R. Godfrey, W. L. Jungers et al. 1998. Ediacara-type fossils in Cambrian sediments. *Nature* 393:567–69.

Jensen, S., B. Z. Saylor, J. G. Gehling, and G. J. B. Germs. 2000. Complex trace fossils from the terminal Proterozoic of Namibia. *Geology* 28:143–46.

Jiang, G., M. J. Kennedy, and N. Christie-Blick. 2003. Stable isotopic evidence for methane seeps in Neoproterozoic postglacial cap carbonates. *Nature* 426:822–26.

Jiang, G. Q., A. J. Kaufman, N. Christie-Blick, S. H. Zhang, and H. C. Wu. 2007. Carbon isotope variability across the Ediacaran Yangtze platform in south China: Implications for a large surface-to-deep ocean $\delta^{13}C$ gradient. *Earth and Planetary Science Letters* 261:303–20.

Johnson, D. T., F. A. Macdonald, B. C. Gill, P. F. Hoffman, and D. P. Schrag. 2012. Uncovering the Neoproterozoic carbon cycle. *Nature* 483:320–23.

Johnston, D. T., S. W. Poulton, C. Dehler, S. Porter, J. Husson et al. 2010. An emerging picture of Neoproterozoic ocean chemistry: Insights from the Chuar Group, Grand Canyon, USA. *Earth and Planetary Science Letters* 290:64–73.

Jones, C. G., J. H. Lawton, and M. Shachak. 1994. Organisms as ecosystem engineers. *Oikos* 69:373–86.

———. 1997. Positive and negative effects of organisms as physical ecosystem engineers. *Ecology* 78:1946–57.

Jones, C. I. 2002. *Introduction to Economic Growth.* New York: Norton.

Jones, D., and I. Thompson. 1977. Echiura from the Pennsylvanian Essex fauna of northern Illinois. *Lethaia* 10:317–25.

Kah, L. C., T. W. Lyons, and T. Frank. 2004. Low marine sulfate and protracted oxygenation of the Proterozoic biosphere. *Nature* 431:834–38.

Kaiser, D. 2001. Building a multicellular organism. *Annual Review of Genetics* 35:103–23.

Kapp, H. 2000. The unique embryology of Chaetognata. *Zoologischer Anzeiger* 239:263–66.

Karlstrom, K. E., S. A. Bowring, C. M. Dehler, A. H. Knoll, S. M. Porter et al. 2000. Chuar Group of the Grand Canyon: Record of breakup of Rodinia, associated change in the global carbon cycle, and ecosystem expansion by 740 Ma. *Geology* 28:619–22.

Kasemann, S. A., A. R. Prave, A. E. Fallick, C. J. Hawkesworth, and K.-H. Hoffmann. 2010. Neoproterozoic ice ages, boron isotopes, and ocean acidification: Implications for a snowball Earth. *Geology* 38:775–78.

Katija, K., and J. O. Dabiri. 2009. A viscosity-enhanced mechanism for biogenic ocean mixing. *Nature* 460:624–26.

Kaufman, A. J., S. B. Jacobsen, and A. H. Knoll. 1993. The Vendian record of Sr and C isotopic variations in seawater: Implications for tectonics and paleoclimate. *Earth and Planetary Science Letters* 120:409–30.

Kaufman, A. J., and A. H. Knoll. 1995. Neoproterozoic variations in the C-isotopic composition of seawater: Stratigraphic and biogeochemical implications. *Precambrian Research* 73:27–49.

Kaufman, A. J., A. H. Knoll, and G. M. Narbonne. 1997. Isotopes, ice ages, and terminal Proterozoic earth history. *Proceedings of the National Academy of Sciences, USA* 94:6600–6605.

Kaufman, A. J., A. H. Knoll, M. A. Semikhatov, J. P. Grotzinger, S. B. Jacobsen et al. 1996. Integrated chronostratigraphy of Proterozoic-Cambrian boundary beds in the western Anabar region, northern Siberia. *Geological Magazine* 133:509–33.

Kendall, B., R. A. Creaser, and D. Selby. 2006. Re-Os geochronology of postglacial black shales in Australia: Constraints on the timing of "Sturtian" glaciation. *Geology* 34:729–32.

Kennedy, M., D. Mrofka, and C. von der Borch. 2008. Snowball Earth termination by destabilization of equatorial permafrost methane clathrate. *Nature* 453:642–45.

Kennedy, M. J., N. Christie-Blick, and L. E. Sohl. 2001. Are Proterozoic cap carbonates and isotopic excursions a record of gas hydrate destabilization following Earth's coldest intervals? *Geology* 29:443–46.

Kennedy, M. J., M. L. Droser, L. M. Mayer, D. Pewear, and D. Mrofka. 2006. Late Precambrian oxygenation; inception of the clay mineral factory. *Science* 311:1446–49.

Keren, H., G. Lev-Maor, and G. Ast. 2010. Alternative splicing and evolution: Diversification, exon definition and function. *Nature Reviews Genetics* 11:345–55.

Khomentovsky, V. V., and G. A. Karlova. 1993. Biostratigraphy of the Vendian-Cambrian beds and the Lower Cambrian boundary in Siberia. *Geological Magazine* 130:29–45.

Kilner, B., C. Niocaill, and M. Brasier. 2005. Low-latitude glaciation in the Neoproterozoic of Oman. *Geology* 33:413–16.

Kimmel, C. B. 1996. Was *Urbilateria* segmented? *Trends in Genetics* 12:329–32.

King, B. L., J. A. Gillis, H. R. Carlisle, and R. D. Dahn. 2011. A natural deletion of the *HoxC* cluster in elasmobranch fishes. *Science* 334:1517.

King, N. 2004. The unicellular ancestry of animal development. *Developmental Cell* 7:313–25.

King, N., and S. B. Carroll. 2001. A receptor tyrosine kinase from choanoflagellates: Molecular insights into early animal evolution. *Proceedings of the National Academy of Sciences, USA* 98:15032–37.

King, N., C. T. Hittinger, and S. B. Carroll. 2003. Evolution of key cell signaling and adhesion protein families predates animal origins. *Science* 301:361–63.

King, N., M. J. Westbrook, S. L. Young, A. Kuo, M. Abedin et al. 2008. The genome of the choanoflagellate *Monosiga brevicollis* and the origin of metazoans. *Nature* 451:783–88.

Kirshvink, J. L. 1992. Late Proterozoic low-latitude glaciation: The Snowball Earth. In *The Proterozoic Biosphere: A Multidisciplinary Study*, edited by J. W. Schopf, C. Klein, and D. Des Marais, 51–52. Cambridge, UK: Cambridge University Press.

Klingenberg, C. P. 2008. Morphological integration and developmental modularity. *Annual Review of Ecology, Evolution and Systematics* 39:115–32.

Knoll, A. H. 2003a. Biomineralization and evolutionary history. *Reviews in Mineralogy and Geochemistry* 54:329–56.

———. 2003b. *Life on a Young Planet: The First Three Billion Years of Evolution on Earth*. Princeton, NJ: Princeton University Press.

———. 2011. The multiple origins of complex multicellularity. *Annual Review of Earth and Planetary Science* 39:217–39.

Knoll, A. H., R. K. Bambach, D. E. Canfield, and J. P. Grotzinger. 1996. Comparative earth history and Late Permian mass extinction. *Science* 273:452–57.

Knoll, A. H., M. R. Walter, G. M. Narbonne, and N. Christie-Blick. 2004. A new period for the geologic time scale. *Science* 305:621–22.

Kobluk, D. R. 1981. Cavity-dwelling biota in Middle Ordovician (Chazy) bryozoan mounds from Quebec. *Canadian Journal of Earth Sciences* 18:42–54.

Kodner, R. B., R. E. Summons, A. Pearson, N. King, and A. H. Knoll. 2008. Sterols in a unicellular relative of the metazoans. *Proceedings of the National Academy of Sciences, USA* 105:9897–902.

Kolata, D. R., J. C. Brower, and T. J. Frest. 1987. Upper Mississippi Valley Champlainian and Cincinnatian echinoderms. *Middle and Late Ordovician Lithostratigraphy and Biostratigraphy of the Upper Mississippi Valley: Minnesota Geological Survey Report of Investigations* 35:179–81.

Koschwanez, J. H., K. R. Foster, and A. W. Murray. 2011. Sucrose utilization in budding yeast as a model for the origin of undifferentiated multicellularity. *PLoS Biology* 9(8):e1001122.

Kosik, K. S. 2009. MicroRNAs tell an evo-devo story. *Nature Reviews Neuroscience* 10:754–59.

Kouchinsky, A., S. Bengtson, W. M. Feng, R. Kutygin, and A. Val'kov. 2009. The Lower Cambrian fossil anabaritids: Affinities, occurrences and systematics. *Journal of Systematic Palaeontology* 7:241–98.

Kouchinsky, A., S. Bengtson, and L. A. Gershwin. 1999. Cnidarian-like embryos associated with the first shelly fossils in Siberia. *Geology* 27:609–12.

Kouchinsky, A., S. Bengtson, B. Runnegar, C. B. Skovstead, M. Steiner et al. 2012. Chronology of Early Cambrian biomineralization. *Geological Magazine* 149:221–51.

Kruse, P. D., A. Gandin, F. Debrenne, and R. Wood. 1996. Early Cambrian bioconstructions in the Zavkhan Basin of western Mongolia. *Geological Magazine* 133:429–44.

Kruse, P. D., A. Y. Zhuravlev, and N. P. James. 1995. Primordial metazoan-calcimicrobial reefs: Tommotian (Early Cambrian) of the Siberian platform. *Palaios* 10:291–321.

Laflamme, M. Forthcoming. Ediacaran clades. Unpublished manuscript.

Laflamme, M., L. I. Flude, and G. M. Narbonne. 2012. Ecological tiering and the evolution of a stem: The oldest stemmed frond from the Ediacaran of Newfoundland, Canada. *Journal of Paleontology* 86:193–200.

Laflamme, M., and G. M. Narbonne. 2008a. Competition in a Precambrian world: Paleoecology of Ediacaran fronds. *Geology Today* 24(5):182–87.

———. 2008b. Ediacaran fronds. *Palaeogeography, Palaeoclimatology, Palaeoecology* 258:162–79.

Laflamme, M., G. M. Narbonne, C. Greentree, and M. M. Anderson. 2007. Morphology and taphonomy of an Ediacaran frond: Charnia from the Avalon Penninsula of Newfoundland. In *The Rise and Fall of the Ediacaran Biota*, edited by P. Vickers-Rich and P. Komarower, 237–57. London: Geological Society.

Laflamme, M., J. D. Schiffbauer, and G. M. Narbonne. 2011. Deep-water microbially induced sedimentary structures (MISS) in deep time: The Ediacaran fossil *Iveshedia*. SEPM Special Publication 101:111–23.

Laflamme, M., S. Xiao, and M. Kowalewski. 2009. Osmotrophy in modular Ediacara organisms. *Proceedings of the National Academy of Sciences, USA* 106:14438–43.

Lake, J. A. 1989. Origin of the multicellular animals. In *The Hierarchy and Evolution of Life*, edited by B. Fernholm, K. Bremer, and H. Jornvall, 273–78. Amsterdam: Excerpta Medica.

Lake, J. A. 1990. Origin of the Metazoa. *Proceedings of the National Academy of Sciences, USA* 87:763–66.

Laland, K. N., and K. Sterelny. 2006. Seven reasons (not) to neglect niche construction. *Evolution* 60:1751–62.

Landing, E. 1994. Precambrian-Cambrian boundary global stratotype ratified and a new perspective of Cambrian time. *Geology* 22:179–82.

Landing, E., S. A. Bowring, K. L. Davidek, A. W. A. Rushton, R. A. Fortey et al. 2000. Cambrian-Ordovician age and duration of the lowest Ordovician Tremadoc series based on U-Pb zircon dates from Avalonian Wales. *Geological Magazine* 137:485–94.

Landing, E., S. A. Bowring, K. L. Davidek, S. R. Westrop, G. Geyer et al. 1998. Duration of the Early Cambrian: U-Pb ages of volcanic ashes from Avalon and Gondwana. *Canadian Journal of Earth Sciences* 35:329–38.

Landing, E., P. M. Myrow, A. P. Benus, and G. M. Narbonne. 1989. The Placentian series: Appearance of the oldest skeletalized faunas in southeastern Newfoundland. *Journal of Paleontology* 63:739–69.

Landing, E. D. 1991. Upper Precambrian through Lower Cambrian of Cape Breton Island: Faunas, paleoenvironments, and stratigraphic revision. *Journal of Paleontology* 65:570–95.

Lang, K. 1963. The relation between the Kinorhyncha and Priapulida and their connection with the Aschelminthes. In *The Lower Metazoa*, edited by E. C. Dougherty, 256–62. Berkeley: University of California Press.

Larroux, C., B. Fahey, D. Liubicich, V. Hinman, M. Guathier et al. 2006. Developmental expression of transcription factor genes in a demosponge: Insights into the origins of metazoan multicellularity. *Evolution and Development* 8:150–73.

Larroux, C., G. N. Luke, P. Koopman, D. Rokhsar, S. M. Shimeld et al. 2008. Genesis and expansion of metazoan transcription factor gene classes. *Molecular Biology and Evolution* 25:980–96.

Lee, M. S., J. B. Jago, D. C. Garcia-Bellido, G. D. Edgecombe, J. G. Gehling et al. 2011. Modern optics in exceptionally preserved eyes of Early Cambrian arthropods from Australia. *Nature* 474:631–34.

Lee, P. N., K. Pang, D. Q. Matus, and M. Q. Martindale. 2006. A WNT of things to come: Evolution of Wnt signaling and polarity in cnidarians. *Seminars in Cell and Developmental Biology* 17:157–67.

Lemons, D., and W. McGinnis. 2006. Genomic evolution of *Hox* gene clusters. *Science* 313:1918–22.

Levine, M., and R. Tjian. 2003. Transcription regulation and animal diversity. *Nature* 424:147–51.

Lewontin, R. C. 2003. Four complications in understanding the evolutionary process. *SFI Bulletin* 18(1) (unpaginated).

Li, C., G. D. Love, T. W. Lyons, D. A. Fike, A. L. Sessions et al. 2010. A stratified redox model for the Ediacaran ocean. *Science* 328:80–83.

Li, D., H. F. Ling, S. Y. Jiang, J. Y. Pan, Y. Q. Chen et al. 2009. New carbon isotope stratigraphy of the Ediacaran-Cambrian boundary interval from SW China: Implications for global correlation. *Geological Magazine* 146:465–84.

Li, G. X., M. Steiner, X. J. Zhu, A. Yang, H. F. Wang et al. 2007. Early Cambrian metazoan fossil record of south China: Generic diversity and radiation patterns. *Palaeogeography, Palaeoclimatology, Palaeoecology* 254:229–49.

Li, X., J. J. Cassidy, C. A. Reinke, S. Fischboeck, and R. W. Carthew. 2009. A microRNA imparts robustness against environmental fluctuation during development. *Cell* 137:273–82.

Li, Z. X., S. V. Bogdanova, A. S. Collins, A. R. Davidson, B. De Waele et al. 2008. Assembly, configuration, and breakup history of Rodinia: A synthesis. *Precambrian Research* 160:179–210.

Lichtneckert, R., and H. Reichert. 2005. Insights into the urbilaterian brain: Conserved genetic patterning mechanisms in insect and vertebrate brain development. *Heredity* 94:465–77.

Lin, J. P., A. C. Scott, C.-W. Li, H. J. Wu, W. I. Ausich et al. 2006. Silicified egg clusters from a Middle Cambrian Burgess Shale–type deposit, Guizhou, south China. *Geology* 34:1037–40.

Lindberg, D. R., and W. F. Ponder. 1996. An evolutionary tree for Mollusca: Branches or roots? In *Origin and Evolutionary Radiation of the Mollusca*, edited by J. D. Taylor, 67–75. Oxford, UK: Oxford University Press.

Linnaeus, C. 1758. *System Naturae*. Stockholm: Laurentii Salvii.

Liu, A. G., D. McIlroy, and M. D. Brasier. 2010. First evidence for locomotion in the Ediacara biota from the 565 Ma Mistaken Point Formation, Newfoundland. *Geology* 38:123–26.

Liu, J., D. G. Shu, J. Han, Z. F. Zhang, and X. L. Zhang. 2008. Origin, diversification and relationships of Cambrian lobopods. *Gondwana Research* 14:277–83.

Liu, J., M. Steiner, J. A. Dunlop, H. Keupp, D. Shu et al. 2011. An armoured Cambrian lobopodian from China with arthropod-like appendages. *Nature* 470:526–30.

Lobo, D., and F. J. Vico. 2010. Evolution of form and function in a model of differentiated multicellular organisms with gene regulatory networks. *BioSystems* 102:112–23.

Logan, G. A., J. M. Hayes, G. B. Heishima, and R. E. Summons. 1995. Terminal Proterozoic reorganization of biogeochemical cycles. *Nature* 376:53–56.

Lohrer, A. M., S. F. Thrush, and M. M. Gibbs. 2004. Bioturbators enhance ecosystem function through complex biogeochemical interactions. *Nature* 431:1092–95.

Loreau, M. 2000. Are communities saturated? On the relationship between α, β and γ diversity. *Ecology Letters* 3:73–76.

Loreau, M., and N. Mouguet. 2000. Immigration and the maintenance of local species diversity. *American Naturalist* 154:427–40.

Losos, J. B. 2009. *Lizards in an Evolutionary Tree: Ecology and Adaptive Radiation of* Anolis. Berkeley: University of California Press.

———. 2010. Adaptive radiation, ecological opportunity, and evolutionary determinism. *American Naturalist* 175:623–39.

Love, G. D., E. Grosjean, C. Stalvies, D. A. Fike, J. P. Grotzinger et al. 2009. Fossil steroids record the appearance of Demospongiae during the Cryogenian period. *Nature* 457:718–21.

Lowe, C. J. 2008. Molecular genetic insights into deuterostome evolution from the direct-developing hemichordate *Saccoglossus kowalevskii*. *Philosophical Transactions of the Royal Society B* 363:1569–78.

Lowe, C. J., M. Terasaki, M. Wu, R. M. Freeman, Jr., L. Runft et al. 2006. Dorsoventral patterning in hemichordates: Insights into early chordate evolution. *PLoS Biology* 4(9):e291.

Lowe, C. J., M. Wu, A. Salic, L. Evans, E. Lander et al. 2003. Anteroposterior patterning in hemichordates and the origin of the chordate nervous system. *Cell* 113:853–65.

Lyell, Charles. 1830–1833. *Principles of Geology, Being an Attempt to Explain the Former Changes of the Earth's Surface, by Reference to Causes Now in Operation*. Vols. 1–3. London: John Murray.

Lynch, M., and J. S. Conery. 2003. The origins of genome complexity. *Science* 302:1401–4.

Lynch, M., and V. Katju. 2004. The altered evolutionary trajectories of gene duplicates. *Trends in Genetics* 20:544–49.

Lynch, M. L. 2006. The origins of eukaryotic gene structure. *Molecular Biology and Evolution* 23:450–68.

———. 2007. *The Origins of Genome Architecture*. Sunderland, MA: Sinauer.

Lynch, V. J., and G. P. Wagner. 2008. Resurrecting the role of transcription factor change in developmental evolution. *Evolution* 62:2131–54.

Ma, X. Y., X. G. Hou, and J. Bergstrom. 2009. Morphology of *Luolishania longicruris* (Lower Cambrian, Chengjiang lagerstätte, southwest China) and the phylogenetic relationships within lobopodians. *Arthropod Structure and Development* 38:271–91.

Maas, A., D. Y. Huang, J. Y. Chen, D. Waloszek, and A. Braun. 2007. Maotianshan-Shale nemathelminths—Morphology, biology, and the phylogeny of Nemathelminthes. *Palaeogeography, Palaeoclimatology, Palaeoecology* 254:288–306.

Maletz, J., M. Steiner, and O. Fatka. 2007. Middle Cambrian pterobranchs and the question: What is a graptolite? *Lethaia* 38:73–85.

Maloof, A. C., S. M. Porter, J. L. Moore, F. O. Dudas, S. A. Bowring et al. 2010. The earliest Cambrian record of animals and ocean geochemical change. *Geological Society of America Bulletin* 122:1731–74.

Maloof, A. C., C. V. Rose, C. C. Calmet, R. Beach, B. M. Samuels et al. 2010. Probable animal body-fossils from pre-Marinoan limestones, South Australia. *Nature Geoscience* 3:653–59.

Maloof, A. C., D. P. Schrag, J. L. Crowley, and S. A. Bowring. 2005. An expanded record of Early Cambrian cycling from the Anti-Atlas margin, Morocco. *Canadian Journal of Earth Sciences* 42:2195–216.

Mangano, M. G., and L. A. Buatois. 2007. Trace fossils in evolutionary paleoecology. In *Trace Fossils: Concepts, Problems, Prospects*, edited by W. F. Miller III, 391–411. Amsterdam: Elsevier.

Margulis, L. 1970. *Origin of Eukaryotic Cells*. New Haven, CT: Yale University Press.

Margulis, L., and K. V. Schwartz. 1982. *Five Kingdoms: An Illustrated Guide to the Phyla of Life on Earth*. San Francisco: Freeman.

Mark-Kurik, E. 1969. Distribtion of vertebrates in the Silurian of Estonia. *Lethaia* 2:145–52.

Marletaz, F., E. Martin, Y. Perez, D. Papillon, X. Caubit et al. 2006. Chaetognath phylogenomics: A protostome with deuterostome-like development. *Current Biology* 16:R577–78.

Marlow, H. Q., D. S. Srivastava, D. Q. Matus, D. Rokhsar, and M. Q. Martindale. 2009. Anatomy and development of the nervous system of *Nematostella vectensis*, an anthazoan cnidarian. *Developmental Neurobiology* 69:235–54.

Marshall, C. R. 2006. Explaining the Cambrian "explosion" of animals. *Annual Review of Ecology, Evolution and Systematics* 34:355–84.

Marshall, C. R., and J. W. Valentine. 2010. The importance of preadapted genomes in the origin of animal bodyplans and the Cambrian explosion. *Evolution* 64:1189–201.

Martin, M. W., D. V. Grazhdankin, S. A. Bowring, D. A. Evans, M. A. Fedonkin et al. 2000. Age of Neoproterozoic bilaterian body and trace fossils, White Sea, Russia: Implications for metazoan evolution. *Science* 288:841–45.

Martindale, M. Q. 2005. The evolution of metazoan axial properties. *Nature Reviews Genetics* 6:917–27.

Martindale, M. Q., J. R. Finnerty, and J. Q. Henry. 2002. The Radiata and the evolutionary origins of the bilaterian body plan. *Molecular Phylogenetics and Evolution* 24:358–65.

Martindale, M. Q., and A. Hejnol. 2009. A developmental perspective: Changes in the position of the blastopore during bilaterian evolution. *Developmental Cell* 17:162–74.

Martindale, M. Q., K. Pang, and J. R. Finnerty. 2004. Investigating the origins of triploblasty: "Mesodermal" gene expression in a diploblastic animal, the sea anemone *Nematostella vectensis* (phylum, Cnidaria; class, Anthozoa). *Development* 131:2463–74.

Mattick, J. S., R. J. Taft, and G. J. Faulkner. 2010. A global view of genomic information—Moving beyond the gene and the master regulator. *Trends in Genetics* 26:21–28.

Matus, D. Q., R. R. Copley, C. W. Dunn, A. Hejnol, H. Eccleston et al. 2006. Broad taxon and gene sampling indicate that chaetognaths are protostomes. *Current Biology* 16:R575–76.

Matus, D. Q., C. R. Magie, K. Pang, M. Q. Martindale, and G. H. Thomsen. 2008. The Hedgehog gene family of the cnidarian, *Nematostella vectensis*, and implications for understanding metazoan Hedgehog pathway evolution. *Developmental Biology* 313:501–18.

Matus, D. Q., K. Pang, H. Marlow, C. W. Dunn, G. H. Thomsen et al. 2006. Molecular evidence for deep evolutionary roots of bilaterality in animal development. *Proceedings of the National Academy of Sciences, USA* 103:11195–200.

Maynard Smith, J., and E. Szathmary. 1995. *The Major Transitions in Evolution.* New York: Freeman.

Mayoral, E., E. Liñán, J. A. Gámez Vintaned, F. Muñiz, and R. Gozalo. 2004. Stranded jellyfish in the lowermost Cambrian (Corduban) of Spain. *Revista española de paleontología* 19:191–98.

Mayr, E. 1960. The emergence of novelty. In *The Evolution of Life*, edited by S. Tax, 349–80. Chicago: University of Chicago Press.

Mazumdar, A., and H. Strauss. 2006. Sulfur and strontium isotope compositions of carbonate and evaporite rocks from the late Neoproterozoic–early Cambrian Bilara Group (Najaur-Ganganagar Basin, India): Constraints on interbasinal correlation and global sulfur cycle. *Precambrian Research* 149:217–30.

McFadden, K. A., J. Huang, X. Chu, G. Jiang, A. J. Kaufman et al. 2008. Pulsed oxidation and biological evolution in the Ediacaran Doushantuo Formation. *Proceedings of the National Academy of Sciences, USA* 105:3197–202.

McGinnis, W., and R. Krumlauf. 1992. Homeotic genes and axial patterning. *Cell* 68:283–302.

McIlroy, D., and G. A. Logan. 1999. The impact of bioturbation on infaunal ecology and evolution during the Proterozoic-Cambrian transition. *Palaios* 14:58–72.

McIntosh, W. C. 1885. Report on the Annelida Polychaeta collected by H.M.S. *Challenger* during the years 1873–76. *Report on the Scientific Results of the Voyage of H.M.S.* Challenger *during the Years 1873–76* (12):1–554.

McKerrow, W. S., C. R. Scotese, and M. D. Brasier. 1992. Early Cambrian continental reconstructions. *Journal of the Geological Society of London* 149:599–606.

McShea, D. W. 2000. Functional complexity in organisms: Parts as proxies. *Biology and Philosophy* 15:641–68.

———. 2002. A complexity drain on cells in the evolution of multicellularity. *Evolution* 56:441–52.

Meert, J. G., and B. S. Lieberman. 2004. A palaeomagnetic and palaeobiological perspective on latest Neoproterozoic and Early Cambrian tectonic events. *Journal of the Geological Society of London* 161:477–87.

———. 2008. The Neoproterozoic assembly of Gondwana and its relationship to the Edicaran-Cambrian radiation. *Gondwana Research* 14:5–21.

Mermillod-Blondin, F., and R. Rosenberg. 2006. Ecosystem engineering: The impact of bioturbation on biogeochemical processes in marine and freshwater benthic habitats. *Aquatic Sciences* 68:434–42.

Meysman, F. J. R., J. J. Middelburg, and C. H. R. Heip. 2006. Bioturbation: A fresh look at Darwin's last idea. *Trends in Ecology and Evolution* 21:688–95.

Michod, R. E. 2007. Evolution of individuality during the transition from unicellular to multicellular life. *Proceedings of the National Academy of Sciences, USA* 104:8613–18.

Miller, A. H. 1949. Some ecologic and morphologic considerations in the evolution of higher taxonomic categories. In *Ornithologie als biologische Wissenschaf*, edited by E. Mayr and E. Schulz, 84–88. Heidelberg, Germany: Carl Winter.

Miller, D. J., E. E. Ball, and U. Technau. 2005. Cnidarians and ancestral gene complexity in the animal kingdom. *Trends in Genetics* 21:536–39.

Minelli, A. 2003. The origin and evolution of appendages. *International Journal of Developmental Biology* 47:573–81.

———. 2009. *Perspectives in Animal Phylogeny and Evolution.* Oxford, UK: Oxford University Press.

Missarzhevsky, V. V. 1989. Drevnejshie skeletnye okamenelosti i stratigrafiya pogranichnykh tolshch dokembriya i kembriya [The oldest skeletal fossils and stratigraphy of the Precambrian-Cambrian boundary beds]. *Trudy Geologicheskogo Instituta AN SSSR* 443:1–237.

Moczydlowska, M., M. Landing, W. L. Zang, and T. Placios. 2011. Proterozoic phytoplankton and timing of Chlorophyte algae origins. *Palaeontology* 54:721–23.

Mounce, R. C. P., M. A. Wills, D. A. Legg, X. Y. Ma, J. M. Wolfe et al. 2011. Phylogenetic position of *Diania* challenged. *Nature* 476:E1–E4.

Muller, G. B., and G. P. Wagner. 1991. Novelty in evolution: Restructuring the concept. *Annual Review of Ecology and Systematics* 22:229–56.

Muller, K. J. 1983. Crustacea with preserved soft parts from the Upper Cambrian of Sweden. *Lethaia* 16:93–109.

Muller, K. J., D. Walossek, and A. Zakharov. 1995. "Orsten" type phosphatized soft-integument preservation and a new record from the Middle Cambrian Kuonamka Formation in Siberia. *Neues Jahrbuch für Geologie und Paläontologie, Abhandlungen* 191:101–18.

Müller, W. E., M. Wiens, T. Adell, V. Gamulin, H. C. Schröder et al. 2004. Bauplan of urmetazoa: Basis for genetic complexity of metazoa. *International Review of Cytology* 235:53–92.

Müller, W. E. G. 2003. The origin of metazoan complexity: Porifera as integrated animals. *Integrative and Comparative Biology* 43:3–10.

Mutvei, H., Y.-B. Zhang, and E. Dunca. 2007. Late Cambrian Plectronocerid nautiloids and their role in cephalopod evolution. *Palaeontology* 50:1327–33.

Narbonne, G. M. 2004. Modular construction of early Ediacaran complex life forms. *Science* 305:1141–44.

———. 2005. The Ediacara biota: Neoproterozoic origin of animals and their ecosystems. *Annual Review of Earth and Planetary Science* 33:421–42.

Narbonne, G. M., M. Laflamme, C. Greentree, and P. Trusler. 2009. Reconstructing a lost world: Ediacaran rangeomorphs from Spaniard's Bay, Newfoundland. *Journal of Paleontology* 83:503–23.

Nei, M. 1969. Gene duplication and nucleotide substitution in evolution. *Nature* 221:40–42.

———. 2005. Selectionism and neutralism in molecular evolution. *Molecular Biology and Evolution* 22:2318–42.

———. 2007. The new mutation theory of phenotypic evolution. *Proceedings of the National Academy of Sciences, USA* 104:12235–42.

Neiber, M. T., T. R. Hartke, T. Stemme, A. Bergmann, J. Rust. et al. 2011. Global biodiversity and phylogenetic evaluation of Remipedia (Crustacea). *PLoS ONE* 6(5):e19627.

Nicholas, C. 1986. The Sr isotopic evolution of the oceans during the "Cambrian explosion." *Journal of the Geological Society of London* 153:243–54.

Nichols, S. A., W. Dirks, J. Pearse, and N. King. 2006. Early evolution of animal signaling and adhesion genes. *Proceedings of the National Academy of Sciences, USA* 103:12451–56.

Nielsen, C. 1987. Structure and function of metazoan ciliary bands and their phylogenetic significance. *Acta Zoologica* 68:205–62.

———. 2001. *Animal Evolution: Interrelationships of the Living Phyla.* 2nd ed. Oxford, UK: Oxford University Press.

———. 2008. Six major steps in animal evolution: Are we derived sponge larvae? *Evolution and Development* 10:241–57.

Nielsen, C., N. Scharff, and D. Eibye-Jacobsen. 1996. Cladistic analyses of the animal kingdom. *Biological Journal of the Linnean Society* 57:385–410.

Niklas, K. J. 1994. Morphological evolution through complex domains of fitness. *Proceedings of the National Academy of Sciences, USA* 91:6772–79.

Nilsen, T. W., and B. R. Graveley. 2010. Expansion of the eukaryotic proteome by alternative splicing. *Nature* 463:457–63.

Nogaro, G., F. Mermillod-Blondin, M. H. Valett, F. Francois-Carcaillet, J. P. Gaudet et al. 2009. Ecosystem engineering at the sediment-water interface: Bioturbation and consumer-substrate interaction. *Oecologia* 161:125–38.

Nomaksteinsky, M., E. Rottinger, H. D. Dufour, Z. Chettouh, C. J. Lowe et al. 2009. Centralization of the deuterostome nervous system predates chordates. *Current Biology* 19:1264–69.

Norris, R. 1989. Cnidarian taphonomy and affinites of the Ediacara biota. *Lethaia* 22:381–94.

Norris, R. E. 1965. Neustonic marine Craspedomonadales (choanoflagellates) from Washington and California. *Journal of Protozoology* 12:589–602.

O'Brien, L. J., and J.-B. Caron. 2012. A new stalked filter-feeder from the Middle Cambrian Burgess Shale, British Columbia, Canada. *PloS One* 7(1):e29233.

Och, L. M., and G. A. Shields-Zhou. 2012. The Neoproterozoic oxygenation event: Environmental perturbations and biogeochemical cycling. *Earth-Science Review* 110:25–57.

Odling-Smee, F. J., K. N. Laland, and M. W. Feldman. 2003. *Niche Construction: The Neglected Process in Evolution*. Princeton, NJ: Princeton University Press.

Ohno, S. 1970. *Evolution by Gene Duplication*. Berlin: Springer.

———. 1996. The notion of the Cambrian pananimalia genome. *Proceedings of the National Academy of Sciences, USA* 93:8475–78.

Oliver, W. A., Jr., and A. G. Coates. 1987. Phylum Cnidaria. In *Fossil Invertebrates*, edited by R. S. Boardman, A. H. Cheetham, and A. J. Rowell, 140–93. Palo Alto, CA: Blackwell Scientific.

Olson, E. N. 2006. Gene regulatory networks in the evolution and development of the heart. *Science* 312:1922–27.

Ou, Q., J. Liu, D. Shu, J. Han, Z. Zhang et al. 2011. A rare onychophoran-like lobopodian from the Lower Cambrian Chengjiang lagerstätte, southwestern China, and its phylogenetic significance. *Journal of Paleontology* 85:587–94.

Page, R. D. M., and E. C. Holmes. 1998. *Molecular Evolution: A Phylogenetic Approach*. Oxford, UK: Blackwell.

Palmer, A. R. 1983. The decade of North American geology 1983: Geologic time scale. *Geology* 11:503–4.

Palmer, D., and B. Rickards. 1991. *Graptolites: Writing in the Rocks*. Woodbridge, UK: Boydell Press.

Pan, Q., O. Shai, L. J. Lee, B. J. Frey, and B. J. Blencowe. 2008. Deep surveying of alternative splicing complexity in the human transcriptome by high-throughput sequencing. *Nature Genetics* 40:1413–15.

Panganiban, G., and J. L. R. Rubenstein. 2002. Developmental functions of the *Distal-less/Dlx* homeobox genes. *Development* 129:4371–86.

Panganiban, G. E. F., S. M. Irvine, C. Lowe, H. Roehl, L. S. Corley et al. 1997. The origin and evolution of animal appendages. *Proceedings of the National Academy of Sciences, USA* 94:5162–66.

Pani, A. M., E. E. Mullarkey, J. Aronowicz, S. Assimacopoulos, E. A. Grove et al. 2012. Ancient deuterostome origins of vertebrate brain signalling centres. *Nature* 483:289–94.

Paps, J., J. Baguña, and M. Ruitort. 2009. Lophotrochozoa internal phylogeny: New insights from an up-to-date analysis of nuclear ribosomal genes. *Proceedings of the Royal Society B* 276:1245–54.

Parkhaev, P. Y. 2004. Malacofauna of the Lower Cambrian Bystraya Formation of eastern Transbaikalia. *Paleontological Journal* 38:590–608.

Paterson, J. R., D. C. Garcia-Bellido, M. S. Y. Lee, G. A. Brock, J. B. Jago et al. 2011. Acute vision in the giant Cambrian predator *Anomalocaris* and the origin of compound eyes. *Nature* 480:237–40.

Paul, C. R. C., and A. B. Smith. 1984. The early radiation and phylogeny of Echinodermata. *Biological Reviews* 59:443–81.

Pearse, V., J. Pearse, M. Bachsbaum, and R. Bachsbaum. 1987. *Living Invertebrates.* Palo Alto, CA: Blackwell Scientific.

Pearson, J. C., D. Lemons, and W. McGinnis. 2005. Modulating *Hox* gene functions during animal patterning. *Nature Reviews Genetics* 6:893–904.

Pecoits, E., K. O. Konhauser, N. R. Aubet, L. M. Heaman, G. Veroslavsky et al. 2012. Bilaterian burrows and grazing behavior at >585 million years ago. *Science* 336:1693–96.

Pedersen, K. J., and L. R. Pedersen. 1988. Ultrastructural observations on the epidermis of *Xenoturbella bocki* Westblad, 1949; with a discussion of epidermal cytoplasmic filament systems of invertebrates. *Acta Zoologica* 69:231–46.

Peel, J. S. 2010. A corset-like fossil from the Cambrian Sirius Passet Lagerstätte of North Greenland and its implications for cycloneuralian evolution. *Journal of Paleontology* 84:332–40.

Peisajovich, S. G., J. E. Garbarino, P. Wei, and W. A. Lim. 2010. Rapid diversification of cell signaling phenotypes by modular domain recombination. *Science* 328:368–72.

Peng, S. C., L. Babcock, R. Robison, H. L. Lin, M. Rees et al. 2004. Global Standard Stratotype-Section and Point (GSSP) of the Furongian Series and Paibian Stage (Cambrian). *Lethaia* 37:365–79.

Peter, I. S., and E. H. Davidson. 2009. Modularity and design principles in the sea urchin embryo gene regulatory network. *FEBS Letters* 583:3948–58.

———. 2010. The endoderm gene regulatory network in sea urchin embryos up to the mid-blastula stage. *Developmental Biology* 340:188–99.

Peter, R. H. 1983. *The Ecological Implications of Body Size.* Cambridge, UK: Cambridge University Press.

Peterson, K. J. 2005. Macroevolutionary interplay between planktic larvae and benthic predators. *Geology* 33:929–32.

Peterson, K. J., and N. J. Butterfield. 2005. Origin of the Eumetazoa: Testing ecological predictions of molecular clocks against the Proterozoic fossil record. *Proceedings of the National Academy of Sciences, USA* 102:9547–52.

Peterson, K. J., J. A. Cotton, J. G. Gehling, and D. Pisani. 2008. The Ediacaran emergence of bilaterians: Congruence between the genetic and the geological fossil records. *Philosophical Transactions of the Royal Society B* 363:1435–43.

Peterson, K. J., J. B. Lyons, K. S. Nowak, C. M. Takacs, M. J. Wargo et al. 2004. Estimating metazoan divergence times with a molecular clock. *Proceedings of the National Academy of Sciences, USA* 101:6536–41.

Peterson, K. J., R. E. Summons, and P. C. J. Donoghue. 2007. Molecular palaeobiology. *Palaeontology* 50:775–809.

Peterson, K. J., B. M. Waggoner, and J. W. Hagadorn. 2003. A fungal analog for Newfoundland Edicaran fossils? *Integrative and Comparative Biology* 43:127–36.

Pflug, H. D. 1972. Sur fauna der Nama-Schichten in Sudwest-Africa. III. Erniettomorpha, und Systematik. *Palaeontographica, Abteilung A* 139:134–70.

Philimore, A. B., and T. D. Price. 2008. Density-dependent cladogenesis in birds. *PLoS Biology* 6:483–89.

———. 2009. Ecological influences on the temporal patterns of speciation. In *Speciation and Patterns of Diversity,* edited by R. K. Butlin, J. R. Brindle, and D. Schluter, 240–56. Cambridge, UK: Cambridge University Press.

Philippe, H., H. Brinkmann, R. R. Copley, L. L. Moroz, H. Nankano et al. 2011. Acoelomorph flatworms are deuterostomes related to *Xenoturbella. Nature* 470:255–58.

Philippe, H., R. Derelle, P. Lopez, K. Pick, C. Borchiellini et al. 2009. Phylogenomics revives traditional views on deep animal relationships. *Current Biology* 19:706–12.

Piatigorsky, J. 2007. *Gene Sharing and Evolution*. Cambridge, MA: Harvard University Press.

Pierrehumbert, R. T., D. S. Abbot, A. Voigt, and D. Koll. 2011. Climate of the Neoproterozoic. *Annual Review of Earth and Planetary Sciences* 39:417–60.

Piper, J. D. A. 2007. The Neoproterozoic supercontinent Palaeopangaea. *Gondwana Research* 12:202–27.

Plachetzki, D. C., and T. H. Oakley. 2007. The origins of novel protein interactions during animal opsin evolution. *PLoS ONE* 2(e1054).

Planavsky, N. J., P. McGoldrick, C. T. Scott, C. Li, C. T. Reinhard et al. 2011. Widespread iron-rich conditions in the mid-Proterozoic ocean. *Nature* 477:448–51.

Planavsky, N. J., O. J. Rouxel, A. Bekker, S. V. Lalonde, K. O. Konhauser et al. 2010. The evolution of the marine phosphate reservoir. *Nature* 467:1088–90.

Poinar, G., Jr. 2003. A rhabdocoel turbellarian (Platyhelminthes, Typhloplanoida) in Baltic amber with a review of fossil and sub-fossil platyhelminths. *Invertebrate Biology* 122:308–12.

Poinar, G. O., Jr., A. Acra, and F. Acra. 1994. Earliest fossil nematode (Mermithidae) in Cretaceous Lebanese amber. *Fundamental and Applied Nematology* 17:475–77.

Porter, S. M. 2004. Closing the phosphatization window: Testing for the influence of taphonomic megabias on the patterns of small shelly fossil decline. *Palaios* 19:178–83.

———. 2007. Seawater chemistry and early carbonate biomineralization. *Science* 316:1302.

———. 2008. Skeletal microstructure indicates chancelloriids and halkeriids are closely related. *Palaeontology* 51:865–79.

———. 2010. Calcite and aragonite seas and the *de novo* acquisition of carbonate skeletons. *Geobiology* 8:256–77.

Post, D. M., and E. P. Palkovacs. 2009. Eco-evolutionary feedbacks in community and ecosystem ecology: Interactions between the ecological theater and the evolutionary play. *Philosophical Transactions of the Royal Society B* 364:1629–40.

Powell, C. M., and S. A. Pisarevsky. 2002. Late Neoproterozoic assembly of East Gondwana. *Geology* 30:3–6.

Putnam, N. H., M. Srivastava, U. Hellsten, B. Dirks, J. Chapman et al. 2007. Sea anemone genome reveals ancestral eumetazoan gene repertoire and genomic organization. *Science* 317:86–94.

Rabosky, D. L. 2009. Ecological limits on clade diversification in higher taxa. *American Naturalist* 173:662–74.

Raff, E. C., J. T. Villinski, F. R. Turner, P. C. J. Donoghue, and R. A. Raff. 2006. Experimental taphonomy shows the feasibility of fossil embryos. *Proceedings of the National Academy of Sciences, USA* 103:5846–61.

Raff, R. A. 1996. *The Shape of Life*. Chicago: University of Chicago Press.

———. 2008. Origins of the other metazoan body plans: The evolution of larval forms. *Philosophical Transactions of the Royal Society B* 363:1473–79.

Rahman, I. A., and S. Zamora. 2009. The oldest cinctan carpoid (stem-group Echinodermata), and the evolution of the water vascular system. *Zoological Journal of the Linnean Society* 157:420–32.

Raikova, O. I., M. Reuter, E. A. Kotikova, and M. K. S. Gustafsson. 1998. A commissural brain! The pattern of 5-HT immunoreactivity in Acoela (Platyhelminthes). *Zoomorphology* 118:69–77.

Ramskold, L., and J. Y. Chen. 1998. Cambrian lobopodians: Morphology and phylogeny. In *Arthropod Fossils and Phylogeny*, edited by G. D. Edgecombe, 107–50. New York: Columbia University Press.

Randell, R. D., B. S. Lieberman, S. T. Hasiotis, and M. C. Pope. 2005. New chancelloriids from the Early Cambrian Sekwi Formation with a comment on chancelloriid affinities. *Journal of Paleontology* 79:987–96.

Ranganayakulu, G., D. A. Elliott, R. P. Harvey, and E. N. Olson. 1998. Divergent roles for NK-2 class homeobox genes in cardiogenesis in flies and mice. *Development* 125:3037–48.

Raymond, J., and D. Segre. 2006. The effect of oxygen on biochemical networks and the evolution of complex life. *Science* 311:1764–67.

Regier, J. C., J. W. Shultz, A. Zwick, A. Hussey, B. Ball et al. 2010. Arthropod relationships revealed by phylogenomic analysis of nuclear protein-coding sequences. *Nature* 463:1079–83.

Reitner, J. 1992. Coralline Spongien: Der Versuch einer phylogenetisch-taxonomischen Analyse. *Berliner Geowissenschaftliche Abhandlungen Reihe E (Paläobiologie)* 1:1–352.

Reitner, J., and G. Wörheide. 2002. Non-lithistid fossil Demospongiae—Origins of their palaeobiodiversity and highlights in history of preservation. In *Systema Porifera: A Guide to the Classification of Sponges,* edited by J. N. A. Hooper and R. W. M. Van Soest, 52–68. New York: Klewer Academic / Plenum.

Retallack, G. J. 1994. Were the Ediacaran fossils lichens? *Paleobiology* 20:523–44.

Richards, G. S., and B. M. Degnan. 2009. The dawn of developmental signaling in the Metazoa. *Cold Spring Harbor Symposia on Quantitative Biology* 74:81–90.

Richardson, T. L., and G. A. Jackson. 2007. Small phytoplankton and carbon export from the surface ocean. *Science* 315:838–40.

Ricklefs, R. E. 2009. Speciation, extinction and diversity. In *Speciation and Patterns of Diversity,* edited by R. K. Butlin, J. R. Brindle, and D. Schluter, 257–77. Cambridge, UK: Cambridge University Press.

Ricklefs, R. E., and D. G. Jenkins. 2011. Biogeography and ecology: Towards the integration of two disciplines. *Philosophical Transactions of the Royal Society of London B* 366:2438–48.

Rigby, J. K., and D. Collins. 2004. Sponges of the Middle Cambrian Burgess Shale and Stephen Formations, British Columbia. *ROM Contributions in Science* 1:1–155.

Ritchie, A., and J. Gilbert-Tomlinson. 1977. First Ordovician vertebrates from the Southern Hemisphere. *Alcheringa* 1:351–68.

Rivera, M. C., and J. A. Lake. 2004. The ring of life provides evidence for a genome fusion origin of eukaryotes. *Nature* 431:152–55.

Robison, R. A., and R. L. Kaesler. 1987. Phylum Arthropoda. In *Fossil Invertebrates,* edited by R. S. Boardman, A. H. Cheetham, and A. J. Rowell, 205–69. Palo Alto, CA: Blackwell Scientific.

Rogov, V., V. Marusin, N. Bykova, Y. Goy, K. Nagovitsin et al. 2012. The oldest evidence of bioturbation on Earth. *Geology* 40:395–98.

Romer, P. M. 1990. Endogenous technological change. *Journal of Political Economy* 98:S71–102.

Rosenzweig, M. L., and R. D. McCord. 1991. Incumbent replacement: Evidence for long-term evolutionary progress. *Paleobiology* 17:202–13.

Rothman, D. H., J. M. Hayes, and R. E. Summons. 2003. Dynamics of the Neoproterozoic carbon cycle. *Proceedings of the National Academy of Sciences, USA* 100:8124–29.

Rouse, G. W., and F. Pleijel. 2001. *Polychaetes.* Oxford, UK: Oxford Universtity Press.

Rowland, S. J. 2001. Archaeocyaths—A history of phylogenetic interpretation. *Journal of Paleontology* 75:1065–78.

Rowland, S. J., and R. S. Shapiro. 2002. Reef patterns and environmental influences in the Cambrian and earliest Ordovician. In *Phanerozoic Reef Patterns,* edited by W. Kiessling and E. Flugel. SEPM Special Publication No. 72, 95–128.

Rozanov, A. Y., V. V. Missarzhevsky, N. A. Volkova, L. G. Voronova, I. N. Krylov et al. 1969. The Tommotian Stage and the Cambrian lower boundary problem. *Transactions of the Geological Institute, Academy of Sciences, USSR* 206:1–380.

Rozanov, A. Y., and A. Y. Zhuravlev. 1992. The Lower Cambrian fossil record of the Soviet Union. In *Origin and Early Evolution of the Metazoa,* edited by J. H. Lipps and P. W. Signor, 205-282. New York: Plenum Press.

Rudwick, M. J. S. 2008. *Worlds Before Adam: The Reconstruction of Geohistory in the Age of Reform.* Chicago: University of Chicago Press.

Ruiz-Trillo, I., J. Paps, M. Loukota, C. Ribera, U. Jondelius et al. 2002. A phylogenetic analysis of myosin heavy chain type II sequences corroborates that Acoela and Nemertodermatida are basal bilaterians. *Proceedings of the National Academy of Sciences, USA* 99:11246–51.

Ruiz-Trillo, I., M. Riutort, D. T. J. Littlewood, E. A. Herniou, and J. Baguña. 1999. Acoel flatworms: Earliest extant bilaterian metazoans, not members of Platyhelminthes. *Science* 283:1919–23.

Runnegar, B. 1995. Vendobionta or metazoa? Developments in understanding the Ediacara "fauna"? *Neues Jahrbuch für Geologie und Paläontologie, Abhandlungen*195:303–18.

———. 2000. Loophole for snowball Earth. *Nature* 405:403–4.

Ruppert, E. E., and R. D. Barnes. 1996. *Invertebrate Zoology.* Orlando, FL: Harcourt.

Ryan, J. F., M. E. Mazza, K. Pang, D. Q. Matus, A. D. Baxevanix et al. 2007. Pre-bilaterian origins of the *Hox* cluster and the *Hox* code: Evidence from the sea anemone, *Nematostella vectensis. PloS ONE* 2:e153.

Salvador, A., ed. 1994. *International Stratigraphic Guide.* 2nd ed. Trondheim, Norway, and Boulder, CO: International Union of Geologic Sciences / Geological Society of America.

Salvini-Plawen, L. V., and G. Steiner. 1996. Synapomorphies and plesiomorphies in higher classification of Mollusca. In *Origin and Evolutionary Radiation of the Mollusca,* edited by J. Taylor, 29–51. Oxford, UK: Oxford University Press.

Sansjofre, P., M. Ader, R. I. Trindade, M. Elie, J. Lyons et al. 2011. A carbon isotope challenge to the snowball Earth. *Nature* 478:93–96.

Sansom, R. S., S. E. Gabbott, and M. A. Purnell. 2010. Non-random decay of chordate characters causes bias in fossil interpretation. *Nature* 463:797–800.

Sappenfield, A., M. L. Droser, and J. G. Gehling. 2011. Problematica, trace fossils, and tubes with the Ediacara Member (South Australia): Redefining the Ediacaran trace fossil record one tube at a time. *Journal of Paleontology* 85:256–65.

Sasai, Y., and E. M. De Robertis. 1997. Ectodermal patterning in vertebrate embryos. *Developmental Biology* 182:5–20.

Savarese, M. 1992. Functional analysis of archaeocyathan skeletal morphology and paleobiological consequences. *Paleobiology* 18:464–80.

———. 1993. Paleobiologic and paleoenvironmental context of coral-bearing Early Cambrian reefs: Implications for Phanerozoic reef development. *Geology* 21:917–20.

Sax, D. F., J. J. Stachowicz, J. H. Brown, J. F. Bruno, M. N. Dawson et al. 2007. Ecological and evolutionary insights from species invasions. *Trends in Ecology and Evolution* 22:465–71.

Schierwater, B., M. Eitel, W. Jakob, H. J. Osigus, H. Hadrys et al. 2009. Concatenated analysis sheds light on early metazoan evolution and fuels a modern "urmetazoon" hypothesis. *PLoS Biology* 7(1):e20.

Schierwater, B., and K. Kuhn. 1998. Homology of *Hox* genes and the zootype concept in early metazoan evolution. *Molecular Phylogenetics and Evolution* 9:375–81.

Schiffbauer, J. D., S. H. Xiao, K. S. Sharma, and G. Wang. 2012. The origin of intracellular structures in Ediacaran metazoan embryos. *Geology* 40:223–26.

Schluter, D. 2000. *The Ecology of Adaptive Radiation.* Oxford, UK: Oxford University Press.

Schmitz, O. J., V. Krivan, and O. Ovadia. 2004. Trophic cascades: The primacy of trait mediated indirect interactions. *Ecology Letters* 7:153–63.

Schmucker, D., J. C. Clemens, H. Shu, C. A. Worby, J. Xiao et al. 2000. *Drosophila* Dscam is an axon guidance receptor exhibiting extraordinary molecular diversity. *Cell* 101:671–84.

Schrag, D. P., R. A. Berner, P. F. Hoffman, and G. P. Halverson. 2002. On the initiation of a snowball Earth. *Geochemistry Geophysics Geosystems* 3(6).

Schram, F. R. 1973. Pseudocoelomates and a nemertine from the Illinois Pennsylvanian. *Journal of Paleontology* 47:985–89.

Schulze, A., E. B. Cutler, and G. Giribet. 2007. Phylogeny of sipunculan worms: A combined analysis of four regions and morphology. *Molecular Phylogenetics and Evolution* 42:171–92.

Schwenk, K. 1995. A utilitarian approach to evolutionary constraint. *Zoology* 98:251–62.

Schwenk, K., and G. P. Wagner. 2001. Function and the evolution of phenotypic stability: Connecting pattern to process. *American Zoologist* 41:552–63.

Scott, C., T. W. Lyons, A. Bekker, Y. Shen, S. W. Poulton et al. 2008. Tracing the stepwise oxygenation of the Proterozoic ocean. *Nature* 452:456–59.

Scott, M. P. 1994. Intimations of a creature. *Cell* 79:1121–24.

Seaver, E. C. 2003. Segmentation: Mono- or polyphyletic? *International Journal of Developmental Biology* 47:583–95.

Sebe-Pedros, A., I. Ruiz-Trillo, A. de Mendoza, B. F. Lang, and B. M. Degnan. 2011. Unexpected repertoire of metazoan transcription factors in the unicellular holozoan *Capsaspora owczarzaki*. *Molecular Biology and Evolution* 28:1241–54.

Seilacher, A. 1956. Der beginn des Kambriums als biologische Wende. *Neues Jahrbuch für Geologie und Paläontologie* 103:155–80.

———. 1967. Fossil behavior. *Scientific Amerian* 217:155–80.

———. 1984. Late Precambrian and Early Cambrian Metazoa: Preservational or real extinction? In *Patterns of Change in Earth Evolution*, edited by H. D. Holland and A. F. Tendall, 159–68. Berlin: Springer-Verlag.

———. 1989. Vendozoa: Organismic construction in the Proterozoic biosphere. *Lethaia* 22:229–39.

———. 1992. Vendobionta and Psammocorallia: Lost constrctions of Precambrian evolution. *Journal of the Geological Society of London* 149:607–13.

———. 1999. Biomat-related lifestyles in the Precambrian. *Palaios* 14:86–93.

———. 2007a. The nature of vendobionts. In *The Rise and Fall of the Ediacaran Biota*, edited by P. Vickers-Rich and P. Komarower, 387–97. London: Geological Society.

———. 2007b. *Trace Fossil Analysis*. Berlin: Springer-Verlag.

Seilacher, A., L. A. Buatois, and M. G. Mangano. 2005. Trace fossils in the Ediacaran-Cambrian transition: Behavioral diversification, ecological turnover and environmental shift. *Palaeogeography, Palaeoclimatology, Palaeoecology* 227:323–56.

Seilacher, A., and J. W. Hagadorn. 2010. Early molluscan evolution: Evidence from the trace fossil record. *Palaios* 25:565–75.

Seilacher, A., and F. Pfluger. 1994. From biomats to benthic agriculture: A biohistoric revolution. In *Biostablization of Sediments*, edited by W. Krumbein, D. M. Paterson, and L. J. Stal, 97–105. Oldenburg, Germany: Bibliotheks-und Informationssystem der Universität Oldenburg.

Seipel, K., and V. Schmid. 2006. Mesodermal anatomies in cnidarian polyps and medusae. *International Journal of Developmental Biology* 50:589–99.

Sempere, L. F., C. N. Cole, M. A. McPeek, and K. J. Peterson. 2006. The phylogenetic distribution of metazoan microRNAs: Insights into evolutionary complexity and constraint. *Journal of Experimental Zoology Part B: Molecular and Developmental Evolution* 306B:2–14.

Sepkoski, J. J., Jr. 2002. A compendium of fossil marine animal genera. *Bulletin of American Paleontology* 363:1–560.

Serezhnikova, E. A., and A. Y. Ivantsov. 2007. *Fedomia mikhaili*—a new spicule-bearing organism of sponge grade from the Vendian of the White Sea, Russia. *Palaeoworld* 16:319–24.

Sergio, F., T. Caro, D. Brown, B. Clucas, J. D. Hunter et al. 2009. Top predators as conservation tools: Ecological rationale, assumptions, and efficiency. *Annual Review of Ecology, Evolution and Systematics* 39:1–19.

Shen, B., L. Dong, S. H. Xiao, and M. Kowalewski. 2008. The Avalon explosion: Evolution of Ediacara morphospace. *Science* 319:81–84.

Shenk, M. A., and M. A. Steel. 1994. A molecular shapshot of the metazoan "Eve." *Trends in Biochemical Science* 18:459–63.

Shick, J. M. 1991. *A Functional Biology of Sea Anemones*. London: Chapman and Hall.

Shields, G. A. 2005. Neoproterozoic cap carbonates: A critical appraisal of existing models and the plumeworld hypothesis. *Terra Nova* 17:299–310.

———. 2007. A normalised seawater strontium curve: Possible implications for Neoproterozoic-Cambrian weathering rates and the further oxygenation of the Earth. *eEarth* 2:35–42.

Shields-Zhou, G., and L. Och. 2011. The case for a Neoproterozoic oxygenation event: Geochemical evidence and biological consequences. *GSA Today* 21(3):4–11.

Shu, D., S. Conway Morris, Z. F. Zhang, J. N. Liu, J. Han et al. 2003. A new species of yunnanozoan with implications for deuterostome evolution. *Science* 299:1380–84.

Shu, D. G., S. Conway Morris, J. Han, L. Chen, X. L. Zhang et al. 2001. Primitive deuterostomes from the Chengjiang lagerstätte (Lower Cambrian, China). *Nature* 414:419–24.

Shu, D. G., S. Conway Morris, and X. L. Zhang. 1996. A *Pikaia*-like chordate from the Lower Cambrian of China. *Nature* 384:157–56.

Shu, D. G., S. Conway Morris, X. L. Zhang, S. X. Hu, L. Chen et al. 1999. Lower Cambrian vertebrates from south China. *Nature* 402:42–46.

Shu, D. G., X. Zhang, and L. Chen. 1996. Reinterpretation of *Yunnanozoon* as the earliest known hemichordate. *Nature* 380:428–30.

Shubin, N. H., C. Tabin, and S. Carroll. 1997. Fossils, genes and the evolution of animal limbs. *Nature* 388:639–48.

Shurin, J. B., and D. S. Srivastava. 2005. New perspectives on local and regional diversity. In *Metacommunities: Spatial Dynamics and Ecological Communities*, edited by M. Holyoak, M. A. Leibold, and R. A. Holt, 399–417. Chicago: University of Chicago Press.

Signor, P. W., and G. J. Vermeij. 1994. The plankton and the benthos: Origins and early history of an evolving relationship. *Paleobiology* 20:297–319.

Signorovitch, A. Y., S. L. Dellaporta, and L. W. Buss. 2005. Molecular signatures for sex in the Placozoa. *Proceedings of the National Academy of Sciences, USA* 102:15518–22.

Sigwart, J. D., and M. D. Sutton. 2007. Deep molluscan phylogeny: Synthesis of palaeontological and neontological data. *Proceedings of the Royal Society B* 274:2413–19.

Sim, M. S., T. Bosak, and S. Ono. 2011. Large sulfur isotope fractionation does not require disproportionation. *Science* 333:74–77.

Simpson, G. G. 1944. *Tempo and Mode in Evolution*. New York: Columbia University Press.

———. 1953. *The Major Features of Evolution*. New York: Columbia University Press.

Simpson, T. L. 1984. *The Cell Biology of Sponges*. New York: Springer-Verlag.

Skovsted, C. B., U. Balthasar, G. A. Brock, and J. R. Paterson. 2009. The tommotiid *Camenella reticulosa* from the Early Cambrian of South Australia: Morphology, scleritome reconstruction, and phylogeny. *Acta Palaeontologica Polonica* 54:525–40.

Skovsted, C. B., G. A. Brock, J. R. Paterson, L. E. Holmer, and G. E. Budd. 2008. The scleritome of *Eccentrotheca* from the Lower Cambrian of South Australia: Lophophorate affinities and implications for tommotiid phylogeny. *Geology* 36:171–74.

Skovsted, C. B., L. E. Holmer, C. M. Larsson, A. E. S. Hogstrom, G. A. Brock et al. 2009. The scleritome of *Paterimitra*: An Early Cambrian stem group brachiopod from South Australia. *Proceedings of the Royal Society B* 276:1651–56.

Skovsted, C. B., and J. S. Peel. 2011. *Hyolithellus* in life position from the Lower Cambrian of north Greenland. *Journal of Paleontology* 85:37–47.

Smith, A. B. 1988. Fossil evidence for the relationships of extant echinoderm classes and their times of divergence. In *Echinoderm Phylogeny and Evolutionary Biology*, edited by C. R. C. Paul and A. B. Smith, 85–97. Oxford, UK: Clarendon Press.

———. 2005. The pre-radial history of echinoderms. *Geological Journal* 40:255–80.

Smith, A. B., and C. Patterson. 1988. The influence of taxonomic method on the perception of patterns of evolution. *Evolutionary Biology* 23:127–216.

Smith, A. B., and K. J. Peterson. 2002. Dating the time of origin of major clades: Molecular clocks and the fossil record. *Annual Review of Earth and Planetary Science* 30:65–88.

Smith, J. P. S. I., S. L. Dellaporta, and L. W. Buss. 1985. The acoel turbellarians: Kingpins of metazoan evolution or a specialized offshoot? In *The Origins and Relationships of Lower Invertebrates*, edited by S. Conway Morris, J. D. George, and H. M. Platt, 123–42. Oxford, UK: Oxford University Press.

Smith, M. P., I. J. Sansom, and K. D. Cochrane. 2001. The Cambrian origin of vertebrates. In *Major Events in Early Vertebrate Evolution—Palaeontology, Phylogeny, Genetics, and Development*, edited by P. E. Ahlberg, 67–84. Systematics Association Special. Vol. 61. London: Taylor and Francis.

Smith, M. R., and J. B. Caron. 2010. Primitive soft-bodied cephalopods from the Cambrian. *Nature* 465:469–72.

Snodgrass, R. E. 1938. Evolution of the Annelida, Onycophora, and Arthropods. *Smithsonian Miscellaneous Collections* 97:1–159.

Sohl, L. E., N. Christie-Blick, and D. V. Kent. 1999. Paleomagnetic polarity reversals in Marinoan (c. 600 Ma) glacial deposits of Australia: Implications for the duration of low-latitude glaciation in Neoproterozoic time. *Geological Society of America Bulletin* 111:1120–39.

Sørensen, M. V., M. B. Hebsgaard, I. Heiner, H. Glener, E. Willerslev et al. 2008. New data from an enigmatic phylum: Evidence from molecular sequence data supports a sister-group relationship between Loricifera and Nematomorpha. *Journal of Zoological Systematics and Evolutionary Research* 46:231–35.

Sperling, E. A., K. J. Peterson, and M. Laflamme. 2011. Rangeomorphs, *Thectardis* (Porifera?) and dissolved organic carbon in the Ediacaran oceans. *Geobiology* 9:24–33.

Sperling, E. A., K. J. Peterson, and D. Pisani. 2009. Phylogenetic-signal dissection of nuclear housekeeping genes supports the paraphyly of sponges and the monophyly of Eumetazoa. *Molecular Biology and Evolution* 26:2261–74.

Sperling, E. A., D. Pisani, and K. J. Peterson. 2007. Poriferan paraphyly and its implications for Precambrian palaeobiology. In *The Rise and Fall of the Ediacaran Biota*, edited by P. Vickers-Rich and P. Komarower, 355–68. London: Geological Society.

Sperling, E. A., J. M. Robinson, D. Pisani, and K. J. Peterson. 2010. Where is the glass? Biomarkers, molecular clocks, and microRNAs suggest a 200-Myr missing Precambrian fossil record of siliceous sponge spicules. *Geobiology* 8:24–36.

Sperling, E. A., and J. Vinther. 2010. A placazoan affinity for *Dickinsonia* and the evolution of late Proterozoic metazoan feeding modes. *Evolution and Development* 12:201–9.

Sprigg, R. G. 1947. Early Cambrian(?) jellyfishes from the Flinders Ranges, South Australia. *Transactions of the Royal Society of South Australia* 71:212–24.

Sprinkle, J. 1973. Morphology and evolution of blastozoan echinoderms. Cambridge, MA: Harvard University Museum of Comparative Zoology, Special Publication.

———. 1980. An overview of the fossil record. In *Echinoderms*, edited by T. W. Broadhead and J. A. Waters, 15–26. Knoxville: University of Tennessee, Department of Geological Sciences, Studies in Geology.

Sprinkle, J., and P. M. Kier. 1987. Phylum Echinodermata. In *Fossil Invertebrates*, edited by R. S. Boardman, A. H. Cheetham, and A. J. Rowell, 550–611. Palo Alto, CA: Blackwell Scientific.

Squire, R. J., I. H. Campbell, C. M. Allen, and C. J. L. Wilson. 2006. Did the transgondwanan supermountain trigger the explosive radiation of animals on Earth? *Earth and Planetary Science Letters* 250:116–33.

Srivastava, M., E. Begovic, J. Chapman, N. H. Putnam, U. Hellsten et al. 2008. The *Trichoplax* genome and the nature of placozoans. *Nature* 454:955–60.

Srivastava, M., O. Simakov, J. Chapman, B. Fahey, M. E. Gauthier et al. 2010. The *Amphimedon queenslandica* genome and the evolution of animal complexity. *Nature* 466:720–26.

Stanley, S. M. 1973. An ecological theory for the sudden origin of multicellular life in the late Precambrian. *Proceedings of the National Academy of Sciences, USA* 70:1486–89.

Stanley, S. M. 1976. Fossil data and the Precambrian-Cambrian evolutionary transition. *American Journal of Science* 276:55–76.

Steiner, M., G. Li, Y. Qian, and M. Zhu. 2004. Lower Cambrian small shelly fossils of northern Sichuan and southern Shaanxi (China), and their biostratigraphic importance. *Geobios* 37:259–75.

Steiner, M., G. Li, Y. Qian, M. Zhu, and B.-D. Erdtmann. 2007. Neoproterozoic to Early Cambrian small shelly fossil assemblages and a revised biostratigraphic correlation of the Yangtze Platform (China). *Palaeogeography, Palaeoclimatology, Palaeoecology* 254:67–99.

Steiner, M., D. Mehl, J. Reitner, and B.-D. Erdtmann. 1993. Oldest entirely preserved sponges and other fossils from the lowermost Cambrian and a new facies reconstruction of the Yangtze Platform (China). *Berlin Geowissen Abhandlungen* 9:293–329.

Steiner, M., and J. Reitner. 2001. Evidence of organic structures in Ediacara-type fossils and associated microbial mats. *Geology* 29:1119–22.

Steiner, M., M. Zhu, Y. Zhao, and B.-D. Erdtmann. 2005. Lower Cambrian Burgess Shale–type fossil associations of south China. *Palaeogeography, Palaeoclimatology, Palaeoecology* 220:129–52.

Steiner, M., M. Y. Zhu, G. X. Li, Y. Qian, and B.-D. Erdtmann. 2004. New Early Cambrian bilaterian embryos and larvae from China. *Geology* 32:833–36.

Stollenwerk, A., M. Schoppmeier, and W. G. M. Damen. 2003. Involvement of *Notch* and *Delta* genes in spider segmentation. *Nature* 423:863–65.

Stoltzfus, A., and L. Y. Yampolsky. 2009. Climbing Mount Probable: Mutation as a cause of nonrandomness in evolution. *Journal of Heredity* 100:637–47.

Struck, T. H., C. Paul, N. Hill, S. Hartmann, C. Hosel et al. 2011. Phylogenomic analyses unravel annelid evolution. *Nature* 471:95–98.

Struck, T. H., N. Schult, T. Kusen, E. Hickman, C. Bleidorn et al. 2007. Annelid phylogeny and the status of Sipuncula and Echiura. *BMC Evolutionary Biology* 7:57.

Sumrall, C. D., and G. A. Wray. 2007. Ontogeny in the fossil record: Diversification of body plans and the evolution of "aberrant" symmetry in Paleozoic echinoderms. *Paleobiology* 33:149–63.

Sun, W., and X. Hou. 1987. Early Cambrian worms from Chengjiang, Yunnan, China: Maotianshania gen. nov. *Acta Palaeontologica Sinica* 26:299–305.

Sutton, M. D., D. E. G. Briggs, D. J. Siveter, and D. J. Siveter. 2004. Computer reconstruction and analysis of the vermiform mollusc *Acaenoplax hayae* from the Herefordshire lagerstätte (Silurian, England), and implications for molluscan phylogeny. *Palaeontology* 47:293–318.

Swadling, K. M, H. J. G. Dartnell, J. A. E. Gibson, E. Saulnier-Talbot, and W. F. Vincent. 2001. Fossil rotifers and the early colonization of an Antarctic lake. *Quaternary Research* 55:380–84.

Swalla, B. J., and A. B. Smith. 2008. Deciphering deuterostome phylogeny: Molecular, morphological and palaeontological perspectives. *Philosophical Transactions of the Royal Society of London B* 363:1557–68.

Swanson-Hysell, N. L., C. V. Rose, C. C. Calmet, G. P. Halverson, M. T. Hurtgen et al. 2010. Cryogenian glaciation and the onset of carbon-isotope decoupling. *Science* 328:608–11.

Szaniawski, H. 1982. Cheotognath grasping spines recognized among Cambrian protoconodonts. *Journal of Paleontology* 56:806–10.

———. 2005. Cambrian chaetognaths recognized in Burgess Shale fossils. *Acta Palaeontologica Polonica* 50:1–8.

Tanaka, M., H. Kasahara, S. Bartunkova, M. Schinke, I. Komuro et al. 1998. Vertebrate homologs of *tinman* and *bagpipe*: Roles of the homeobox genes in cardiovascular development. *Developmental Genetics* 22:239–49.

Tang, F., S. Bengtson, Y. Wang, X. L. Wang, and C. Y. Yin. 2011. *Eoandromeda* and the origin of Ctenophora. *Evolution and Development* 13:408–14.

Tautz, D. 2004. Segmentation. *Developmental Cell* 7:301–12.

Telford, M. J. 2008. Xenoturbellida: The fourth deuterostome phylum and the diet of worms. *Genesis* 46:580–96.

Telford, M. J., S. J. Bourlat, A. Economou, D. Papillon, and O. Rota-Stabelli. 2008. The evolution of the Ecdysozoa. *Philosophical Transactions of the Royal Society B* 363:1529–37.

Thompson, M. D., and S. A. Bowring. 2000. Age of the Squantum "tillite" Boston Basin, Massachusetts: U-Pb zircon constraints on terminal Neoproterozoic glaciation. *American Journal of Science* 300:630–55.

Todd, J. A., and P. D. Taylor. 1992. The first fossil entoproct. *Naturwissenschaften* 79:311–14.

Traylor-Knowles, N., U. Hansen, T. Q. Dubuc, M. Q. Martindale, L. Kaufman et al. 2010. The evolutionary diversification of LSF and *Grainyhead* transcription factors preceded the radiation of basal animal lineages. *BMC Evolutionary Biology* 10:101.

Tseng, W. F., T. H. Jang, C. B. Huang, and C. H. Yuh. 2011. An evolutionarily conserved kernel of *gata5, gata6, otx2* and *prdm1a* operates in the formation of endoderm in zebrafish. *Developmental Biology* 357:541–57.

Turner, J. T. 2002. Zooplankton fecal pellets, marine snow and sinking phytoplankton blooms. *Aquatic Microbial Ecology* 27:57–102.

Tyler, S. 2001. The early worm: Origins and relationships of the lower flatworms. In *Interrelationships of the Platyhelminthes*, edited by D. T. Littlewood and R. A. Bray, 3–12. London: Taylor and Francis.

Tziperman, E., I. Halevy, D. T. Johnston, A. H. Knoll, and D. P. Schrag. 2011. Biologically induced initiation of Neoproterozoic snowball-Earth events. *Proceedings of the National Academy of Sciences, USA* 108:15091–96.

Ubaghs, G., and K. E. Caster. 1967. Homalozoans. In *Treatise on Invertebrate Paleontology, Part S*, edited by H. H. Beaver, K. E. Caster, J. W. Durham, R. O. Fay, H. B. Fell et al., S495–627. Lawrence: Geological Society of America / University Press of Kansas.

Ubaghs, G., and R. A. Robison. 1985. A new homoiostelean and a new eocrinoid from the middle Cambrian of Utah. *University of Kansas Paleontological Contributions* 115:1–24.

Ubaghs, G., and D. Vizcaíno. 1990. A new eocrinoid from the Lower Cambrian of Spain. *Palaeontology* 33:249–56.

Ushatinskaya, G. T. 1987. Unusual inarticulate brachiopods from the Lower Cambrian sequence of Mongolia. *Paleontological Journal* 21:59–66.

Vacelet, J., and N. Boury-Esnault. 1995. Carnivorous sponges. *Nature* 373:333–35.

Valentine, J. W. 1973. *Evolutionary Paleoecology of the Marine Biosphere*. Englewood Cliffs: Prentice-Hall.

———. 1980. Determinants of diversity in higher taxonomic catagories. *Paleobiology* 6:444–50.

———. 1992. *Dickinsonia* as a polypoid organism. *Paleobiology* 18:378–82.

———. 1997. Cleavage patterns and the topology of the metazoan tree of life. *Proceedings of the National Academy of Sciences, USA* 94:8001–5.
———. 2004. *On the Origin of Phyla*. Chicago: University of Chicago Press.
Valentine, J. W., A. G. Collins, and C. P. Meyer. 1994. Morphological complexity increase in metazoans. *Paleobiology* 20:131–42.
Valentine, J. W., D. Jablonski, and D. H. Erwin. 1999. Fossils, molecules and embryos: New perspectives on the Cambrian explosion. *Development* 126:851–59.
Valentine, J. W., D. Jablonski, A. Z. Krug, and K. Roy. 2008. Incumbancy, diversity and latitudinal gradients. *Paleobiology* 34:169–78.
Valentine, J. W., and E. M. Moores. 1970. Plate-tectonic regulation of faunal diversity and sea level: A model. *Nature* 228:657–59.
Valentine, J. W., and T. D. Walker. 1986. Diversity trends within a model taxonomic hierarchy and evolution. *Physica* 22D:31–42.
———. 1987. Extinction in a model taxonomic hierarchy. *Paleobiology* 13:107–93.
Van de Peer, Y., S. Maere, and A. Meyer. 2009. The evolutionary significance of ancient genome duplications. *Nature Reviews Genetics* 10:725–32.
Van Roy, P., and D. E. Briggs. 2011. A giant Ordovician anomalocaridid. *Nature* 473:510–13.
Van Roy, P., P. J. Orr, J. P. Botting, L. A. Muir, J. Vinther et al. 2010. Ordovician faunas of Burgess Shale type. *Nature* 465:215–18.
Vannier, J., I. Calandra, C. Gaillard, and A. Zylinska. 2010. Priapulid worms: Pioneer horizontal burrowers at the Precambrian-Cambrian boundary. *Geology* 38:711–14.
Vannier, J., J.-Y. Chen, D.-Y. Huang, S. Charbonnier, and X.-Q. Wang. 2006. The Early Cambrian origin of the thylacocephalan arthropods. *Acta Palaeontologica Polonica* 51:201–14.
Vendrasco, M. J., G. Li, S. M. Porter, and C. Z. Fernandez. 2009. New data on the enigmatic *Ocruranus-Eohalobia* group of Early Cambrian small skeletal fossils. *Palaeontology* 56:1373–96.
Verdel, C., B. P. Wernicke, and S. A. Bowring. 2011. The Shuram and subsequent Ediacaran carbon isotope excursions from southwest Laurentia, and implications for environmental stability during the metazoan radiation. *Geological Society of America Bulletin* 123:1539–59.
Vermeij, G. J. 1974. Adaptation, versatility and evolution. *Systematic Zoology* 22:466–77.
———. 1989. The origin of skeletons. *Palaios* 4:585–90.
Vickaryous, M. K., and B. K. Hall. 2006. Human cell type diversity, evoluion, development, and classification with special reference to cells derived from the neural crest. *Biological Reviews* 81:425–55.
Vickers-Rich, P., and P. Komarower, eds. 2007. *The Rise and Fall of the Ediacaran Biota*. London: Geological Society.
Vinther, J. 2009. The canal system in sclerites of Lower Cambrian *Sinosachites* (Halkieriidea: Sachitida): Significance for the molluscan affinities of the sachitids. *Palaeontology* 52:689–712.
Vinther, J., and C. Nielsen. 2005. The Early Cambrian *Halkieria* is a mollusc. *Zoologica Scripta* 34:81–89.
Vinther, J., P. Van Roy, and D. E. G. Briggs. 2008. Machaeridians are Palaeozoic armoured annelids. *Nature* 451:185–88.
Vogel, C., M. Bashton, N. D. Kerrison, C. Chothia, and S. A. Teichmann. 2004. Structure, function and evolution of multidomain proteins. *Current Opinion in Structural Biology* 14:208–16.
Vogel, S. 1994. *Life in Moving Fluids: The Physical Biology of Flow*. Princeton, NJ: Princeton University Press.
Vopalensky, P., and Z. Kozmik. 2009. Eye evolution: Common use and independent recruitment of genetic components. *Philosophical Transactions of the Royal Society B* 364:2819–32.

Wade, M. 1968. Preservation of soft-bodied animals in Precambrian sandstones at Ediacara, South Australia. *Lethaia* 84:183–88.

Waggoner, B. M. 2003. The Ediacaran biotas in space and time. *Integrative and Comparative Biology* 43:104–13.

Waggoner, B. M., and G. O. Poinar Jr. 1993. Fossil habrotrochid rotifers in Dominican amber. *Cellular and Molecular Life Sciences* 49:354–57.

Wagner, A. 2008. Neutralism and selectionism: A network-based reconciliation. *Nature Reviews Genetics* 9:965–74.

Wagner, G. P., C. Amemiya, and F. Ruddle. 2003. *Hox* cluster duplications and the opportunity for evolutionary novelties. *Proceedings of the National Academy of Sciences, USA* 100:14603–6.

Wagner, G. P., J. Mezey, and R. Calabretta. 2005. Natural selection and the origin of modules. In *Modularity: Understanding the Development and Evolution of Natural Complex Systems*, edited by W. Calebaut and D. Rasskin-Gutman, 33–60. Cambridge, MA: MIT Press.

Wagner, G. P., M. Pavlicev, and J. M. Cheverud. 2007. The road to modularity. *Nature Reviews Genetics* 8:921–31.

Walker, J. D., and J. W. Geissman. 2009. 2009 GSA geologic time scale. *GSA Today* 19:60–61.

Walossek, D. 1995. The Upper Cambrian *Rehbachiella*, its larval development, morphology and significance for the phylogeny of Branchiopoda and Crustacea. *Hydrobiologia* 298:1–13.

Walossek, D., J. E. Repetski, and K. J. Muller. 1994. An exceptionally preserved parasitic arthropod, *Heymonsicambria taylori* n. sp. (Arthropoda incertae cedis: Pentastomida), from Cambrian-Ordovician boundary beds of Newfoundland, Canada. *Canadian Journal of Earth Sciences* 31:1664–71.

Waloszek, D., and J. A. Dunlop. 2002. A larval sea spider (Arthropoda: Pycnogonida) from the Upper Cambrian "Orsten" of Sweden, and the phylogenetic position of pycnogonids. *Palaeontology* 45:421–46.

Waloszek, D., A. Maas, J. Y. Chen, and M. Stein. 2007. Evolution of cephalic feeding structures and the phylogeny of Arthropoda. *Palaeogeography, Palaeoclimatology, Palaeoecology* 254:273–87.

Waloszek, D., and K. J. Muller. 1990. Stem-lineage crustaceans from the Upper Cambrian of Sweden and their bearing upon the position of *Agnostus*. *Lethaia* 23:409–27.

Wang, E. T., R. Sandberg, S. Luo, I. Khrebtukova, L. Zhang et al. 2008. Alternative isoform regulation in human tissue transcriptomes. *Nature* 456:470–76.

Wang, Y., and X. L. Wang. 2011. New observations on *Cucullus* Steiner from the Neoproterozoic Doushantuo Formation of Guizhou, south China. *Lethaia* 44:275–86.

Warn, J. 1974. Presumed Myzostomid infestation of an Ordovician crinoid. *Journal of Paleontology* 48:506–13.

Warsh, D. 2006. *Knowledge and the Wealth of Nations*. New York: Norton.

Watters, W. A., and J. P. Grotzinger. 2001. Digital reconstruction of calcified early metazoans, terminal Proterozoic Nama Group, Namibia. *Paleobiology* 27:159–71.

Webster, B. L., R. R. Copley, R. A. Jenner, J. A. Mackenzie-Dodds, S. J. Bourlat et al. 2006. Mitogenomics and phylogenomics reveal priapulid worms as extant models of the ancestral Ecdysozoan. *Evolution and Development* 8:502–10.

Wennberg, S. A., R. Janssen, and G. E. Budd. 2008. Early embryonic development of the priapulid worm *Priapulus caudatus*. *Evolution and Development* 10:326–38.

Wesley-Hunt, G. D. 2005. The morphological diversification of carnivores in North America. *Paleobiology* 31:35–55.

West, G. B., and J. H. Brown. 2005. The origin of allometric scaling laws in biology from genomes to ecosystems: Towards a quantitative unifying theory of biological structure and organization. *Journal of Experimental Biology* 208:1575–892.

Westblad, E. 1949. *Xenoturbella bocki* n. g. n. sp., a peculiar, primitive turbellarian type. *Arkiv för Zoologi* 1:3–29.

Wheeler, B. M., A. M. Heimberg, V. N. Moy, E. A. Sperling, T. W. Holstein et al. 2009. The deep evolution of microRNAs. *Evolution and Development* 11:50–68.

Whittington, H. B. 1978. The lobopod animal *Aysheaia pedunculata* Walcott, Middle Cambrian, Burgess Shale, British Columbia. *Philosophical Transactions of the Royal Society B* 284:165–97.

Wilbur, B. C. 2006. Reduction in the number of Early Cambrian helicoplacoid species. *Palaeoworld* 15:283–93.

Wilkins, A. S. 2002. *The Evolution of Developmental Pathways*. Sunderland, MA: Sinauer.

Wille, M., T. F. Nagler, B. Lehmann, S. Schroder, and J. D. Kramers. 2008. Hydrogen sulphide release to surface waters at the Precambrian/Cambrian boundary. *Nature* 453:767–69.

Williams, A., and A. J. Rowell. 1965. Brachiopod anatomy. In *Treatise on Invertebrate Paleontology. Part H. Brachiopoda,* edited by R. C. Moore, H6–57. Lawrence: Geological Society of America / University of Kansas.

Williams, D. M., J. F. Kasting, and L. A. Frakes. 1998. Low-latitude glaciation and rapid changes in the Earth's obliquity explained by obliquity-oblateness feedback. *Nature* 396:453–55.

Williams, G. C. 1966. *Adaptation and Natural Selection*. Princeton, NJ: Princeton University Press.

Williams, M., D. J. Siveter, M. J. Salas, J. Vannier, L. E. Popov et al. 2008. The earliest ostracods: The geological evidence. *Palaeobiodiversity and Palaeoenvironments* 88:11–21.

Willman, S., M. Moczydlowska, and K. Grey. 2006. Neoproterozoic (Ediacaran) diversification of acritarchs—A new record from the Murnaroo 1 drillcore, eastern Officer Basin, Australia. *Review of Palaeobotany and Palynology* 139:17–39.

Wills, M. A. 1993. Annelida. In *The Fossil Record 2*, edited by M. J. Benton, 271–78. London: Chapman and Hall.

———. 1998. Cambrian and recent disparity: The picture from priapulids. *Paleobiology* 24:177–99.

Wilmer, P. 1990. *Invertebrate Relationships: Patterns in Animal Evolution*. Cambridge, UK: Cambridge University Press.

Wittebolle, L., M. Marzorati, L. Clement, A. Balloi, D. Daffonchio et al. 2009. Initial community evenness favours functionality under selective stress. *Nature* 458:623–26.

Wittkopp, P. J., and G. Kalay. 2012. *Cis*-regulatory elements: Molecular mechanisms and evolutionary processes underlying divergence. *Nature Reviews Genetics* 13:59–69.

Wood, R. 1999. *Reef Evolution*. Oxford, UK: Oxford University Press.

Wood, R., A. Y. Zhuravlev, and C. T. Anaaz. 1993. The ecology of Lower Cambrian buildups from Zuune Arts, Mongolia: Implications for early metazoan reef evolution. *Sedimentology* 40:829–58.

Wood, R. A. 2011. Paleoecology of the earliest skeletal metazoan communities: Implications for early biomineralization. *Earth-Science Reviews* 106:184–90.

Wood, R. A., J. P. Grotzinger, and J. A. Dickson. 2002. Proterozoic modular biomineralized metazoan from the Nama Group, Namibia. *Science* 296:2383–86.

Wootton, J. T. 1994. The nature and consequences of indirect effects in ecological communities. *Annual Review of Ecology and Systematics* 25:443–66.

Worm, B., E. B. Barbier, N. Beaumont, J. E. Duffy, C. Folke et al. 2006. Impacts of biodiversity loss on ocean ecosystem services. *Science* 314:787–90.

Wray, G. A., M. W. Hahn, E. Abouheif, J. P. Balhoff, M. Pizer et al. 2003. The evolution of transcriptional regulation in eukaryotes. *Molecular Biology and Evolution* 20:1377–419.

Wright, J. P., and C. G. Jones. 2006. The concept of organisms as ecosystem engineers ten years on: Progress, limitations, and challenges. *BioScience* 56:203–9.

Xia, F. S., S. G. Zhang, and Z. Z. Wang. 2007. The oldest bryozoans: New evidence from the late Tremadocian (Early Ordovician) of East Yangtze Gorges in China. *Journal of Paleontology* 81:1308–26.

Xiao, S. H., H. M. Bao, H. F. Wang, A. J. Kaufman, C. M. Zhou et al. 2004. The Neoproterozoic Quruqtagh Group in eastern Chinese Tianshan: Evidence for a post-Marinoan glaciation. *Precambrian Research* 130:1–26.

Xiao, S. H., J. W. Hagadorn, C. Zhou, and X. L. Yuan. 2007. Rare helical spheroidal fossils from the Doushantuo lagerstätte: Ediacaran animal embryos come of age? *Geology* 35:115–18.

Xiao, S. H., J. Hu, X. L. Yuan, R. L. Parsley, and R. J. Cao. 2005. Articulated sponges from the Lower Cambrian Hetang Formation in southern Anhui, south China: Their age and implications for the early evolution of sponges. *Palaeogeography, Palaeoclimatology, Palaeoecology* 220:89–117.

Xiao, S. H., and A. H. Knoll. 1999. Fossil preservation in the Neoproterozoic Douchantuo phosphorite lagerstätte, south China. *Lethaia* 32:219–40.

Xiao, S. H., and A. H. Knoll. 2000. Phosphatized animal embryos from the Neoproterozoic Doushantuo Formation at Weng'an, Guizhou, south China. *Journal of Paleontology* 74:767–88.

Xiao, S. H., A. H. Knoll, X. L. Yuan, and C. M. Pueschel. 2004. Phosphotized multicellular algae in the Neoproterozoici Doushantuo Formation, China, and the early evolution of florideophyte red algae. *American Journal of Botany* 91:214–27.

Xiao, S. H., and M. Laflamme. 2009. On the eve of animal radiation: Phylogeny, ecology and evolution of the Ediacara biota. *Trends in Ecology and Evolution* 24:31–40.

Xiao, S. H., B. Shen, C. M. Zhou, and X. L. Yuan. 2005. A uniquely preserved Ediacaran fossil with direct evidence for a quilted bodyplan. *Proceedings of the National Academy of Sciences, USA* 102:10227–332.

Xiao, S. H., Y. Yun, A. H. Knoll, and J. K. Bartley. 1998. Three-dimensional preservation of algae and animal embryos in a Neoproterozoic phosphorite. *Nature* 391:553–58.

Yin, L., M. Zhu, A. H. Knoll, X. Yuan, J. Zhang et al. 2007. Doushantuo embryos preserved inside diapause egg cysts. *Nature* 446:661–63.

Yin, Z. J., M. Y. Zhu, P. Tafforeau, J. Y. Chen, P. J. Liu et al. Forthcoming. Early embryogenesis of potential bilaterian animals with polar lobe formation from the Ediacaran Weng'an biota, south China. *Precambrian Research.*

Yoder, J. B., E. Clancey, S. des Roches, J. M. Eastman, L. Gentry et al. 2010. Ecological opportunity and the origin of adaptive radiations. *Journal of Evolutionary Biology* 23:1581–96.

Young, G. A., and J. W. Hagadorn. 2010. The fossil record of cnidarian medusae. *Palaeoworld* 19:212–21.

Young, G. C. 1997. Ordovician microvertebrate remains from the Amadeus Basin, central Australia. *Journal of Vertebrate Paleontology* 17:1-25

Young, J. Z. 1981. *The Life of Vertebrates*. Oxford, UK: Clarendon Press.

Yuan, X. L., J. Li, and R. J. Cao. 1999. A diverse metaphyte assemblage from the Neoproterozoic black shales of south China. *Lethaia* 32:143–55.

Zamora, S. 2010. Middle Cambrian echinoderms from north Spain show echinoderms diversified earlier in Gondwana. *Geology* 38:507–10.

Zamora, S., E. Liñán, P. Domínguez Alonso, R. Gozalo, and J. A. Gámez Vintaned. 2007. A Middle Cambrian edrioasteroid from the Murero biota (NE Spain) with Australian affinities. *Annales de Paléontologie* 93:249–60.

Zaret, K. 1999. Developmental competence of the gut ectoderm: Genetic potentiation by GATA and HNF3/Forkhead proteins. *Developmental Biology* 209:1–10.

Zhang, S. H., G. Q. Jiang, J. M. Zhang, B. Song, M. J. Kennedy et al. 2005. U-Pb sensitive high-resolution ion microprobe ages from the Doushantuo Formation in south China: Constraints on late Neoproterozoic glaciations. *Geology* 33:473–76.

Zhang, X. G., and X.-G. Hou. 2004. Evidence for a single median fin-fold and tail in the Lower Cambrian vertebrate, *Haikouichthys ercaicunensis*. *Journal of Evolutionary Biology* 17:1162–66.

Zhang, X. G., A. Maas, J. T. Haug, D. J. Siveter, and D. Waloszek. 2010. A eucrustacean metanauplius from the Lower Cambrian. *Current Biology* 20:1075–79.

Zhang, X. G., B. R. Pratt, and C. Shen. 2011. Embryonic development of a Middle Cambrian (500 myr old) scalidophoran worm. *Journal of Paleontology* 85:898–903.

Zhang, X. G., D. J. Siveter, D. Waloszek, and A. Maas. 2007. An epipodite-bearing crown-group crustacean from the Lower Cambrian. *Nature* 449:595–98.

Zhang, X. L., and D. E. G. Briggs. 2007. The nature and significance of the appendages of *Opabinia* from the Middle Cambrian Burgess Shale. *Lethaia* 40:161–73.

Zhang, X. L., and J. Reitner. 2006. A fresh look at *Dickinsonia*: Removing it from the Vendobionta. *Acta Geological Sinica* 80:636–42.

Zhang, X. L., and D. G. Shu. 2005. A new arthropod from the Chengjiang lagerstätte, Early Cambrian, southern China. *Alcheringa* 29:185–94.

Zhou, C., M. D. Brasier, and Y. Xue. 2001. Three-dimensional phosphatic preservation of giant acritarchs from the terminal Proterozoic Doushantuo Formation in Guizhou and Hubei provinces, south China. *Palaeontology* 44:1157–78.

Zhou, C. M., G. W. Xie, K. McFadden, and S. H. Xiao. 2007. The diversification and extinction of Doushantuo-Pertatataka acritarchs in south China: Causes and biostratigraphic significance. *Geological Journal* 42:229–62.

Zhu, M. Y., L. E. Babcock, and S. C. Peng. 2007. Advances in Cambrian stratigraphy and paleontology: Integrating correlation techniques, paleobiology, taphonomy and paleoenvironmental reconstruction. *Paleoworld* 15:217–12.

Zhu, M. Y., J. G. Gehling, S. H. Xiao, Y. L. Zhao, and M. L. Droser. 2008. Eight-armed Ediacara fossil preserved in countrasting taphonomic windows from China and Australia. *Geology* 36:867–70.

Zhu, M. Y., J. Vannier, H. Van Iten, and Y. L. Zhao. 2004. Direct evidence for predation on trilobites in the Cambrian. *Proceedings of the Royal Society London B* 271(Suppl. 5):S277–80.

Zhu, M. Y., J. M. Zhang, and A. Yang. 2007. Integrated Ediacaran (Sinian) chronostratigraphy of south China. *Palaeogeography, Palaeoclimatology, Palaeoecology* 254:7–62.

Zhu, M. Y., Y. Zhao, and J. Chen. 2002. Revision of the Cambrian discoidal animals *Stellostomites eumorphus* and *Pararotadiscus guizhouensis* from south China. *Geobios* 35:165–85.

Zhuravlev, A. Y. 1993. Were Ediacaran Vendobionta multicellulars? *Neues Jahrbuch für Geologie und Paläontologie, Abhandlungen* 190:299–314.

Zhuravlev, A. Y., E. Linan, J. A. Gamez-Vintaned, F. Debrenne, and A. B. Fedorov. 2012. New finds of skeletal fossils in the terminal Neoproterozoic of the Siberian Platform and Spain. *Acta Palaeontologica Polonica* 57:205–24.

Zhuravlev, A. Y., and R. A. Wood. 1996. Anoxia as a cause of the mid-Early Cambrian (Botomian) extinction event. *Geology* 24:311–14.

———. 2008. Eve of biomineralization: Controls on skeletal mineralogy. *Geology* 36:923–26.

Zuckerkandl, E., and L. Pauling. 1965. Evolutionary divergence and convergence in proteins. In *Evolving Genes and Proteins*, edited by V. Bryson and H. J. Vogel, 97–181. New York: Academic Press.

Index

Boldface indicates an illustration.